Systems Biology

Philosophical Foundations

Systems Biology
Philosophical Foundations

Edited by

Fred C. Boogerd
and
Frank J. Bruggeman
Department of Molecular Cell Physiology
Faculty of Earth and Life Sciences
Vrije Universiteit, Amsterdam, The Netherlands

Jan-Hendrik S. Hofmeyr
Department of Biochemistry, Faculty of Science
University of Stellenbosch, Stellenbosch, South Africa

Hans V. Westerhoff
Department of Molecular Cell Physiology
Faculty of Earth and Life Sciences
Vrije Universiteit, Amsterdam, The Netherlands
and
Manchester Centre for Integrative Systems Biology
The University of Manchester, Manchester, United Kingdom

ELSEVIER

Amsterdam • Boston • Heidelberg • London • New York • Oxford
Paris • San Diego • San Francisco • Singapore • Sydney • Tokyo

Elsevier
Radarweg 29, PO Box 211, 1000 AE Amsterdam, The Netherlands
The Boulevard, Langford Lane, Kidlington, Oxford OX5 1GB, UK

First edition 2007

Library of Congress Cataloging-in-Publication Data
A catalog record for this book is available from the Library of Congress

British Library Cataloguing in Publication Data
A catalogue record for this book is available from the British Library

ISBN-13: 978-0-444-52085-2
ISBN-10: 0-444-52085-6

For information on all Elsevier publications
visit our website at books.elsevier.com

07 08 09 10 11 10 9 8 7 6 5 4 3 2 1

Transferred to digital printing in 2007.

Contents

List of Contributors

William Bechtel, Department of Philosophy, University of California, San Diego, USA

Fred C. Boogerd, Department of Molecular Cell Physiology, Faculty of Earth and Life Sciences, Vrije Universiteit, Amsterdam, The Netherlands

Frank J. Bruggeman, Department of Molecular Cell Physiology, Faculty of Earth and Life Sciences, Vrije Universiteit, Amsterdam, The Netherlands

Werner Callebaut, Konrad Lorenz Institute for Evolution and Cognition Research, Altenberg, Austria and Universiteit Hasselt Diepenbeek, Belgium

David A. Fell, School of Biological and Molecular Sciences, Oxford Brookes University, Oxford, UK

Evelyn Fox Keller, Program in Science, Technology, and Society, Massachusetts Institute of Technology Cambridge, USA

Jan-Hendrik S. Hofmeyr, Department of Biochemistry, Faculty of Science, University of Stellenbosch, Stellenbosch, South Africa

Douglas B. Kell, School of Chemistry, The Manchester Interdisciplinary Biocentre, The University of Manchester, Manchester, UK

Ulrich Krohs, Department of Philosophy, University of Hamburg, Hamburg, Germany and Konrad Lorenz Institute for Evolution and Cognition Research, Altenberg, Austria

Alvaro Moreno, Department of Logic and Philosophy of Science, University of the Basque Country, San Sebastian-Donostia, Spain

Robert C. Richardson, Department of Philosophy, University of Cincinnati, Cincinnati, USA

Kenneth F. Schaffner, Department of History and Philosophy of Science, University of Pittsburgh, Pittsburgh, USA

Robert G. Shulman, Departments of Molecular Biophysics and Biochemistry, Yale University, New Haven, USA

Achim Stephan, Institute of Cognitive Science, University of Osnabrück, Osnabrück, Germany

Mukhtar Ullah, Systems Biology and Bioinformatics Group, Department of Computer Science, University of Rostock, Rostock, Germany

Hans V. Westerhoff, Department of Molecular Cell Physiology, Faculty of Earth and Life Sciences, Vrije Universiteit, Amsterdam, The Netherlands and Manchester Centre for Integrative Systems Biology, The University of Manchester, Manchester, UK

William C. Wimsatt, Department of Philosophy and Committee on Evolutionary Biology, The University of Chicago, Chicago, USA

Olaf Wolkenhauer, Systems Biology and Bioinformatics Group, Department of Computer Science, University of Rostock, Rostock, Germany

Contributor Biographies

William Bechtel is a Professor in the Department of Philosophy and the interdisciplinary programs in Cognitive Science and Science Studies at the University of California, San Diego. Much of his recent research addresses the nature of mechanisms and mechanistic explanation as they figure in biology and cognitive science. His most recent book, *Discovering Cell Mechanisms* (Cambridge, 2006), examines the development of cell biology as a distinct biological discipline in the mid twentieth century from the perspective of the challenge of developing accounts of mechanisms. He is currently completing a book entitled *Mental Mechanisms* to be published by Lawrence Erlbaum that applies the framework of mechanistic explanation to contemporary research in cognitive science and cognitive neuroscience. He is also the coauthor of *Discovering Complexity* (with Robert Richardson, Princeton, 1993) and *Connectionism and the Mind* (with Adele Abrahamsen, Blackwell, 2002). In addition, he is the coeditor of *A Companion to Cognitive Science* (Blackwell, 1998) and *Philosophy and the Neurosciences* (Blackwell, 2001) as well as the journal *Philosophical Psychology*.

Fred C. Boogerd is an Assistant Professor in the Department of Molecular Cell Physiology at the Vrije Universiteit of Amsterdam, The Netherlands. He obtained his doctoral (MSc) in Chemistry (cum laude) in 1979 and his PhD in 1984 at the Vrije Universiteit in Amsterdam. He worked for several years as a post-doc at Leiden University and Delft University of Technology, both in The Netherlands, and he was on sabbatical at the Technical University of Denmark in Copenhagen in Denmark. He is interested in physiology in general and has studied in particular the microbial processes of denitrification, manganese oxidation, desulphurization of coal, nitrogen fixation, succinate transport, and ammonium assimilation. Recently, he has developed a strong interest in systems biology and in the philosophy of (systems) biology, especially in the topics of reductionism, emergence, explanation and mechanism in Biology.

Frank J. Bruggeman is a Lecturer in the Systems Biology Group within the Manchester Interdisciplinary Biocentre (School for Chemical Engineering and Analytical Science, Faculty for Physical Sciences, University of Manchester, UK) and post-doc in the Department of Molecular Cell Physiology at the Vrije Universiteit in Amsterdam, The Netherlands. He obtained his doctoral (MSc) in Biology (cum laude) in 1999 at the University of Leiden and his PhD with Honors in 2005 at the Vrije Universiteit of Amsterdam. He is interested in the philosophy of biology (complex systems, emergence, reductionism, (mechanistic) explanation, modularity), regulation and adaptation of cellular physiology, and application of (dynamical systems and control) theory to the (experimental) analysis of cellular phenomena. He has worked on philosophy (mechanistic explanation and emergence) and systems biology (modular response analysis, modelling of ammonium assimilation in *Escherichia coli*, MAPK signalling, pH homeostasis in muscle, robustness, phytohormone transport).

Werner Callebaut studied philosophy at the University of Ghent, Belgium (PhD, 1983). He is the scientific manager of the Konrad Lorenz Institute for Evolution and Cognition Research (KLI) in Altenberg, Austria, and the editor-in-chief of the new journal *Biological Theory: Integrating Development, Evolution, and Cognition*, published by The MIT Press. Callebaut also teaches philosophy of science at Hasselt University, Belgium and theoretical biology at the University of Vienna. He was previously affiliated to Maastricht University, the Netherlands and is a past president of the Belgian Society for Logic and Philosophy of Science. His main research interests concern the philosophy of biology, general philosophy of science in evolutionary, naturalistic, and systems-theoretic perspectives, evolutionary economics and bounded rationality, evolutionary epistemology, and interdisciplinary science studies. His major publications are *Taking the Naturalistic Turn, or How Real Philosophy of Science is Done* (University of Chicago Press, 1993) and *Modularity: Understanding the Development and Evolution of Natural Complex Systems* (coeditor, with Diego Rasskin-Gutman; University of Chicago Press, 2005).

David Fell is a Professor of Biochemistry and Assistant Dean in the School of Biological and Molecular Sciences at Oxford Brookes University. After studying biochemistry at Oxford University, followed by a DPhil on the physical biochemistry of yeast pyruvate kinase, he started lecturing at Oxford Polytechnic, now known as Oxford Brookes University. His research moved from enzymology into computer simulation and theoretical analysis of metabolic control, and he has written the only textbook on metabolic control analysis, *Understanding the Control of Metabolism*. More recently, he has been analysing the structure of metabolic networks and extending his computer modelling into signal transduction pathways and the cell cycle. Applications of his work include metabolic engineering and modelling of drug action. In 2001, he became part-time Scientific Director of the Oxford company Physiomics plc, which is using computer simulation of cellular systems for the development and analysis of therapeutic strategies for the pharmaceutical industry.

Jan-Hendrik S. Hofmeyr is a Professor in the Department of Biochemistry at the University of Stellenbosch, South Africa. He obtained his PhD in 1986 at the University of Stellenbosch after collaborating with Henrik Kacser (one of the founders of metabolic control analysis) and the enzymologist Athel Cornish-Bowden. Hofmeyr and his colleagues Jacky Snoep and Johann Rohwer form the Triple-J Group for Molecular Cell Physiology, a research group that studies the control and regulation of cellular processes using theoretical, computer modelling and experimental approaches. He has made numerous fundamental contributions to the development of metabolic control analysis and computational cell biology, and with Athel Cornish-Bowden developed both coresponse analysis and supply–demand analysis as a basis for understanding metabolic regulation. A recent interest is to seek a way of expressing formally the functional organization of the cell in terms of a theory of molecular fabrication, which could form a theoretical foundation for both systems biology and nanotechnology. He has a National Research Foundation A-rating and is a member of the Academy of Science of South Africa and a Fellow of the Royal Society of South Africa. He currently chairs BTK-ISSB (International Study Group for Systems Biology). He won the Harry Oppenheimer Fellowship Award for 2002 and the Beckman Gold Medal of the South African Biochemical Society in 2003.

Douglas B. Kell Top Scholar, Bradfield College, Berks (1996–1970). BA (Hons) Biochemistry at St John's College, Oxford (1975) (Class 2–1 with Distinction in Chemical Pharmacology). Senior Scholar of St John's College, Oxford, MA (Oxon), DPhil (Oxon) 1978. SRC Postdoctoral Research Fellow (1978–1980), Postdoctoral Research Assistant (1980–1981) and SERC Advanced Fellow (1981–1983) and 'New Blood' lecturer in Microbial Physiology (1983–1988), all at the Department of Botany & Microbiology, University College of Wales, Aberystwyth. Reader in Microbiology, Dept of Biological Sciences, UCW, Aberystwyth, 1988–1992, and Founding Director, Aber Instruments Ltd, Science Park, Aberystwyth. Personal Chair, The University of Wales, 1992. Director of Research, Institute of Biological Sciences, UWA 1997–2002. Founding Director, Aber Genomic Computing, 2001–. EPSRC/RSC Research Professor of Bioanalytical Science, UMIST 2002–, The University of Manchester 2004–. 2005–Director, BBSRC Manchester Centre for Integrative Systems Biology. He has published over 350 scientific papers, 11 of which have been cited over 100 times (plus 5 over 90). His H-index is 46. 1986 Recipient of the Fleming Award of the Society for General Microbiology. 1998 Aber Instruments received a

Queen's Award for Export Achievement. 2004 Royal Society of Chemistry Interdisciplinary Science Award. 2005 FEBS-IUBMB Theodor Bücher prize. 2005 Royal Society/Wolfson Merit Award. 2000–2006 Member, BBSRC Council, BBSRC Strategy Board, NERC Environmental Genomic Committee.

Evelyn Fox Keller received her PhD in theoretical physics at Harvard University, worked for a number of years at the interface of physics and biology, and is now Professor of History and Philosophy of Science in the Program in Science, Technology and Society at MIT. She is the author of many articles and books, including: *A Feeling for the Organism: The Life and Work of Barbara McClintock; Reflections on Gender and Science; Secrets of Life, Secrets of Death: Essays on Language, Gender and Science; Refiguring Life: Metaphors of Twentieth Century Biology; The Century of the Gene;* and *Making Sense of Life: Explaining Biological Development with Models, Metaphors, and Machines.* This year, she is in Paris (at REHSEIS) as a Chaire Blaise Pascal.

Ulrich Krohs is "Privatdozent" of philosophy at the University of Hamburg, Germany, and research fellow at the Konrad Lorenz Institute for Evolution and Cognition Research in Altenberg, Austria. Having studied biochemistry and philosophy at the universities of Tübingen, Aachen, Brighton and Hamburg, he holds a PhD in biochemistry from the Technical University of Aachen and the post doctoral "Habilitation" in philosophy from Hamburg University. His current research interests include the structure of biological and of technological theories, and the epistemic justification of function ascriptions in biology, technology and the social sciences. Ulrich Krohs has authored a book on biological theories and co-edited an Introduction into the Philosophy of Biology (both in German). Papers on function and design, on the concept of Darwinian fitness, and on the structure of functional theories are forthcoming.

Alvaro Moreno is currently Professor of Philosophy of Science at the University of the Basque Country, in Spain, where he leads a research group specialized in Complex Systems, Philosophy of Biology, Cognitive Science and the methodological and epistemological aspects of Artificial Life. His work in Philosophy of Biology has been mainly focused on the structure, nature and origin of living organization, an area in which he has been working for 20 years. He is also interested in the problem of the origin of cognition, where he has developed a biologically focused approach. He is the author of many publications and organizer of several international workshops.

Robert C. Richardson is currently the Charles Phelps Taft Professor of Philosophy at the University of Cincinnati. He received his BA *magna cum laude* from the University of Colorado in 1971; he received his MA in 1972 and

PhD with Honors in 1977, both from the University of Chicago. He has been a visiting professor within the Department of Molecular Cell Physiology at the Free University of Amsterdam during 2003 and 2004, and was the Mercator Guest Professor at the University of Osnabrück during 2005. His research specialities are History and Philosophy of Science, Philosophy of Biology, and Cognitive Science. He has published broadly in journals including *Philosophy of Science, Mind, Biology and Philosophy, Philosophical Psychology, Erkenntnis, and Synthese.* With William Bechtel, he published *Discovering Complexity: Decomposition and Localization as Strategies in Scientific Research* (Princeton: Princeton University Press, 1993), and has a forthcoming book, *Maladapted Psychology: Infelicities of Evolutionary Psychology* (Cambridge: MIT Press, 2006).

Kenneth F. Schaffner (PhD, Columbia, 1967; MD, University of Pittsburgh, 1986) is University Professor of History and Philosophy of Science and University Professor of Philosophy at the University of Pittsburgh, and also University Professor of Medical Humanities Emeritus at the George Washington University. He has been a Guggenheim Fellow and is a Fellow of both the Hastings Center and the American Association for the Advancement of Science. He has published extensively in philosophical and medical journals on ethical and conceptual issues in science, medicine, and psychiatry, and is a member of the World Psychiatric Association–World Health Organization (WHO) workgroup, which advises on the approach and content of the Mental Health Section of the eleventh version of the International Classification of Diseases (ICD-11), due out in 2011. He is the author of *Discovery and Explanation in Biology and Medicine*, 1993, and is completing a book on *Behaving: What's Genetic and What's Not, and Why Should We Care?* He currently serves on a number of journal editorial boards and is a former editor-in-chief of *Philosophy of Science*.

Robert Shulman has been the leader in the application of NMR methods to biochemical and biomedical problems for 45 years. At Bell Laboratories, where he was head of biophysical research, starting in 1961 his group led the way into NMR studies of proteins and nucleic acids. In $\simeq$ 1974, his NMR interests shifted from macromolecular studies to NMR studies of metabolites *in vivo*. Since then he has been a leader in the field of Magnetic Resonance Spectroscopy *in vivo* where his ^{13}C NMR studies have focused on muscle and brain in rats and humans. Dr Shulman holds bachelor's and PhD degrees from Columbia University, and has been a member of the Yale faculty since 1979. He came to Yale from Bell Laboratories in New Jersey, where he had been head of the Biophysics Research Department for 26 years. He is a member of the National Academy of Sciences and the Institute of Medicine.

Achim Stephan (1955) is a Professor of *Philosophy of Mind and Cognition* at the Institute of Cognitive Science at the University of Osnabrück, Germany. He has studied Philosophy, Mathematics, Psychotherapy & Psychosomatics, and received his PhD from the University of Göttingen (1988) with a dissertation on Freud's implicit theory of meaning. In 1998, he habilitated at the University of Karlsruhe with a comprehensive book on emergence, prediction, and self-organization. He had visiting positions at the VU Amsterdam and the University of Ulm, and has held research fellowships at both the Hanse Institute of Advanced Study (Delmenhorst) and the Konrad Lorenz Institute (Altenberg/Vienna). Recently, he was a member of two ZiF Research groups at the Center for Interdisciplinary Research (Bielefeld) on *Emotions as Bio-Cultural Processes* and *Embodied Communication in Humans and Machines*, respectively. He has written extensively on the topic of emergence, and published and edited several books (e.g. *Emergenz. Von der Unvorhersagbarkeit zur Selbstorganisation*, 2005).

Mukhtar Ullah received his BSc in Electrical Engineering from N-W.F.P University of Engineering and Technology Peshawar, N-W.F.P, Pakistan in 1999 and his MSc in Advanced Control and Systems Engineering from the Department of Electrical Engineering and Electronics, at the University of Manchester Institute of Science and Technology (UMIST), Manchester, UK in 2002. He is now a research assistant in the Systems Biology and Bioinformatics group in the Department of Computer Science, University of Rostock, Germany. His research is concerned with the identification of dynamic models of intra-cellular processes.

Hans V. Westerhoff has been active in many preludes to systems or integrative biology, from biological non-equilibrium thermodynamics to hierarchical control and regulation analysis. He integrates theoretical and experimental aspects, presently as bottom-up systems biology. Hans Westerhoff is a driver of the Siliconcell programme (www.siliconcell.net), which makes and collects computer replica of pathways in living organisms. He sees these as laboratories for philosophers of modern biology and views systems biology as a new type of science.

After staff positions at the (US) National Institutes of Health and the Netherlands Cancer Institute, Hans Westerhoff became Professor of Microbial Physiology at the Free University Amsterdam, Professor of Mathematical Biochemistry at the University of Amsterdam, and also AstraZeneca Professor of Systems Biology at the University of Manchester. He is on boards of systems biology programs in various countries and organizer of Courses and Conferences on systems biology (www.systembiology.net).

William C. Wimsatt professes Philosophy, Evolutionary Biology, Conceptual and Historical Studies of Science, and directs the Big Problems program at the University of Chicago, where he went on a post-doc in 1969 (after a PhD from the University of Pittsburgh) to study population biology with Richard Lewontin and Richard Levins. He studied Engineering Physics and Philosophy at Cornell University and spent a year designing adding machines. His interests include history of genetics, mathematical modelling, the evolution and development of and heuristics for studying complex systems. He has worked on the strengths and limits of methodological reduction and reductionism, and teaches philosophy of the inexact sciences, which he asserts includes much of physics. For some years, he has tried to do more justice to the richness and complexity of cultural evolution by seeing how the interactive structure of individual development and our constructed scaffolding at multiple social and cultural levels enables cumulative cultural change and differentiation.

Olaf Wolkenhauer received degrees in control engineering from the University of Applied Sciences, Hamburg, Germany and the University of Portsmouth, UK in 1994. He received his PhD for work on possibility theory applied to data analysis in 1997 from the University of Manchester Institute of Science and Technology (UMIST) in Manchester, UK. From 1997 to 2000 he was a lecturer in the Control Systems Centre, UMIST and held a joint senior lectureship between the Department of Biomolecular Sciences and the Department of Electrical Engineering & Electronics until 2003. He is now a full professor (Chair in Systems Biology and Bioinformatics) at the University of Rostock, Germany. His research interest is in data analysis and mathematical modelling with applications to molecular and cell biology. The research focus is on dynamic modelling of gene expression and cell signalling.

Preface

This book, Systems Biology: Philosophical Foundations, represents the culmination of our studies on a number of philosophical issues related to systems biology we have undertaken in the recent years. In 1999, Hans Westerhoff, Frank Bruggeman, and Fred Boogerd from the Department of Molecular Cell Physiology (Faculty of Earth and Life Sciences, Vrije Universiteit, Amsterdam, The Netherlands) embarked on a project called 'Living reductionism', supported by the Vrije Universiteit through a so-called USF-grant. We gratefully acknowledge the contribution of Willem Drees (Leiden University) in the initial phase of the project.

In the beginning, our department collaborated with the Departments of Artifical Intelligence (Jan Treur) and Theoretical Psychology (Huib Looren-de Jong) of the Vrije Universiteit to study the concept of reductionism and antireductionism in Biology, Computer Science and Theoretical Psychology. This work resulted in a collection of four papers which was jointly published as a symposium section in Philosophical Psychology (vol. 14, no. 4, December 2002).

Thereafter, our department set out to focus more specifically on the idea of enriching Systems Biology and Philosophy of Biology by the mutual application of insights of both disciplines. To be able to accomplish this ambitious task, we thought it worthwhile to seek support from philosophers to elaborate on a variety of philosophical issues that naturally surfaced when thinking about the foundations of systems biology and what exactly made it so different from molecular biology. We are very grateful to Robert Richardson (Cincinnati, USA) and Achim Stephan (Osnabrück, Germany) whom we met in 2002, and with whom we started thinking about central philosophical notions in systems biology. We have had many inspiring and fruitful discussions and their contributions and support were essential to arrive at our current ideas on the foundations of systems biology.

In June 2005, our department organized a symposium: Towards a Philosophy of Systems Biology together with Jan-Hendrik Hofmeyr (Department of Biochemistry, University of Stellenbosch, South Africa). It was attended by some 120 persons with various scientific backgrounds. Thirteen Systems Biologists and Philosophers (William Bechtel, David Fell, Jan-Hendrik Hofmeyr,

Douglas Kell, Evelyn Fox Keller, Alvaro Moreno, Robert Richardson, Kenneth Schaffner, Robert Shulman, Achim Stephan, Hans Westerhoff, William Wimsatt, and Olaf Wolkenhauer) shed light on themes related to the subject of the symposium. All of them contributed to this book as an author. Werner Callebaut, Ulrich Krohs, and Mukhtar Ullah, who were not present at the symposium, also contributed as authors.

We thank the editors at Elsevier, Joke Zwetsloot and Anne Russum, for their help and support.

"Systems Biology: Philosophical Foundations" is the first book to be published on philosophical foundations of Systems Biology and we are grateful to have had the opportunity to edit and publish this work.

Fred C. Boogerd
Frank J. Bruggeman
Jan-Hendrik S. Hofmeyr
Hans V. Westerhoff

SECTION I

Introduction

1

Towards philosophical foundations of Systems Biology: introduction

Fred C. Boogerd, Frank J. Bruggeman, Jan-Hendrik S. Hofmeyr and Hans V. Westerhoff

1. SYSTEMS BIOLOGY: A NEW SCIENCE IN SEARCH OF METHODOLOGIES AND PHILOSOPHICAL FOUNDATIONS

The aim of systems biology is to understand how functional properties and behaviour of living organisms are brought about by the interactions of their constituents (Alberghina & Westerhoff, 2005). For new properties to arise in interactions, the latter must effect or affect processes that are nonlinear in terms of their kinetics or inhomogeneous in terms of their organization. Consequently, precise and comprehensive, quantitative experimental analyses of living systems at levels between the system and its molecules are a requirement for systems biology, as is the accurate interpretation of the resulting experimental data. Both need to be able to address systems that are complex enough to effect functionality of life.

Contemporary systems biology is a vigorous and expanding discipline, in many ways a successor to molecular biology and genomics on the one hand and mathematical biology and biophysics on the other (Westerhoff & Palsson, 2004). It is perhaps unprecedented in its combination of biology with a great many other sciences, from physics to ecology, mathematics to medicine and linguistics to chemistry. Because systems biology is at the interface of so many different scientific disciplines, because it tends to overstretch these disciplines when implementing them, and because it bears on an issue that transcends all other disciplines and perhaps even science as such ('What is life?'), the question arises as to the nature of its philosophical foundations: Are they just a mixture of

Systems Biology
F.C. Boogerd, F.J. Bruggeman, J.-H.S. Hofmeyr and H.V. Westerhoff (Editors)
Isbn: 978-0-444-52085-2

the foundations of all contributing disciplines, or are unique foundations required for this unique discipline? In simpler terms, is systems biology just the combined application of the above-mentioned disciplines to living systems, or does it have its own unique foundations and methodology, enabling this science to emerge from the other disciplines? Answering this question may also resolve a long standing issue, i.e., the extent to which biology is entitled to its own scientific foundations rather than being dominated by existing science philosophies such as that of physics. In this sense, systems biology may be the culmination of biology.

In contemporary experimental (molecular) biology, philosophy is not an issue. In broad terms, the aim of molecular biology is to characterize the molecular constituents of living organisms. Its agenda is noncomplex; it measures the properties of each component and, in case of molecular cell biology, its localization in the living cell. Although its methods and results are breathtaking and highly important, they are straightforward and do not require any philosophy that extends the philosophical foundations of physics (Carnap, 1966). Short of vitalism, no science denies that living systems consist solely of molecules. Since molecular biology characterizes all the molecules of these systems, it must also characterize those living systems themselves, at least according to some molecular biologists. What else could there be? Hence, what else other than the Philosophical Foundations of Physics could be needed?

By contrast, systems biology is concerned with the relationship between molecules and cells; it treats cells as organized, or organizing, molecular systems having both molecular and cellular properties. It is concerned with how life or the functional properties thereof that are not yet in the molecules, emerge from the particular organization of and interactions between its molecular processes. It uses models to describe particular cells and generalizes over various cell types and organisms to arrive at new theories of cells as molecular systems. It is concerned with explaining and predicting cellular behaviour on the basis of molecular behaviour. It refers to function in ways that would not be permitted in physics. It addresses an essential minimum complexity exceeding that of any physical chemical system understood until now. It shies away from reduction of the system under study to a collection of elementary particles. Indeed, it seems to violate many of the philosophical foundations of physics, often in ways unprecedented even by modern physics.

The premise of systems biology is that there is something to be discovered, i.e. living systems do have functional properties that cannot be discovered and understood by molecular biology alone; functional properties that are not in the molecules themselves. Because living systems are composed of nothing but molecules, this is at least a paradox, and indeed a contradiction to some molecular biologists. Scientific developments have, however, shown that there was and is something to be discovered. Systems biology has led to many new scientific insights, some of which will be reviewed below.

If essential, the paradox and the emancipation from the philosophical foundations of physics may warrant the development of philosophical foundations that are unique to systems biology – in contrast with those of molecular biology and physics and perhaps all other disciplines mentioned above. Because systems biology may well become main-stream biology and medicine in the coming years, and because practicing systems biologists are often hindered by paradigm battles with molecular biologists, we found it important that this issue be discussed intensely and openly by experts in fields ranging from molecular and systems biology to the philosophy of science. This book constitutes such a discussion and possibly some of the philosophical foundations of systems biology.

2. SYSTEMS BIOLOGY

2.1. History of systems biology

Systems biology did not come out of the blue (Westerhoff & Palsson, 2004; Alberghina & Westerhoff, 2005). It has two roots in scientific history: a 'components' root and a 'systems' root (see also Chapter 9). The emphasis in the former root was on individual macromolecules. Mainstream molecular biology was strongly associated with it; recently, it culminated in a variety of new high-throughput techniques (X-omics, bioinformatics). The latter root, with its emphasis on formal analysis of functional behaviour that arises when many molecules interact simultaneously, is represented by nonequilibrium thermodynamics, mathematical modelling, control theory and related disciplines, and it was cherished by a minority of systems-oriented scientists comprising control theoreticians, mathematical biologists and biologists engaged in quantitative analysis of metabolism. We would agree that these two distinct lines of inquiry in molecular biology are converging to form the systems biology of the twenty-first century (Westerhoff & Palsson, 2004). This convergence was not possible any earlier, because sufficiently large, comprehensive and precise sets of experimental data on molecules in living cells had been lacking. The theoretical approaches to metabolism in the 1960s and 1970s involving kinetic modelling and control theory had not become part of mainstream biology, because there was not enough data to construct and validate models comprehensively enough to explain functional phenomena completely. Because the new methodologies are genome wide, they must comprise (virtually) all molecules that are relevant. On the contrary, the convergence was not much solicited before it became evident that by themselves the resulting large data sets did not lead to the understanding of how living organisms function, expliciting the limitations to molecular biology to be addressed by systems biology (cf. above).

2.2. What is contemporary systems biology?

Development of the various high-throughput technologies used in genome sequencing, transcriptomics, proteomics and metabolomics have enabled the comprehensive analysis of complete living systems in terms of the identity and concentration of all their components (Joyce & Palsson, 2006). However, on their own these methodologies will not lead to the understanding of the living cell, hence of life, for they do not study the interactions between those molecules and their organization within the cell. Reaching such understanding will require a systems biology that is defined as the science that deciphers how biological functions arise from the interactions between components of living organisms. Because the simplest systems that are alive are unicellular organisms, systems biology studies the gap between ('dead') molecules and life, and it is on this new terrain that the contents of this book focus.

2.3. Approaches to systems biology

Two approaches to systems biology can be distinguished: Top-down and bottom-up systems biology (see Chapters 2 and 9). Top-down systems biology starts with experimental data on the behaviour of molecules in living systems as a whole. Nowadays, this is often done using high-throughput approaches to measure types and levels of (macro)molecules in the cell on a large scale, e.g. through metabolomics, transcriptomics, proteomics or fluxomics, together called functional genomics (Joyce & Palsson, 2006). In the analysis of such data, new hypotheses on the molecular organization and functioning of the organism may be induced on the basis of correlations in the behaviour of the concentrations of the molecules (Kell, 2004). In contrast, bottom-up systems biology starts from the interactive properties of the molecules and determines how these interactions lead to functional behaviour. The interactions affect or effect processes that enable a living system to develop in time or maintain its state through processes that repair damage or compensate for dissipation. In some systems, the molecular constituents are sufficiently understood to allow the construction of detailed kinetic models of reaction networks ('silicon cells') (Bakker et al., 1997; Kholodenko et al., 1999; Rohwer et al., 2000; Teusink et al., 2000; Bruggeman et al., 2005). The emergent properties are predicted by calculating how the model behaves *in silico* and compared to observations made on the system level. The lack of correspondence leads to the discovery of interactive or organizational properties that are important for biological function (e.g. Teusink et al., 1998; Bakker et al., 2000). Such properties are then inserted in new generations of the model, and eventually detailed and accurate models should be obtained. These can be used to design drugs *in silico* (Bakker et al., 2002) to engineer strains for biotechnology (Hoefnagel et al., 2002) or to better understand how molecules jointly bring about cellular behaviour (Teusink et al., 1998). Top-down and bottom-up systems biological studies are not mutually exclusive. Ultimately they should be synergistic. The top-down systems biologist exhibits the pragmatism of an engineer for which phenomenological explanations suffice, whereas the bottom-up systems biologist is more guided by a desire to understand mechanism or systems-theory principles (cf. O'Malley & Dupré, 2005). A recent book on systems biology provides more information on these two approaches to systems biology (Alberghina & Westerhoff, 2005).

3. TOWARDS A PHILOSOPHY OF SYSTEMS BIOLOGY

3.1. The philosophy of molecular biology itself needs no further elaboration

After the rise of biochemistry and molecular biology in the 1930s and 1950s and their spectacular successes throughout the second half of the twentieth century, the underlying reductionist approach has been almost beyond dispute among practitioners of these disciplines. Studying (ensembles of) individual (macro)molecules had become a rewarding enterprise. The reigning paradigm became that by breaking up the system and studying the properties of the resulting parts one should be able to understand the system fully, simply because the parts would constitute and hence determine the whole. This reductionist mindset provided a powerful research methodology, a clear epistemological guiding principle for acquiring knowledge, an effective strategy for judging the quality of manuscripts and grant applications and a consistent view of the world. Strategy and success were so clear that no further philosophizing appeared necessary.

One should put this in the perspective of those times. It had been noted that dissecting living organisms or even living cells to even the slightest extent removed virtually all properties that one associated with life (as we now know because of the interference with energy metabolism and communication, the depletion of enzymes and coenzymes, or in fact the dependence of anything on virtually everything in the organism). Therefore, it seemed obvious that looking at molecules was not going to help to understand life. The alternative of the holistic physiology that refused to look at anything other than the intact organism was also limited however. Leaving the living organism intact, one could only describe experimental results and observed behaviour in terms of phenomenological models, the components of which (e.g. growth rate) had no physical/material entity. Consequently, there was no unique dependence of behaviour on the activity of those components, not even under well-defined experimental conditions. An example would be the dependence of the energy state of the cell on growth rate. On the basis of an empirical observation, a phenomenological model would describe that dependence in a certain way. According to the philosophical foundations of physics, one should then be able to verify or falsify that description. As we now know with our molecular and systems biology understanding, one should expect this 'dependence' to be positive if a catabolic enzyme is stimulated, but negative if the ribosomes are stimulated to increase growth rate. However, knowing which of the two was activating was not an option for the holist physiological approach. Consequently, models would be falsified and nonmolecular biology obtained a bit of a reputation of an unpredictable science if a science at all.

By contrast, after complete dissection (reduction), one could define the system in terms of physical entities that one could then modulate in only one way (e.g. the concentration of an mRNA), *in vitro*. Then a reproducible scientific strategy was possible, thanks to this reductionist strategy. Who would question the reductionist mood of a winning team of molecular biologists? With this book, this is exactly what we intend to do. To be sure, we fully agree that the reductionist approach worked extremely well for some five decades, but to maintain the winning mood in the twenty-first century, it is not clear that this approach will suffice. At the turn of the twentieth century, an ever-growing number of biological scientists mention limits to reductionism (e.g. Bock & Goode, 1998; Gallagher & Appenzeller, 1999; Kiberstis & Roberts, 2002; Chong & Ray, 2002). The molecular biological revolution led to a characterization of the molecular constitution of organisms. Systems biology aims to decipher how the molecules jointly bring about cellular behaviour. The fact that the molecules are supposed to do this jointly suggests that studying them only individually without a focus on their interactions may not work. On the other hand, it is clear that a return to the holist physiology strategy will not work either. Perhaps some new strategy is needed, with unique philosophical foundations.

3.2. Philosophers focus on philosophy of evolutionary biology

Up to the 1960s' the philosophy of science hardly explored the nature of sciences other than physics (Machamer, 2002). Philosophy of biology came of age only in the latest three decades (Callebaut, 2005). To date, philosophers of biology have mainly engaged in discussions about (i) the autonomy of biology (ii) evolutionary biology and (iii) molecular (functional) biology. As Mayr (1996) noted, the position of biology among the sciences has been the most prominent and controversial issue of the philosophy of biology; some consider biology an extension of chemistry and physics, others claim the autonomy of biology. The position of biology among the sciences as an autonomous science is by and large safeguarded by the peculiar characteristics of the living world, as encountered in functional biology, but particularly in evolutionary biology (Mayr, 1961; Mayr, 1996). The potential reduction of classical (Mendelian) genetics to molecular genetics has been the subject of philosophical discussions on reduction in molecular biology (Sarkar, 1998). And then, philosophers of biology have argued about biological functions, and about causal and functional explanations (e.g. Mahner & Bunge, 1997; Wouters, 2005). But the philosophy of biology is still incomplete; we cannot as yet deal satisfactorily with the relationship between a living system and its functional organization, i.e. how molecules in action determine the characteristics of living systems. Evolutionary biology studies how living systems came to be, whereas systems biology studies how living systems are; a biology of becoming versus a biology of being. This is a profound difference.

3.3. A philosophy of systems biology is lacking but needed

Experimental biologists may not care much about philosophy, while philosophers of biology deal predominantly with evolutionary biology and, to a much lesser extent, with molecular biology and, importantly, not at all with systems approaches to biology. Philosophy of biology is thereby concerned with a biology which is descriptive rather than predictive, a biology which asserts that an understanding of the living world will come from descriptions of the histories of organisms, in particular their genetic history, and from a catalogue of the molecules of which organisms are composed. In view of what seems to have become the mainstream of biology since the advent of biochemistry, molecular biology and functional genomics, this philosophy of biology may be incomplete and therefore unsatisfactory. Indeed, the existing philosophy of biology fails to address the rather profound issue of what distinguishes the living from the nonliving, except to say that something lives because its ancestors lived. In its current state, the philosophy of biology does not have the wherewithal to consider a single organism as an integrated, functionally organized system that can be understood per se, independent of its evolutionary history.

The issue has become acute only recently. Physiology aimed at understanding living organisms as wholes, but not in terms of their molecules. Molecular biology aimed at understanding living organisms as sums of the properties of their individual components without allowing for the emergence of additional properties in their interactions. Only systems biology takes up the challenge of understanding living organisms as wholes, in terms of integrals of their interacting and organizing constituents. It aims to predict systemic behaviour of organisms from their constituent processes by merging the molecular and cellular levels, rather than by describing them independently as is done in evolutionary and molecular biology. This surely has profound philosophical implications for biology as a whole.

The science of systems biology addresses the issue of what life is, irrespective of its origin, but for this the philosophical foundations appear to be lacking. We believe this challenge should be taken up.

4. INTRODUCTION OF A NUMBER OF PHILOSOPHICAL ASPECTS OF SYSTEMS BIOLOGY

4.1. Two types of reductionism

Most biologists use reductionist strategies in some way or another. Often they are indifferent to the philosophical aspects of their methodologies, theories and explanations. Quinn highlighted the attitude of most scientists towards reductionism by paraphrasing Mark Twain's comment on the weather:

> "Mark Twain once said about the weather 'everybody talks about it, but nobody does anything about it'. It seems to me that the inverse applies to reductionism: everybody does it, but nobody wants to talk about it!"
>
> (Bock & Goode, 1998, p. 218)

The notion of reductionism in connection to biology has attracted the attention of at least some biologically oriented scientists and of philosophers of science for several decades (e.g. The Alpbach Symposium in 1968 and the Novartis Foundation Symposium in 1998 (Koestler & Smythies, 1969; Bock & Goode, 1998)).

The term 'reductionism' is not unambiguous; it is used to refer to (i) the research strategy of trying to understand complex systems by studying their parts and (ii) the claim that a particular scientific theory has been reduced to another theory according to a particular paradigm of how theories can be reduced to each other. Examples of the latter claim are the reduction of thermodynamics to statistical mechanics and the present efforts to reduce the four interactions (forces) of physics to a single underlying one. Reductionism in the former sense is quite distinct from that in the latter. The context will determine which type of reductionism is being referred to in a particular instance. In both cases though, the natural world is thought to be somehow stratified in higher and lower levels. Living systems (wholes) are invariably composed of components (parts), where the parts in their turn can each be considered as wholes themselves that consist of even smaller parts, and so on (e.g. cells are composed mainly of macromolecules such as enzymes and enzymes are composed of amino acids). Entities of higher levels are thought to be composed of entities that populate lower levels. Although their notions of what constitutes a level differ, most scientists and philosophers agree that the implementation of some concept of level or hierarchy is indispensable to understanding the natural world. Given this basic organization of the world, some have argued that it should be possible to reduce entire disciplines at higher levels to disciplines at lower levels (e.g. to reduce biology to chemistry which is then reduced to physics, or to reduce physiology to biochemistry or to reduce Mendelian genetics to molecular genetics). Eventually this should lead to a 'theory of everything' (Nagel, 1979). Theory reduction would also result in the growth of scientific knowledge. The discussion whether intertheoretic reduction can be a criterion for the progress of biological sciences lasted for several decades of the last century. It led to the consensus that the existing paradigms for intertheoretic reduction are inadequate, especially for biology.

The nested whole-part relationship of living systems immediately suggests a reductionist research strategy and this refers back to the first type of reductionism mentioned above: break the whole apart and study the parts in isolation. Radical reductionists are convinced that a complete knowledge of the parts will give the behaviour of the whole without further ado. This view on reality can be illustrated with some hard-boiled reductionist adages, such as 'an organism is essentially nothing but a collection of atoms and molecules' (Crick, 1966), 'there is only one science, physics, all the rest is social work' (Watson in conversation with Rose, 1994), and 'all science is either physics or stamp collecting' (attributed to Rutherford).

In the alternative antireductionistic or holistic approach, complex wholes are considered not to be understandable from the mere knowledge of the behaviour of the parts in isolation. A holist rather takes the opposite stance; only properties of the system as a whole may offer understanding. In many cases, the antireductionist argues that the context of the whole entirely determines the behaviour of the parts and that the behaviour of the parts within the system is qualitatively different from their behaviour in isolation. A crucial aspect of the context of the organism is that the parts almost invariably engage in nonlinear interactions. In nonlinear interactions, qualitatively new properties can arise, depending on the state the system is in, as the strength of the interactions vary with that state. Consequently, the emergence of new properties from the interaction of two components, depends on the activity of other components of the system, possibly many other components if not the entire genome. These new properties and the conditionality of when they arise compromise the statement that the properties of all parts in isolation may provide a sufficient base for prediction of the properties of the whole. Such a base is only provided if the properties of the parts include their interactive properties and the state of the system is given (Bruggeman et al., 2002; Boogerd et al., 2005). Here it is important that the state a living system is in is not an equilibrium state, but is maintained by the continued activity of a number of dissipative processes. The state is a manifestation of the functioning whole. Synthesis is not simply the reverse of analysis, and the whole should be taken into account, if only because it sets the condition of the system including the type, number and relative position of its components.

Antireductionists emphasize that the whole should be taken into account, but usually fail to explicate precisely how (this point has been made formally by Rosen (1991) in *Life Itself*). The above implies that all levels of organization have their value. It also suggests that there is use for autonomous theories at each level. At its extreme, radical antireductionism asserts a priori that the whole can only be meaningfully studied at the level of the whole: any reference to parts is useless. Breaking up the whole and studying its parts would never return systemic behaviour and only distracts the scientists from engaging in proper science. This view on reality can be illustrated with the hard-boiled holistic adage; the whole is always qualitatively, irreducibly, different from the parts. The issue of the limits of reductionism and of antireductionism is central to systems biology (Bruggeman et al., 2002). Living systems, being nonlinear dynamical systems, have properties different from their constituents in isolation, properties which emerge from the interactions among the molecular constituents; accordingly, it is the organization of these intermolecular processes in organisms that underlies their characteristic living properties. A reductionist or antireductionist strategy alone does not do justice to this claim. A new strategy seems needed, and this is one of the main themes elaborated on in this book.

4.2. A continuum of reductionism to antireductionism

Most biologists will not adhere to the extreme versions of reductionism (microscopic 'nothing but' statements) or antireductionism (macroscopic 'all or nothing' statements). The radical views represent the two extremes of a continuum of positions (see Philosophical Psychology Symposium, 2002; Laughlin et al., 2000). Systems biology does not make an a priori choice for an exclusively reductionist or holistic view on life, but it strongly advocates taking a middle position to be dictated empirically by the particular case in question (Bruggeman et al., 2002).

Such a systems biological approach, an integrative interlevel approach, that combines both molecular and systems properties and that attempts to explain the systemic behaviour of organisms in terms of their functional organization appears promising. Within the evolving discipline of biology, it makes sense for systems biology to succeed molecular biology. But again, the precise methodologies to be followed and the status of the scientific theories that ensue are unclear. They will be discussed in this book.

4.3. Types of explanation

Traditionally, within the logical positivists' framework, scientific theories were taken to consist of coherent networks of laws, which were universal, general and necessary. A particular behaviour was considered explained if it could be derived from general laws plus the initial conditions of the particular case (Hempel & Oppenheim, 1948; Nagel, 1979). However, the received view at present is that the formal deductive nomological way of explanation (the DN-model) is not applicable to biology because of several reasons: (i) the logical-positivists' interpretation was mainly developed for physical theories, without considering the idiosyncracies of the animate world, (ii) biological laws, if there are any, do not necessarily apply throughout the universe (since they are domain-specific), they are not exceptionless, and they are contingent, (iii) the DN-model appears to refer to a finished and ideal science (Railton, 1978), whereas biological theories are almost always 'in progress'.

In response to the inadequacy of the DN-model, alternative modes of explanations were constructed (e.g. Railton, 1978; Salmon, 1989). Currently, in the philosophy of science, there are two mainstream ideas about what counts as a scientific explanation: the unificationist and the causal/mechanical type of explanation. The first view, which is in many ways a descendant of the DN-model, also endeavours to unify disparate phenomena observed at different levels of organization by finding laws at the lower level that together with initial conditions yields an explanation of the phenomenon at the higher level (e.g. Kitcher, 1985). It seeks the unity of science as much as possible, but at the same time it avoids ending up with a 'theory of everything'. Systems theoretic-oriented systems biologists in particular emphasize that finding more general principles, if not laws, is of utmost importance in systems biology; in this sense they are unificationists (e.g. Hornberg et al., 2005). The second view is the causal/mechanical type of explanation (Wimsatt, 1974; Railton, 1978; Salmon, 1984; Bechtel & Richardson, 1993; Machamer et al., 2000; Craver, 2001; Woodward, 2003). This view is much more widespread in biology. In the next section, we will introduce this type of explanation. There is no a priori reason though why systems biology should not engage in both types of explanation, but again this is an issue to be discussed in this book.

4.4. Mechanistic explanation

In biology, a particular type of a reductive interlevel explanation, which is related to the causal/mechanical view of reality, is called mechanistic explanation and it deserves some special attention as it seems to capture many explanations found in systems biology. This kind of explanation has long been neglected by philosophers of biology, but as from 1990 the philosophical interest in mechanisms and mechanistic explanations is on the increase. The growing interest also shows up in this book, since several authors extensively deal with this issue (see chapters of Bechtel; Westerhoff & Kell; Richardson & Stephan; Schaffner; Shulman). The essential difference between a mechanistic explanation and a reductionistic explanation lies at the heart of systems biology. In the sciences, reductionism is mostly understood as a means to explain phenomena generated by systems in terms of the properties of their parts, often when considered in isolation. If complex systems that have systemic properties brought about by the interactions of their parts are considered, where none of the parts in isolation display similar properties, then reductionism is a not a fruitful strategy to explain phenomena. It is only satisfactory for a description of the system's parts and their properties; it yields only a parts list, an inventory of the parts. To explain behaviour, the parts have to be considered in their natural context, i.e. as components in the functioning system, and at the same time, the system has to be studied as a whole. Often the kind of explanation sought for is a mechanistic explanation which tries to make intelligible how the joint behaviours of the parts, while embedded in the system, bring about the behaviour of the system that is to be explained. Mechanistic explanation, therefore, is an interlevel activity that does not cherish the system more than the parts or vice versa; that is to say, it is neither holistic nor reductionistic. In the end, a mechanistic explanation of a cellular behaviour amounts to a quantitative, mechanistic model that captures those molecular phenomena in the cell responsible for the cellular behaviour that is be explained.

The search for and articulation of mechanisms also has methodological implications. Various experimental strategies can be followed to go from 'possible' to 'plausible' to 'actual' mechanisms (Darden & Tabery, 2005). Such strategies have three basic elements: (i) an experimental set-up in which the mechanism is running, (ii) an intervention (perturbation) technique and (iii) a detection technique (Craver, 2002). Basically, intervention strategies may involve activating the normal working of the mechanisms or modifying the working of the mechanism, and in both cases, detecting some downstream effect. Systems biology stresses the idea that besides the qualitative aspects of mechanisms, represented by cartoons of mechanism, quantification of the behaviour of mechanisms is of utmost importance. It is noted that many scientists do not pay much attention to experimental quantification as a means to find plausible and actual mechanisms. On average, molecular biology has a tendency to articulate mechanisms only in qualitative terms. It is stressed here that combined experimental–theoretical approaches as carried out in systems biology allow for such a quantitative analysis of the systemic behaviour upon perturbations made to the system (e.g. Jensen et al., 1993; Bakker et al., 1997; Hoefnagel et al, 2002; Snoep et al., 2002; Hornberg et al., 2005; Bruggeman et al., 2005). Also, most philosophers of science do not engage in discussions about quantitative, computational aspects of mechanisms. They almost invariably refer to mechanisms in qualitative terms. Systems biology, however, strongly focuses on this quantitative aspect and is therefore also fundamentally different from molecular biology in this respect. The underlying reason is that in nonlinear systems (i) the actual operation of a mechanism depends on the precise state the system is in and (ii) more than one mechanism may operate at the same time.

4.5. Systems biology and models

Living organisms are complex systems. Cellular behaviour arises though the action and interaction of thousands of molecules and macromolecules.

An interesting problem then arises: 'How should one speak about causality in complex systems involving many interacting factors?'. Also because cells are complex systems acting as integrated wholes, linear chains of causes and effects are virtually absent and 'circular' causality abounds (this volume Hofmeyr; Westerhoff & Hofmeyr in Alberghina & Westerhoff, 2005). Our intuitive understanding of phenomena cannot cope with the overwhelming complexity of nonlinear interactions among cellular components. The use of mathematical models is then an indispensable tool to tackle this complexity systematically (Westerhoff & Kell, this volume; Wolkenhauer & Ullah, this volume; Bruggeman et al. in Kriete & Eils, 2006). Such models may also be a much more appropriate domain of an investigation into the nature of causality than description in everyday language (e.g. Wagner, 1999). When computations are done with *in silico* models of living systems, integrating all the known molecular properties of such systems, surprising counterintuitive results appear, results that would not have been anticipated without the integration of the molecular properties into mathematical models (Noble, 2002; Westerhoff & Kell, this volume). Examples of such models can be found in the silicon cell's live model base on the Internet (Snoep & Olivier, www.siliconcell.net). A similar statement can be made for the counterintuitive finding of metabolic control analysis that there is not necessarily a 'rate-limiting' step in a metabolic pathway as was generally believed, but that control tends to be distributed among various steps (enzymes) of the pathway (Fell, this volume). Reconstruction of system's behaviour from the behaviours of the constituent parts can, in principle, be done in two ways: *in vitro* reconstitution and *in silico* modelling. Although impressive results have been obtained, *in vitro* reconstruction of system's behaviour is experimentally rather difficult and laborious. *In silico* modelling through computational integration of the properties of and interactions among parts is more promising (Westerhoff & Kell, this volume; Wolkenhauer & Ullah, this volume). The crux here is that for integrative computation, binding and kinetic data of all the parts should be determined such that their values equal, or at least approach, their values *in vivo*, at the operating state (see also above). Experimental data that are obtained *in vitro* are not necessarily the same as those that are valid *in vivo*, unless the conditions and the states the components are in, are identical. Models may be used as a hypothesis generator or predictor for desired perturbations, but become much more powerful after validation through combined experimental–theoretical approaches.

4.6. What is life?

Biology is the science of living systems and many, if not all, scientists and philosophers will agree to the statement that 'the cell is the smallest unit of life'. Although molecules constitute living systems, they are by themselves not alive. No matter how thin the dividing line is, there is qualitative jump between a living and a nonliving system (cf. Mahner & Bunge, 1997). Therefore, much of biology and systems biology starts with cell biology.

In a classical paper, Mayr argued that "the word biology is a label for two largely separate fields which differ greatly in methods, Fragestellung and basic concepts . . . which may be designated functional biology and evolutionary biology" (Mayr, 1961). This book sets out to highlight the systems biological approach of functional biology, and, as a consequence, comparatively little attention is given to evolutionary biology (but see Chapter 5). In other words, the book mainly focuses on the living processes occurring in extant life forms. However, the quest for minimal life, i.e. the smallest unit of life among autonomous cells, is an important and interesting experimental and conceptual issue within systems biology (see chapters by Westerhoff & Kell; Keller; Bechtel). In this respect, also the transition from inanimate to animate matter, i.e. the origin of life, is a topic that is highly relevant to systems biology.

Most often, in the literature as well in as in biology textbooks, life is described in terms of a list of purported characteristics (e.g. Koshland, 2002; Duve de, 1991). Such lists differ among each other with respect to the number and the nature of the characteristics. It seems always possible to find exceptions, i.e. systems that appear to be living but do not exhibit all of the characteristics mentioned in the list. These lists also invariably display appreciable overlap and it seems very well possible to distill a 'consensus' list of the essential features of the living state. Giving an unambiguous definition of life, however, is a much more difficult task and has proven to be an enigmatic problem. Most authors refrain from attempting this and confine themselves to presenting a list of vital characteristics. In this book, the enigmatic problem of the question 'What is life and how did it originate' is addressed by Hofmeyr (Chapter 10), Moreno (Chapter 11), and Bechtel (Chapter 12).

5. AIM AND OVERVIEW OF THE BOOK

This book attempts to provide discussion material that should lead to a better definition of the ways in which systems biology should be carried out. Thus it should provide philosophical foundations to this new discipline. The book is divided into this introductory chapter, three major sections and a concluding Chapter 14. The sections are each dedicated to a major theme. The first sets the scene in terms of describing systems biology research programs (Chapters 2–5). The second discusses theory and models (Chapters 6–9). The third of these sections deals with organization in biological systems (Chapters 10–13). All authors except Callebaut and Krohs presented a first form of their contribution at a symposium that was held at the Vrije Universiteit (the Netherlands, Amsterdam) from June 2–5 in 2005. The chapters in this book are the result of the symposium and contain the adjustments the authors made as a result of those discussions.

REFERENCES

Alberghina L & Westerhoff HV (Eds.) *Systems Biology: Definitions and Perspectives* (Topics in Current Genetics), Springer-Verlag Berlin and Heidelberg GmbH & Co, 2005.

Bakker BM, Assmus HE, Bruggeman FJ, Haanstra FJ, Klipp E & Westerhoff HV. *Network-based selectivity of antiparasitic inhibitors.* Molecular Biology Reports: 29, 1–5, 2002.

Bakker BM, Mensonides FIC, Teusink B, van Hoek P, Michels PAM & Westerhoff HV. *Compartmentation protects trypanosomes from the dangerous design of glycolysis.* Proceedings of the National Acadamy of Sciences USA: 97(5), 2087–2092, 2000.

Bakker BM, Michels PAM, Opperdoes FR & Westerhoff HV. *Glycolysis in bloodstream form* Trypanosoma brucei *can be understood in terms of the kinetics of the glycolytic enzymes.* Journal of Biological Chemistry: 272(6), 3207–3215, 1997.

Bechtel W & Richardson RC. *Discovering complexity. Decomposition and localization as strategies in scientific research.* Princeton University Press, Princeton, New Jersey, 1993.

Bock GR & Goode JA (Eds.) *The Limits of Reductionism in Biology.* (Novartis Foundation Symposium 213, John Wiley, London, 1998.

Boogerd FC, Bruggeman FJ, Richardson RC, Stephan A & Westerhoff HV. *Emergence and its place in nature: A case study of biochemical networks.* Synthese: 145, 131–164, 2005.

Bruggeman FJ, Bakker BM, Hornberg JJ & Westerhoff HV. *Introduction to Computational Models of Biochemical Reaction Networks.* In: Computational Systems Biology, Kriete A & Eils R (Eds.), Chapter 7, Elsevier Academic Press, Burlington, MA, 2006.

Bruggeman FJ, Boogerd FC & Westerhoff HV. *The multifarious short-term regulation of ammonium assimilation of* Escherichia coli*: dissection using an in silico replica.* FEBS Journal: 272, 1965–1985, 2005.

Bruggeman FJ, Westerhoff HV & Boogerd FC. *BioComplexity: a pluralist research strategy is necessary for a mechanistic explanation of the "live" state.* Philosophical Psychology: 15(4), 411–440, 2002.

Callebaut W. *Again, what the philosophy of biology is not.* Acta Biotheoretica: 53, 93–122, 2005.

Carnap R. *Philosophical Foundations of Physics*, Basic books, New York, 1966.

Chong L & Ray LB. Whole-istic Biology. Science: 295(5560), 1661, 2002.

Craver CF. *Role Functions, Mechanisms and Hierarchy.* Philosophy of Science: 68, 31–55, 2001.

Craver CF. *Interlevel Experiments, Multilevel Mechanisms in the Neuroscience of Memory.* Philosophy of Science: 69 (*Supplement*), S83–S97, 2002.

Crick FHC. *Of molecules and man.* Washington University Press, Seattle, WA, 1966.

Darden L & Tabery J. *Molecular Biology.* The Stanford Encyclopedia of Philosophy (Spring 2005 Edition), Zalta EN (ed.), URL=http://plato.stanford.edu/archives/spr2005/ entries/molecular-biology.

Duve de C. *Blueprint for a Cell: The Nature and Origin of Life.* Neil Patterson Publishers, Carolina Biological Supply Company, Burlington, 1991.

Gallagher R & Appenzeller T. *Beyond reductionism.* Science: 284(5411), 79, 1999.

Hempel CG & Oppenheim P. *Studies in the Logic of Explanation.* Philosophy of Science: 15, 135–175, 1948.

Hoefnagel MHN, Starrenburg MJC, Martens DE, Hugenholtz J, Kleerebezem M, Van Swam II, Bongers R, Westerhoff HV & Snoep JL. *Metabolic engineering of lactic acid bacteria, the combined approach: kinetic modelling, metabolic control and experimental analysis.* Microbiology: 148, 1003–1013, 2002.

Hornberg J, Binder B, Bruggeman FJ, Schoeberl, Heinrich BR & Westerhoff HV. *Control of MAPK signalling: from complexity to what really matters.* Oncogene: 24, 5533–5542, 2005.

Jensen PR, Michelsen O & Westerhoff HV. *Control analysis of the dependence of* Escherichia coli *physiology on the H^+-ATPase.* Proceedings of the National Acadamy of Sciences USA: 90, 8068–8072, 1993.

Joyce AR & Palsson BO. *The model organism as a system: integrating 'omics' data sets.* Nature Reviews on Molecular Cell Biology: 7(3), 198–210, 2006.

Kell DB. *Metabolomics and systems biology: making sense of the soup.* Current Opinion in Microbiology: 7(3), 296–307, 2004.

Kholodenko BN, Demin OV, Moehren G & Hoek JB. *Quantification of short term signaling by the epidermal growth factor receptor.* Journal of Biological Chemistry: 274(42), 30169–81, 1999.

Kiberstis P & Roberts L. It's not just the genes. Science: 296(5568), 685, 2002.

Kitcher P. *Two approaches to explanation.* The Journal of Philosophy LXXXII: 632–639, 1985.

Koshland Jr DE. *The seven pillars of life.* Science: 295, 2215–2216, 2002.

Koestler A & Smythies JR (Eds.) *Beyond Reductionism – New Perspectives in the Life Sciences.* The Alpbach Symposium 1968, Hutchinson, London, 1969.

Laughlin RB, Pines D, Schmalian J, Stojkovic BP & Wolynes P. *The middle way.* Proceedings of the National Acadamy of Sciences USA: 97(1), 32–37, 2000.

Machamer P. *A brief historical introduction to the Philosophy of Science.* In: Blackwell Guide to the Philosophy of Science (Eds. Machamer P & Silberstein M), pp. 1–17, 2002.

Machamer P, Darden L & Craver CF. *Thinking About Mechanisms.* Philosophy of Science: 67, 1–25, 2000.

Mahner M & Bunge M. *Foundations of Biophilosophy.* Springer-Verlag, Berlin, 1997.

Mayr E. *Cause and effect in biology: Kinds of causes, predictability, and teleology are viewed by a practicing biologist.* Science: 134, 1501–1506, 1961.

Mayr E. *The autonomy of Biology: The position of biology among the sciences.* Quarterly Review of Biology: 71, 97–106, 1996.

Nagel E. *The Structure of Science: Problems in the Logic of Scientific Explanation.* Cambridge, Hackett publishing company, 1979.

Noble D. *Modeling the heart—from genes to cells to the whole organ.* Nature: 295, 1678–1682, 2002.

O'Malley MA & Dupré P. *Fundamental issues in systems biology.* BioEssays: 27(12), 1270–1276, 2005.

Philosophical Psychology Symposium. *Inter-level relations in computer science, biology , and psychology*, volume 15 (4) 381–473, 2002.

Railton P. *A Deductive-Nomological Model of Probabilistic Explanation.* Philosophy of Science: 45, 206–226, 1978.

Railton P. *Probability, Explanation, and Information.* Synthese: 48, 233–256, 1981.

Rohwer JM, Meadow ND, Roseman S, Westerhoff HV & Postma PW. *Understanding glucose transport by the bacterial phosphoenolpyruvate:glycose phosphotransferase system on the basis of kinetic measurements in vitro.* Journal of Biological Chemistry: 275(45), 34909–21, 2000.

Rosen R. *Life Itself: A Comprehensive Inquiry into the Nature, Origin, and Fabrication of Life.* Columbia University Press, New York, 1991.

Salmon W. *Scientific Explanation and the Causal Structure of the World.* Princeton University Press, Princeton, 1984.

Salmon W. *Four Decades of Scientific Explanation*. University of Minnesota Press, Minneapolis, 1989.

Sarkar S. *Genetics and reductionism*. Cambridge Studies in Philosophy and Biology, Cambridge University Press, Cambridge, 1998.

Snoep JL, van der Weijden CC, Andersen HW, Westerhoff HV & Jensen PR. *DNA supercoiling in* Escherichia coli *is under tight and subtle homeostatic control, involving gene-expression and metabolic regulation of both topoisomerase I and DNA gyrase*. European Journal of Biochemistry: 269, 1662–1669, 2002.

Teusink B, Passarge J, Reijenga CA, Esgalhado E, van der Weijden CC, Schepper M, Walsh MC, Bakker BM, van Dam K, Westerhoff HV & Snoep JL. *Can yeast glycolysis be understood in terms of in vitro kinetics of the constituent enzymes? Testing biochemistry*. European Journal of Biochemistry: 267(17), 5313–29, 2000.

Teusink B, Walsh MC, van Dam K & Westerhoff HV. *The danger of metabolic pathways with turbo design*. Trends in Biochemical Sciences: 23, 162–169, 1998.

Wagner A. *Causality in complex systems*. Biology and Philosophy: 14, 83–101, 1999.

Watson JD, in conversation with Rose, Steven, 1994. Quoted by Rose S in *Biology Beyond Reductionism*. p.vi., UK: Open University Press, 1996.

Westerhoff HV & Palsson BO. *The evolution of molecular biology into systems biology*. Nature Biotechnology: 22, 1249–1252, 2004.

Wimsatt WC. Reductive explanation: A functional account. In: Michalos AC, Hooker, CA. Pearce G & Cohen, RS (Eds.) PSA-1974 (Boston Studies in the Philosophy of Science: 30, 671–710, 1974.

Woodward J. *Making things happen: A theory of causal explanation*. Oxford University Press, Oxford, 2003.

Wouters A. *The function debate in philosophy*. Acta Biotheoretica: 53, 123–151, 2005.

SECTION II

Research programs of Systems Biology

2

The methodologies of systems biology

Hans V. Westerhoff and Douglas B. Kell

SUMMARY

In this book on philosophical aspects of systems biology, this chapter summarizes the philosophical status of a variety of sciences. Biology, physics and molecular biology offer particular contrast here. It is contended that philosophy and methodology should be determined substantially by the degree of complexity of the system under study. Some of the new experimental methods that have made systems biology possible are summarized. Research strategies that claim to be systems biology yet approach the topic in different ways are described. Inductive reasoning and the development and exploitation of suitable technologies are important parts of the systems biology agenda but are not themselves hypothesis-dependent science. A new methodology for systems biology is sketched that spirals in an iterative manner between experiments and theory but makes inherent use of mathematics in ways that are new to the life sciences. It is shown that the construction of a computer replica of parts of living systems has become possible and that the 'silicon cell' strategy enables the calculation of emergent properties. This may then serve as a basis for subsequent discussions with philosophers of science about how new and unique the philosophical foundations of systems biology are or should be.

1. THE METHODOLOGY AND PHILOSOPHICAL FOUNDATIONS OF THE VARIOUS SCIENCES

1.1. Physics

According to classical philosophy of science (e.g. Carnap, 1966; Nagel, 1961), science advances by an iteration between the world of mental constructs (ideas,

Systems Biology
F.C. Boogerd, F.J. Bruggeman, J.-H.S. Hofmeyr and H.V. Westerhoff (Editors)
Isbn: 978-0-444-52085-2

background knowledge, hypotheses) and the world of sense data (experimental observations). Laws (theories, hypotheses) are induced from empirical findings (Carnap, 1966). Consequences deduced by combining hypotheses with established underlying principles (such as fundamental laws of chemistry and physics) are examined experimentally to test the new hypotheses (see also Fig. 3). Given sufficient positive testing, they are transformed to underlying principles through theorization. For testing, theories should be quantitative (Carnap, 1966). It is seen as a great asset when laws and theories can also be *reduced* to underlying theories of greater validity and generality. Here thermodynamics has always served as an example; its first and second laws were first determined empirically (Nagel, 1961). The former was then elevated to a general scientific law that is also valid at the more microscopic level. The latter was deduced from the underlying principle of large numbers of substates and evolution towards increased probability with time. Quantum mechanics has also served as such an example: Schrödinger's equation and wave functions were 'induced' so as to be able to explain observations, such as the periodicity in the Table of Chemical Elements. Modern elementary particle physics appears to continue along these lines, ever inducing new phenomena and properties such as quarks, charms and colours. More generally, physics aims to explain multiple phenomena on the basis of simpler and fewer principles. Indeed, the first law of thermodynamics is much simpler than the 100% efficient conversion between all sorts of energy that it prescribes. In the classical philosophy of science, explanation by simple underlying principles is important (cf. Nagel, 1961, p. 321).

Of course, this philosophy of science is incomplete. It is very often too simplistic to state that science deduces predictions from hypotheses that can be verified. Indeed, it is seen in most quarters as much more important to try to make predictions that can then be used to falsify hypotheses (Popper, 1992). Then in practice, the sociology of science also comes in, where hypotheses are not actually falsified by their originators, but rather by competing, younger researchers, albeit only after the proponents of the original hypothesis have become less active or passed on (cf. Kuhn, 1996; Lakatos, 1978; Primas, 1981). However, this is not the issue we would like to discuss here, as we shall focus on the extent to which classical, molecular and systems biology do conform to what used to be defined as science by the main philosophers of science, or more specifically physics (Carnap, 1966).

1.2. Biology

While theoretical physics is both respectable and a major part of the activities of physicists, theoretical biology is a minor part of modern biology and is treated largely with disdain by most experimentalists (Kell, 2006). Not all of classical biology conformed strictly to the scientific methods recalled above, as it was largely observational (Brent, 1999). Much of that science of biology accepted the diversity that appeared to inhabit the biosphere: organisms were classified and compared, and their behavior was studied in the sense of establishing correlations between properties. These correlations were rarely put to the test in the sense of falsification or even verification; observations were dominant; laws, even phenomenological ones, were rare. Classificatory concepts sufficed (Carnap, 1966).

Physicists were much stricter; they expected their codifications to produce immutable laws. Thus, the type of biology being studied caused many physicists to disdain biology, which would then be seen as an 'other science' if a science at all. Biology was 'stamp collecting', and it was claimed that physics was superior. 'Science is either physics or stamp collecting' is a statement attributed to Rutherford. Those who have witnessed field biologists efficiently recognizing birds in complex ecosystems, and predicting with an 80% success rate what the individual birds would do next, are perhaps less convinced of the truth of the dictum of the physicists. After all, the complexity of the prediction made by the biologist and what one might consider the total success of that prediction (i.e. success rate multiplied by complexity) was many times higher than that of the physicist predicting the probability of the location of an electron on the basis of a wavefunction. Interestingly, chemistry and biochemistry have always been middlemen; although chemistry was claimed to be a science conforming to the principles proclaimed by the philosophers of science, it often was not; organic chemistry, for instance, was rule-based rather than theory-based, albeit fairly successful in predicting possible chemical reactions and reaction mechanisms. Chemistry warrants its own philosophy of science, distinct from that of physics (Primas, 1981).

We suggest that the basic problem of biologists at that time, and to some degree now, which distinguished it from the objects of study surveyed by physicists, was that the object of their study, i.e. life, was too complex to be amenable to the 'Physics' of Rutherford. The number of unobserved and in fact unobservable degrees of freedom was virtually unlimited. Every possible hypothesis would always be falsifiable, as there could always be exceptions, or additional unknown components of the system that would perturb the rule (the 'hidden variables' of certain approaches to understanding the behaviour of quantum systems). Even Mendel's 'laws' were subject to many exceptions, and it is now all too easy to scorn Mendel for overemphasizing the overall principles and for down playing the aberrations (it is also widely accepted that Mendel's data were 'too good to be true'). What would have happened with Newton discovering the laws of classical mechanics had the velocity of light been 0.1 m/s? Then Newton would have been plagued by apparent exceptions (because of relativistic corrections). Or what would have happened if all the objects around us had had substantial Coulombic charge, so as to perturb the observation of $F = ma$, in those days at least?

Classical (Organismal (Nagel, 1961)) biology was (and is) a science in that it obeyed strict methods, was devoid of unfounded predictions and aimed for reproducibility. It was, however, seen as incomplete in that its predictions were often perturbed by unexpected variations. On the contrary, it did not shy away from studying the complex and the most interesting phenomena in existence, i.e. life.

Much (though not all (Primas, 1981)) of physics did conform to the scientific methods delineated by the classical philosophers of science. How could it? Well, first of all it studied objects that happened to be simpler than the objects studied by biology; billiard balls, protons and electrons are inherently simpler than haemoglobins, monkeys and tumor cells. Certainly, it has been an extreme challenge to mankind to understand the circling of electrons around conglomerates of protons and neutrons, but the scientific achievements have been enormous. However, the number of degrees of freedom involved in the explanations of physics has been much smaller than the number of degrees of freedom in the objects of biology. Physicists (and engineers) sought this simplicity; they preferred to study single objects or systems with very few degrees of freedom, and preferably linear interactions. This enabled the discovery of simple principles and their codification by analytical mathematics. Physics could be physics and not stamp collecting, precisely because physicists selected a particular subset of stamps rather than the most beautiful and extensive stamp collections as objects of study.

This focus on simpler systems and the emphasis on simple principles, often enforced by first- and perhaps second-order linear approximations, have been very good for the development of science. Enormous progress was made for those objects of study that were simple in the above sense. Doubts arose when others noted that many problems in the environment around us were not being solved by physics. These included the weather, the behavior of the stock market, the behavior of the majority of (nonideal) gases, and life and disease.

When confronted with those issues, some physicists reversed the argumentation. It was not physics itself that was unfit to study those systems that were more complex. Rather, those objects of studies were unfit for pure physics; they might perhaps be studied by applied, less pure physics, perhaps through simulation of all the special cases. Nonequilibrium thermodynamics of the Westerhoff (Westerhoff & van Dam, 1987) type, nonequilibrium statistical mechanics of the Keizer type (Keizer, 1987) and later the discovery of deterministic chaos (e.g. Gleick, 1988) were such 'impure' physics. On the contrary, they demonstrated that many aspects of reality may be beyond the understanding of simpler physical theory. Prigogine was a case in point, searching for a general principle of nonequilibrium steady states in arbitrary systems, which does not exist (Nicolis & Prigogine, 1977). Some physicists moved towards biology, accepting that physics itself should change and adopt complexity. Terrell Hill is one of these, being attracted to biology because its phenomena were inherently interesting and developing physics methods so as to be able to deal optimally with its complexity (Hill, 1977). Much of modern physics of course does accept the complexity and is subject to the limitations of nongenerality and nonlinearity plaguing biology (Fröhlich & Kremer, 1983; Primas, 1981). In this sense, we admit that we here caricature physics to serve as a contrast in a description of the essence of systems biology.

1.3. Biochemistry and molecular biology

Whilst it was welcome that physics was able to deal so elegantly with a number of phenomena, the problem for science was that much of what is inherently interesting to mankind appeared to be left intractable. Life itself, in the sense of understanding the material basis of the functioning of living organisms, therewith eluded the science that followed the methodology of physics (Rosen, 1991). There could be only two ways out of this dilemma: either physics adapted to life as an object of study, or the object of study, 'life' was adapted to the methodology of physics (perhaps with new, superphysical laws to be added, as in Schrödinger's agenda (Schrödinger, 1944, p. 80)). The latter strategy has been the basis of yet another success story, i.e. that of biochemistry, biophysics and molecular biology. It was indeed set in motion by physical scientists such as Michaelis and Menten, Franklin, Watson and Crick. Michaelis and Menten set out to study the reaction catalyzed by a single protein, while Franklin, Watson and Crick looked at a piece of a double-stranded DNA molecule. The molecular processes carried out by macromolecules in living organisms were characterized in this manner. In addition, simple and qualitative schemes of how they function together were drawn as cartoons (such cartoon-based modelling was and is a significant part of these sciences (Kell & Knowles, 2006)). This includes the one showing that a piece of DNA contains the inheritable information, which can be expressed through mRNA into proteins, which then carry out function by catalyzing metabolic conversions, signalling and work. In these three disciplines of biochemistry, biophysics and molecular biology hypotheses were proposed and verified experimentally.

However, although they tried and claimed to operate in accordance with the methodology of physics, as time proceeded, biochemistry and molecular biology became less and less anchored on the principles expounded by chemistry and physics. The hypotheses and the activities of molecular biology became intentionally largely qualitative, and the concepts comparative (Carnap, 1966), so that their tests (verifications/falsifications) could give a digital yes/no answer. With this and with a strong tendency to empirical-rather than hypothesis-driven science, biochemistry and molecular biology became immensely successful. It is now possible to purify many or most of the water-soluble macromolecules that are active in living cells and determine their structure by X-ray crystallography. For membrane proteins, this is still a challenge, but progress is being made. The mechanism of quite a few enzymes is now considered to be understood reasonably well (although fundamental issues remain (Scrutton et al., 1999; Sutcliffe & Scrutton, 2000)) and so are regulatory mechanisms in the sense of which molecule might bind to which other molecule and regulate the activity of the latter. Pathways and networks of metabolism, gene expression and signal transduction have been mapped.

1.4. Cell Biology: The living cell

Near the end of the twentieth century genomics revolutionized this landscape. This revolution was preceded by a long and ever accelerating progress in biochemistry, molecular biology and the related disciplines of microbiology and biophysics and led to a combined discipline: cell biology. It defined the organization of life at the cellular level in qualitative terms of its molecules. With apologies for the readers who know their cell biology, but with due respect to the philosophers who may not quite do so but are interested in systems biology, we shall now describe the essence of this definition.

Early on, biochemistry had shown that all (most) chemical conversions carried out by living organisms occurred in a number of simpler steps such as dehydration, transfer of phosphate from ATP, dehydrogenation and isomerization. Each of these is catalysed by a protein, called an enzyme, which consists of one or a few chains of amino acids and sometimes an additional organic or inorganic chemical molecule or ion, folded into a complex structure. The amino acids are virtually limited to a set of 20 types. The protein is different for every type of molecule that needs to be converted. This led to the concept of metabolism consisting of large networks of chemical reactions through which mass flows, with a correspondence of every step to a protein (Beadle & Tatum, 1941). The metabolic networks are extremely powerful chemically, being able to convert many types of molecule into many other types, and many thousands of metabolites are known (Kell, 2004). The former correspond to almost anything that occurs in the environment of living organisms and is useful to them as food. The latter are suitable building blocks for the organism. The pluripotency of metabolism appears limited only by impossibilities stemming from a number of fundamental laws, such as the impossibility to create chemical elements from other chemical elements and the impossibility to generate Gibbs' free energy (Westerhoff & van Dam, 1987). The consequence is that there are metabolic networks ensuring that sufficient of each of these commodities is harvested from the food and supplied to biosynthesis. Metabolism is a network that makes biomass from food, although it does not seem to have evolved to be efficient in the thermodynamic sense (Kell et al., 2005; Westerhoff et al., 1983).

The question of how the proteins are synthesized led to the discovery of a network that is orthogonal to this metabolic network at each step of the latter (see Fig. 1). Each protein is synthesized from amino acids by a complex machinery, called the ribosome, which consists of protein and a second main type of macromolecule, i.e. ribosomal ribonucleic acid (rRNA). The diversity of the proteins stems from the fact that the sequence at which amino acids are attached to its nascent chain is specified by a specific messenger RNA (mRNA) molecule. RNA molecules are chains of four types of nucleotide, which are referred to by a mnemonic of the name of the corresponding 'bases', i.e. as A, U, G and C. Each of the 64 triplets of such bases corresponds to an amino acid, with just a few exceptions that deal with the regulation of protein synthesis itself. Each mRNA molecule is a copy of part of single stranded DNA, i.e. a very long chain of nucleotides referred to as dA, dT, dG and dC (the 'd's are often omitted). It occurs in combination with a complementary single stranded DNA molecule which has a T, A, C or G, respectively, where the other strand has an A, T, G and C, respectively. This double strandedness makes the DNA a robust way of storing the information. Damage that can be recognized as such can be repaired by referring to the sequence of the complementary strand. The part of the chain that is copied into an mRNA and is ultimately translated into protein is often called a gene (although this word actually refers to a concept that predates the discovery of DNA). The copying, which is called transcription, is

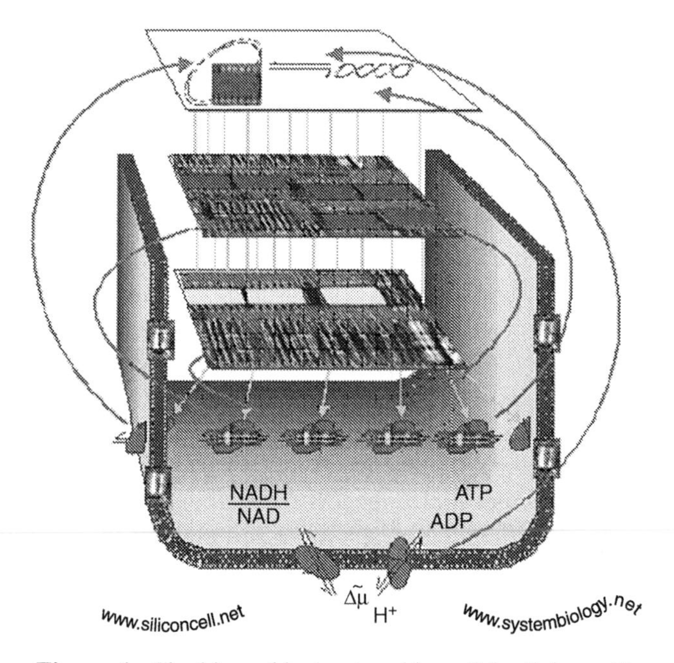

Figure 1 The hierarchical networking of the living cell.

carried out by a large enzyme complex called RNA polymerase. Preceding cell division, the DNA is copied, and the original and the copy end up in different daughter cells.

This set of networks that drive the synthesis of proteins on the basis of information of nucleic acids and information concerning the status of the cell and its environment is one that is often summarized as 'DNA makes RNA makes protein' (see Fig. 1). It is the domain of molecular biology.

Two aspects are of additional importance here: (i) DNA is not converted into RNA, nor is RNA converted into protein. This is a difference with a metabolic pathway where material parts ('mass') of the first molecule ends up in the last molecule. The gene-expression pathways only transfer information. (ii) Where the scheme suggests a hierarchy, DNA directing RNA, which directs enzymes, which then catalyse and hence also direct metabolism, this 'hierarchy' is not dictatorial but 'democratic' (Westerhoff et al., 1990): The rate at which transcription occurs depends on the binding of other proteins (called transcription factors) to parts of the DNA close to or relating to the gene. That binding in turns depends on the concentrations of metabolites that may bind to these, depending on whether the transcription factors are in the proximity of the DNA or depending on whether they have been modified chemically.

The chemical modification of transcription factors responds to the status of intracellular metabolism and to the presence of extracellular signals, such as light, and the presence of food. This response is achieved by yet another set of networks. These networks specialize in this signal transduction and again consist of pathways in which each step is catalysed by proteins. In most of these pathways however, there is no transfer of mass from the beginning to the end. Only information about the conditions measured at the beginning of the pathway is reflected by the state elsewhere in the pathway.

Metabolism, gene-expression and signal-transduction constitute networks in the dimensions of time, information and chemistry. The living cell also depends on other networks that address the dimensions of chemistry, structure and space. The cell itself is a membrane-bounded compartment. In eukaryotes such as mammals, the cell also contains many membrane bounded subcompartments, which house networks that can be incompatible with networks in other subcompartments. Without catalysis, transport across most of the membranes is impossible, and the transport of some macromolecules through compartments is also catalysed. The DNA is folded into a complex structure with proteins called chromatin. These networks of structure and transport through and around structures have been well characterized. In recent years, more and more of these structures have been shown to be displaced from equilibrium, being maintained continuously by regulated active networks. Examples include the DNA structure, certainly in bacteria (Snoep et al., 2002), the asymmetric lipid distribution in membranes and the microtubular and actin networks in the cell sap.

Molecular biology became a further success story when it joined forces with biochemistry and microbiology and became modern biotechnology. First, it was discovered that many organisms make enzymes that cut DNA with specific nucleotide sequences. By not having those base sequences themselves, those organisms could protect themselves against invading viruses. These 'restriction enzymes' were used by scientists to put genes of interest into organisms that did contain those sequences. By growing these organisms and then again applying the restriction enzyme to isolated DNA, pieces of DNA corresponding to genes could be 'cloned', i.e. purified and their amounts greatly amplified. The resulting material could then be introduced into other living cells which would then express that DNA into protein. If those cells altered their functioning this helped in establishing the function of the gene. The amplified amount of DNA also enabled that DNA to be sequenced, first by tedious methodology, but in a demand-driven mode this led to the development of new and rapid sequencing methodology. (The methodology to amplify DNA also became much more effective when the polymerase chain reaction (PCR) was developed, allowing for the amplification to occur *in vitro*.) The result was that the nucleotide sequence of each gene of interest could be determined. Because of the 64-to-20 mapping of DNA sequence to protein sequence, this implied that the amino acid sequence of the corresponding protein was also determined. Through the above cloning procedure, larger amounts of proteins could be obtained enabling structure determination through X-ray crystallography and NMR. At present the structure of almost any soluble protein can be determined, albeit at relatively low throughput.

It also became possible to determine whether any given gene was expressed in an organism. Here the base-pairing phenomenon that underpins DNA and mRNA function served molecular biology. Tagged DNA or RNA molecules that were complementary in terms of nucleotide sequence were synthesized and made to react ('hybridize') with mRNA isolated from living organisms. If a certain mRNA was expressed then the hybridization would betray this. Because so many genes are expressed in any organism and because of background reactivity, the mRNAs first had to be separated from each other, which was accomplished by gel electrophoresis. A corresponding methodology was developed for the measurement of expression at the level of protein, by using specific antibodies for the proteins. The separation power of these methods is however limited, and therefore they were not suitable for genome wide measuring of gene expression.

Another powerful tool came from genetics applied to rapidly growing microorganisms. Mutations were made in the DNA of these organisms and the consequences for their functioning was determined. Through the above methodologies, mutations could be related to proteins. Deleting genes and observing the consequences, pathways could be constructed that should be responsible for certain aspects of cellular behaviour. When this was done for different organisms and combined with nucleotide sequencing, an astonishing phenomenon turned up.

This was the extensive homology of organisms in terms of their intracellular organization, as well as in terms of the amino acid sequences of their corresponding proteins. In principle, the major food substance glucose could be oxidized in many ways to carbon dioxide with the harvest of much of the corresponding free energy. Virtually all organisms, however, possess the glycolytic pathway and the tricarboxylic acid cycle, and many contain the membrane-associated electron-transfer chain, which comprise one way of accomplishing this overall process. *A fortiori*, the enzyme that catalyses the phosphorylation of glucose by ATP, is sufficiently homologous also in terms of its amino acid sequence, for its sequence to be identified in many newly sequenced genomes, through the sophisticated techniques of bioinformatics. Even more strongly so, functional domains of proteins (such as ATP binding sites) have been sufficiently conserved through evolution to be recognized between genomes. On another planet with perhaps much higher rates of net mutagenesis, and much lower selection pressure, this may be different, but for our planet this phenomenon of extensive homology has been an enormous asset to molecular biology. To many newly sequenced genes, a function is assigned simply on the basis of homology of sequence, and in many cases this assignment turns out to be correct, qualitatively. An important consequence is also that the phrase 'understanding life' does have a meaning. It could have been such that molecules, mechanisms and pathways differed immensely between organisms and that each organism had solved the problem of how to stay alive in its own, entirely different way. It is quite clear now that this is not the case; life as we know it in a broad sense is probably maintained in just one single way, with 'minor' variations on the theme. This is not to say that this variation, which is minor in terms of principle and quality, is not vast in terms of quantity. Biological diversity especially in the microbial realm is enormous. Accordingly, life is able to maintain itself under a very wide variety of conditions on this planet, but again, essentially through extensive variation on a single theme. Of course, this greatly motivates the scientific question of what constitutes this essentially uniform molecular basis of life.

The maps and structures of living cells, i.e. the field that may be called cell biology, were considered known in the 1980s in their essence. What was lacking was the completeness. Although for each type of network, a number of examples had been well documented, many actual networks had not yet been identified. More disturbingly, however, every now and then a cellular component was discovered that was strongly involved in the already 'known' pathways, most often in their regulation, but often even in their mechanism. Examples included fructose 2,6 bisphosphate in glycolysis, the chaperonins in the protein synthesis pathway and ubiquitinylation in signal transduction. In addition, although some cellular behaviour could be explained qualitatively on the basis of the known networks, much other behaviour was in conflict with what was known, or simply not explained by it. The conflicts could not be used constructively as falsifications

(Popper, 1992), because it was well recognized that there were many unknown components and regulatory mechanisms in the cell that could affect the pathway that was under investigation. For similar reasons verifications were limited in value. What often resulted was an escape of biochemistry and molecular biology to well defined *in vitro* systems, where at least the mechanisms of the proposed pathway or molecules could be established, even though the relevance for their operation *in vivo* became unclear.

2. LIMITATIONS TO THE SCIENTIFIC STATUS OF BIOCHEMISTRY AND MOLECULAR BIOLOGY

Notwithstanding their success concerning the understanding of single types of macromolecules, classical biochemistry and molecular biology face limitations when compared to the science aimed at by the philosophers of classical physics. These limitations are

(1) Inaccuracy: no quantitative, i.e. accurate, testing of hypotheses
(2) Inability to deal with emergent properties: because of lack of quantization it is impossible to test a number of qualitative hypotheses that are highly important for the emergent properties in living systems
(3) Irreducibility: biochemistry and molecular biology theories cannot be reduced to physical chemical theories
(4) Impotency, i.e. inability to address Life itself and lack of connection to organismal Biology
(5) Undefinedness: not all factors that play important roles are known and consequently hypotheses cannot be tested
(6) Inaccessibility to experimentation: the systems under study cannot be experimented on through a sufficient number of degrees of freedom
(7) Lack of analyzability

We now discuss these limitations, one at a time.

2.1. Inaccuracy

The first limitation is that the cartoon-type hypotheses were not quantitative and thereby unfit for the strictest possible quantitative testing, a procedure desired by the philosophy of physics (Carnap, 1966). Being quantitative enables tests to be more stringent (Laughlin, 2005). If the temperature of a closed vessel with an ideal gas rises by 10% then the qualitative test of the law of Boyle asks if the pressure goes up, whilst the quantitative test asks whether the pressure goes up by precisely 10%. Clearly, the qualitative test has a 50% chance of being passed by coincidence, whereas the quantitative test has a much smaller such chance, depending on the experimental accuracy.

2.2. Inability to deal with emergence

A second limitation also derives from the lack of being quantitative but, paradoxically, pertains to failure to test the prediction of qualitative phenomena. The behaviour of systems of independent components is nothing but the simple addition of the behaviour of those components. In sufficiently nonlinear systems and even in linear systems with certain networking (for simplicity we shall here call the latter also 'nonlinear'), qualitatively new behaviour may emerge, which is often important for biological function. In fact for survival of living organisms, a number of properties is essential that are absent from the individual molecules in those organisms. They must emerge from certain nonlinear interactions. We shall refer to those nonlinear interactions as 'essential' nonlinearities. Examples include oscillations in networks of components that would themselves never oscillate (Goldbeter et al., 2001), and free-energy transduction between components that would themselves only dissipate free energy (Westerhoff & van Dam, 1987). For biological macromolecules, the nonlinearity varies between conditions, as it depends on their environment. We briefly illustrate this by considering what may be the rate equation of an enzyme in an intracellular network:

$$v = \frac{[S] \cdot V}{K_m + [S]} \tag{1}$$

where v, [S], K_m and V refer to the actual reaction rate, the concentration of the substrate of the reaction, the Michaelis–Menten 'constant' and the 'maximum' reaction rate, respectively. The way in which the enzyme affects the behaviour (both in the qualitative and in the quantitative sense) of the network is fairly well described by the elasticity coefficients for the metabolites with which it interacts, in this simplest case, the substrate S. This elasticity coefficient corresponds to the log–log derivative of the rate with respect to the concentration of the substrate, i.e.

$$\varepsilon_S = \frac{\partial \ln v}{\partial \ln[S]} = \frac{K_m}{K_m + [S]} \tag{2}$$

The equation shows that the role of the enzyme in the system is not only determined by that enzyme itself (through K_m) but also by its environment (i.e. by $[S]$) and by how it interacts with that environment (in terms of S/K_m).

The new behaviour that emerges depends on the type of nonlinearity that reigns in the network, e.g. on the value of the above elasticity coefficient (Westerhoff & van Dam, 1987). Consequently any theory explaining the occurrence of oscillations will only predict oscillations for certain states of the system (i.e. certain magnitudes of $[S]$) and not for others (as is observed, e.g., Ihekwaba et al., 2004; Nelson et al., 2004), and their nature can depend even qualitatively on multiple enzymes in the system (e.g. Ihekwaba et al., 2005). Testing whether the theory indeed explains oscillations that occur in a living cell will first have to determine what the state of the system is, in a quantitative sense (i.e. how high $[S]$ is, and not just whether there is some S or not), then to ask whether for that state the theory predicts oscillations, and then to test whether under those conditions oscillations are indeed observed experimentally. The implication is not only that theory and experiments need to be quantitative but also that they need to pertain to the conditions of the living state, i.e. they need to be performed under conditions as close as possible to those that are considered to pertain *in vivo*, preferably in the living organism itself.

An actual example is the following. If one observes synchronous glycolytic oscillations in intact yeast cells (Davey et al., 1996; Richard et al., 1993), and one proposes that the stimulation of the enzyme phosphofructokinase by AMP is 'responsible', one can test this hypothesis by mutating the enzyme and removing that stimulation. However, any alteration that alters the system such that its state is no longer in the oscillatory domain, will do away with the oscillations. In fact the proposed mutation of phosphofructokinase could well do away with the oscillations by simply shifting the system to a different operating point even if this product stimulation were responsible for the oscillations. A proper test of the hypothesis thus removes the AMP effect whilst simultaneously modulating the system so as to keep it at its operational state. Better, one removes the AMP effect gradually and asks if the frequency or amplitude of the oscillations changes (Reijenga et al., 2005b). In nonlinear systems, even qualitative statements therefore need quantitative tests (Ihekwaba et al., 2005).

How important is this issue? Well, the rate and equilibrium equations for most biological processes are nonlinear or at least nonproportional (Hill, 1977; Westerhoff & van Dam, 1987). Moreover, many of the biological processes that are important for function exhibit properties that one would not see in individual molecules and that therefore require nonlinear interactions between those molecules. These processes include differentiation, development, the cell cycle, robust signal transduction and most transport processes. Their theories can only be tested if they are quantitative, and strictly only by quantitative experimentation that is performed inside the living cell.

2.3. Frustrated aspiration of biochemistry and molecular biology to . . . biology

Another type of limitation to biochemistry and molecular biology is that they do not by themselves produce the overlying science, i.e. biology. In principle, biochemistry and molecular biology study all the molecules that occur in organisms, but they refrain from addressing the life that is embodied by all those molecules. Although this claim of insufficiency of biochemistry and molecular biology has often been made by physiologists and other organismal biologists, it is not immediately appreciated by all, and certainly its remedy is not. There is indeed a paradox: if biochemistry and molecular biology were to continue to study and establish the structure and the mechanisms of action of every macromolecule of a living organism, then they should ultimately understand that whole living cell. For what else is there in a living cell than its molecules? This contention is the most common version of the reductionist agenda: dissect any system into its elements, study all those elements individually, and then just understand the system. Technically, the 'just understand the system', implies that system's behaviour can be understood as a superposition of how all its components behave individually. The organismal biologists often observe that when a living system is taken apart, it loses much of the essential behaviour of living systems. This makes some of them turn to the holist agenda, which studies only intact systems. This then makes them subject to much of the limitations noted above for organismal biology, and more importantly, it implies that those limitations will stay forever, independently of the progress of science.

The reductionist and holist paradigms seem to be irreconcilable, but below we shall propose that through systems biology and the silicon-cell approach they may not be. Here we shall first indicate why the 'just understand the system' methodology does not work, i.e. why by themselves biochemistry and molecular biology cannot produce biology. The reason is again the essential nonlinearities of biological systems. Much of biology depends on dynamic phenomena that emerge in nonlinear interactions. These cannot be understood by the simple addition of the behaviour of the components in isolation. This is one reason why biology lies outside the realm of biochemistry and molecular biology *sensu stricto*. In other words, what makes a system different from its parts list is the non-linear interactions between those parts, and these are changed or lost upon disassembly.

2.4. Irreducibility

A third limitation is again related to the cartoon aspect of biochemistry and molecular biology: in these new disciplines molecules are not drawn in terms of their structure or chemical equation, but by coloured balls with short, non-chemical names, such as hexokinase, HXK, Ras or wnt. These names serve reasonably well as mnemonics. Attempts to give enzymes systematic names produced names that referred to their activity rather than to their chemical formula or structure. The reason was that for many enzymes the chemical structure could not be established, whereas at least some of the catalytic activities could be. The names and concepts of biochemistry were not reduced to the underlying physical chemistry (in the sense of reduction of theories to underlying more general theories, cf. above). Similarly, 'the' structure of nucleic acids and proteins was determined by X-ray crystallography, but the question of whether that structure was stable with respect to the physical forces between amino acids and between bases, was not addressed. This was in part because it could not be addressed effectively. Virtually none of these structures can presently be calculated *ab initio* (see (Popelier & Joubert, 2002) for an example), precisely because the interactions are nonlinear, and with many interactions depending on other interactions. Likewise, electric field effects on transmembrane movements of ions cannot be vested in physics and chemistry because too much of the details of the transport matters and is in fact unknown. Although there has been some progress in the calculation of enzyme catalysis in terms of physical–chemical interactions, most such reaction mechanisms cannot be verified in terms of precise physics and chemistry. The same is true for the pathways of processes that make living cells operate. The fluxes through them cannot be calculated *ab initio* either, but only from direct physical–chemical interactions and atomic structures. In biochemical textbooks, pathways are therefore drawn as roadmaps running through many towns and connecting major cities or hubs (Barabási & Oltvai, 2004). Indeed, reduction of molecular biology and biochemistry to the underlying physics and chemistry is rare, and not even an aim of these disciplines anymore; both disciplines are entirely successful on the basis of their own concepts and laws, immaterial whether these are reducible to physics and chemistry or not. However, this general problem of intractability in terms of the underlying physics and chemistry caused reluctance among many physicists and chemists to consider biochemistry and molecular biology as serious sciences. The biology of entire living systems was observed to be too complex and ill defined for the hypotheses to be strict, testable and falsifiable. To some, this made molecular biology and biochemistry appear to remain stamp collecting.

Indeed, the above limitations suggest that neither biochemistry nor molecular biology connect to physics. They fail to meet the criteria of classical physics that were once proposed to be the criteria of proper science (Carnap, 1966). Looking at chemistry beyond quantum chemistry, this may not be a novelty among the experimental sciences; chemistry may not connect to physics either (Primas, 1981).

2.5. Lack of testability because of undefinedness

Another important limitation of biochemistry and molecular biology relates more literally to holism. Returning to Eqn (1), we realize that the Michaelis 'constant' is independent of the concentration of S but not necessarily constant otherwise. Agents binding to the enzyme catalysing the reaction may influence this Michaelis constant, and certainly the concentration of the product of the reaction changes the (effective) K_m, i.e.

$$K_{\text{m, apparent}} = K_{m'} \cdot \left(1 + [P] \big/ K_p\right) \tag{3}$$

All components of the same living cell may therewith affect the role the enzyme plays in the cell's behaviour, including the components that are not yet known.... This pinpoints one of the arguments of holism, in that to understand the role of one of the molecules in a system with the type of nonlinearities found in cell biology, one must look at the whole. We do not think that one should necessarily be able to look at the whole all the time, but certainly one should be able to look at all the possible molecular factors that play a role. Until recently, not all molecules of the living cell were known or even knowable, making it impossible for biochemistry and molecular biology even to determine with certainty the role a molecule of choice might play in determining the nonlinear behaviour of the living system, simply because unknown factors could well be clouding any issue. Post-genomics is beginning to change this.

2.6. Lack of experimental accessibility

As emphasized by Carnap (1966) for physics, it is important that hypotheses are tested under all relevant conditions and in terms of all relevant degrees of freedom. In living systems, many factors may exert an influence and it should therefore be mandatory that proposed mechanisms are tested by modulation of all those factors individually. For as long as not all those factors were known, it was difficult for biology to carry out these tests; the living system was not accessible enough for such testing.

2.7. Lack of analysability

Because many factors are likely to be involved in the sustenance of the living state, hypotheses concerning mechanisms are likely to be multifactorial. Accordingly, many of these factors should be monitored simultaneously in tests. Although quite a few factors can be measured individually by biochemistry and molecular biology, until recently it was impossible to monitor many components simultaneously.

Summarizing, we see a landscape where biochemistry and molecular biology could extend neither to physics nor to organismal biology because of at least these seven types of limitation. We shall now discuss recent changes in molecular biology that would seem to do away with some of these limitations.

3. RISING ABOVE THE LIMITATIONS

3.1. Genomics

A major cause of the above limitations was that there existed no complete understanding of inventory of all the components of a living cell, even though such an inventory had been identified in principle, i.e. the DNA: the DNA contains the information for all the proteins in the cell and the proteins catalyse all the reactions. It was thought that in principle, the sequence of the DNA should determine everything that happens in the living cell, under any given set of environmental conditions. It became quite important therefore to sequence all the DNA of a living organism, and in the 1990s of the previous century, large consortia of researchers embarked on accomplishing this aim in activities referred to as 'genomics'. It may seem that genomics was not much different from the molecular biology that preceded it. Indeed, many of the most active scientists in genomics continued to be molecular biologists as well. Yet, for our discussion here, the transition between molecular biology and genomics has been quintessential. Genomics went after the determination of the complete DNA sequence of an organism, rather than of DNA sequence of many of its components, i.e. genomics went for the system rather than for its components. By 1995, the first complete sequences of the genomes of free-living organisms (cf mitochondria in 1981 (Anderson et al., 1981)) became available (Fleischmann et al., 1995), and importantly also the sequences of the two best-known model organisms soon followed, i.e. the eukaryote yeast (Goffeau et al., 1996) and the bacterium *Escherichia coli* (*E. coli*) (Blattner et al., 1997). By 2001, the DNA sequence of humans was nominally established and sequences of many organisms have become known as we write this. In essence, the DNA sequence of any organism can now be determined. Because of the homology discussed above and thanks to bioinformatics, the function of many genes can be proposed with appreciable success rates when the homology to genes of known function is close. Although for half of all sequenced genes (this fraction differs between organisms), the function is uncertain or unclear, this fraction is considered to be on the decrease. (We would stress, of course, that many genes with some 'known' functions will turn out to have other functions that are as yet unknown.)

Knowing most of the genes of an organism provided a strong motivation for what has been called 'functional genomics', i.e. for determining whether those genes function in terms of being expressed and what their role is. Because of the strong tendency of nucleic acids of complementary sequence to react with each other, this was possible in principle by making populations of small RNA molecules each of which was complementary to part of one of all the genes in the genome. A breakthrough came when those probe molecules could be spotted as an array onto a slide and could be provided with a fluorescent tag that lights up when an mRNA molecule hybridized. This nucleotide array technology is now used to determine the expression of all genes at the level of mRNA, at accuracies beyond 30%.

No similar hybridization chemistry exists at the level of a chain of amino acids (yet). Using immunological techniques however, antibody-like molecules are now spotted onto arrays, and the abundance of proteins in extracts from cells is determined (Walter et al., 2000). Alternative modes of genome-wide detection of protein abundances include a methodology in which all proteins are separated in a highly reproducible way through two-dimensional (2D) gel electrophoresis, such that each location in 2D corresponds to a specific protein. The mapping of spot location to the identity of the gene is a slow process, but for smaller genomes this methodology is getting close to the possibility of genome wide detection of gene expression at the level of protein. This methodology is inherently limited in three important ways. First, the resolution of 2D gel electrophoresis is insufficient to separate all proteins of genomes larger than a few thousand genes; though useful for bacteria, the methodology is still of more limited value for human biology. Second, the method is not quantitative yet, and indeed many proteins, especially membrane proteins, are missed entirely. And third, it is difficult to identify the individual proteins. The latter problem is now being alleviated by the implementation of mass spectrometry. By extracting protein from a specific location on the 2D gel, subjecting that to limited proteolytic digestion, determining the precise mass and/or sequence of the resulting peptides and combining the resulting information with the known sequence of the genome, the protein spots can now often be attributed to specific proteins.

Mass spectrometry also offers methods that may analyse genome-wide expression at the protein level. The gel-electrophoresis step can be replaced by capillary chromatography, a separation by mass spectrometry on the basis of the total mass of the protein (or fragments thereof), fission of the protein/peptide in the gas phase and then a second mass spectrometry step to determine what the resulting fractions are. Again the availability of the genome sequence enables one to identify the protein. For mass spectrometry, molecules have to be brought into the gas phase as electrically charged molecules. However, existence in the gas phase is far from the thermodynamically most favourable mode of existence for most of the molecules that constitute the living cell. The effectiveness at which the entry into the gas phase is achieved is low therefore more importantly, it depends much on the presence and properties of the other molecules in the mixture. Other molecules with electric charge can affect the tendency of a given molecule to enter the gas phase. Consequently, the mass spectrometry method is inherently irreproducible in the quantitative sense; it is hard to determine expression levels accurately with this method (although this is improving both by changing conditions in the mass spectrometer (Vaidyanathan et al., 2003) and by isotope-based quantification. This is because isotopes behave essentially identically with respect to the above problems, yet can be discriminated readily by the mass spectrometer. Spiking samples with known amounts of an isotope of the substance of which the quantity needs to be determined, therefore enables quantitative determination of amounts of proteins (more often in relative terms but occasionally absolutely (e.g. Beynon et al., 2005)).

The genome-wide determination of gene expression at the levels of mRNA and protein are called transcriptomics and proteomics, respectively. Genome-wide analysis of the expression at the level of metabolism, which is often closest to function, is called metabolomics. Genome-wide metabolomics has not yet been developed to the same extent as transcriptomics (Dunn, Bailey & Johnson, 2005; Dunn & Ellis, 2005; Goodacre et al., 2004). Mass spectrometry methods akin to the ones described above for proteins are being developed for metabolomics. Again it is a problem to get the metabolites into the gas phase and to determine their level quantitatively; isotope methodology can again solve this problem (though one needs an isotope for each determinand, and the larger problem resides in the fact that we do not know what most of these molecules are . . .).

Cell function is determined not only by the expression levels of proteins but also by where they are expressed. Here three developments are highly important. One is that of high-resolution microscopy. The second is the development of many fluorescent probes for important molecules and ions in living cells. And third is the possibility of fluorescence- or luminescence-based reporter proteins, which are either fused to proteins of interest or are put under the control of the gene-expression control elements that normally drive the expression of proteins of interest. Thanks to these methodologies, the timing of expression and the dynamic localization of many molecules in the living cell can now be determined.

Another less profound, yet highly important advance in technology is that of robotization and automation for high throughput experimentation. By using plates with many reaction vessels and robots doing the pipetting, many experiments can be performed in parallel and at much enhanced reproducibility.

At present one can determine for all genes in a genome simultaneously whether they are expressed at the level of mRNA. Soon this will also be possible at the level of protein and in terms of their relationship to further levels of functionality, e.g. at the level of metabolites. Through functional genomics, therefore, everything will potentially soon be knowable and known about living cells. For unicellular organisms this should imply that everything will be known about a living organism, albeit that collections of such cells remain highly heterogeneous (Davey & Kell 1996). Every component can be manipulated by expressing the corresponding gene in the organism under the control of a regulatory element that can be steered by the experimenter. Everything will come to be known therefore and systems of Life will come under complete experimental control. The limitations of the 'undefinedness' and inaccessibility to falsification–verification experiments of biology, will soon be gone. Finally biology can stop collecting stamps and become 'proper Physics', or so it would seem.

3.2. Soon everything will be known . . . : Will biology become physics, at last?

Indeed, the vast increase in power of molecular biology, and the ability to experiment and analyse genome wide, should get biology much closer to the ideal of constructing completely verifiable and falsifiable theories. Of the above list of seven limitations, it would seem that the ones regarding undefinedness, inaccessibility and lack of analysability have disappeared with the advent of functional and post-genomics. These three criteria come close to the criteria that proper physics should adhere to, e.g. according to Carnap (1966). Provided that the analyses of functional genomics are made quantitative, it would seem that the first criterion (accuracy) will also be met. It would seem therefore that with functional genomics Biology would all but graduate to become proper physics.

From the point of view that science should be one and indivisible, the reduction of biology to just another physical chemical science with 'just' the same methodologies and quality criteria, would seem to be a great good. Whether this should actually happen is the fundamental issue that is the subject matter of this book. We shall now indicate why we think that this reduction is not to be expected.

3.3. Observing or understanding?

Functional genomics will enable us to observe virtually everything that happens in living organisms. The aim of the sciences, however, is also to understand the observations. Such understanding can consist of the possibility of deducing what is observed from pre-existing theories. It can also amount to the understanding on the basis of theories that are being generated as many more observations are made, i.e., through induction, principled hypothesis formulation and hypothesis testing through verification/falsification procedures.

We shall first address the former basis of understanding. It turns out that functional genomics has not removed the limitation of irreducibility from biochemistry and molecular biology, and that it will not do this in any foreseeable future. When it was proposed to sequence the whole genome of organisms, one of the underlying arguments might have been that this should automatically lead to the understanding of the functioning of living cells and organisms in molecular terms. Folding of a protein was perhaps thought to be determined by it finding the structure with the lowest free energy. Because that free energy is determined by the interactions of all its amino acids and the sequence of these in the chain, it was perhaps thought that one should be able to calculate that structure *ab initio*. For all but the simplest proteins, the calculation of the structure with the lowest free energy from the amino acid sequence is still impossible. The problem is strongly nonlinear and hence much too complex to be carried out by existing computers. In fact, the calculations of protein structure that are being done with some success are not truly *ab initio* but use phenomenological force fields and/or knowledge of existing structures. At present structure predictions of proteins on the basis of their sequence are occasionally fairly successful, but such predictions are virtually only based on comparison with homologous structures. The next step, i.e. the calculation of catalytic action from the protein structure is equally difficult. Here too, success is based on comparison of homologous series. The *ab initio* calculation of kinetic properties of entire pathways might all be possible in principle, but it is impracticable at present and in fact for any foreseeable future, due to the sheer complexity and nonlinearities of the interactions that are involved (see also Westerhoff & Kell, 1987).

In the living cell there are also catalysts of correct protein folding, i.e. chaperonins or by the action of the Ribosome. Because both these assisting proteins couple this process to a reaction consuming free energy, it is quite possible that they put their target protein in a structural state with a free-energy that is higher than minimal. Indeed, the structure of proteins may not even correspond to the free energy minimum but be determined by the mechanism of folding. After all, the spontaneous conversion between native and denatured states of proteins is rarely effective.

A lingering feature of biology could well be important here. This is its inherent hysteresis. The concept of biology as straightforward though complicated physical chemistry, should be most consistent with the following picture of the genesis of a new living cell: in an existing living cell, all the components of a daughter cell might be synthesized independently *de novo*, inclusive of the lipids necessary for its membrane and its DNA. Then a closed spherical lipid bilayer would be formed around all the newly synthesized components, and the newly formed cell that sat inside the mother cell would be extruded by that mother cell. After their synthesis, all components for the new cell would assume their minimum free energy structure independent of the activities of the mother cell. The state of the daughter cell would then be determined entirely by free-energy minima, hence by the physical chemistry of its molecules. This mechanism of generating new cells might be entirely possible and would in fact be consistent with what Van Leeuwenhoek expected to see in terms of *homunculi* through his microscope. But it is not what actually happens. Instead, the membrane of the daughter cell is formed by splitting off a part of the membrane of the mother cell; the DNA of the daughter cell is the result of a semiconservative replication of the mother cell, i.e. the mother and the daughter cell receive both one strand of the DNA of the mother cell, the other strand having been synthesized *de novo*. According to our current knowledge, the proteins that end up in the daughter cell are not all proteins that have been synthesized *de novo*. Newly synthesized proteins and pre-existing proteins and even newly synthesized organelles and pre-existing ones end up in both the daughter cell and the mother cell. In many organisms the mother cell after division and the daughter cell are effectively the same; division yields two daughters and the mother ceases to exist. In other organisms such as *Saccharomyces cerevisiae*, division is asymmetric, and the mother differs from the daughter, yet appreciable mixing has occurred. Importantly also, the DNA, mRNA and proteins of the young daughter cell have been synthesized by the DNA polymerase, RNA polymerase and ribosomes of the mother cell. Consequently, rather than that each cell is an entirely new physical-chemical phenomena, all cells are in fact continuous with each other. If there were a process of excessively slow relaxation in a cell, the same process would be in the same state in all daughter cells. That this not in part is reflected by observations of epigenetic phenomena.

The extent to which this possible hysteresis is actually important is unclear at the moment. For molecules of low molecular weight and complexity, it is unimportant because relaxation to an equilibrium structure is fast enough. For macromolecules and for the regulatory state of networks it might be important. This issue simply has not been looked at sufficiently yet. In some cases of regulation, such as for instance with the *lac* operon in *E. coli*, the regulatory state is effectively inheritable through this type of mechanisms, which has the effect of zonation of its colonies. In its ultimate form the point of hysteresis appears obvious. All amino acids in proteins have the *L*-stereoisomeric constellation. The mirror world with all *R* amino acids should be energetically equally probable. Yet new cells with all their proteins in the *R* form do not arise, because the enzymes that make their amino acids make the *L* form.

The conclusion is that the feature that it is too difficult to calculate structures of proteins on the basis of physical–chemical principles may not even be too relevant. It is quite possible that most of the structures that exist in living cells are determined by more than the straightforward physical chemistry of those molecules themselves. They may also depend on pre-existing structures of other molecules with which they interacted during synthesis. The fact then that biochemistry and molecular biology do not start from underlying physical–chemical principles but with their own elementary objects such as enzymes and genes, may be an asset rather than a disadvantage. The corollary is that also the irreducibility of biochemistry and molecular biology to physics is much more fundamental than technical. Any molecule-based biology may therewith be a science that is fundamentally different from physics.

Evolution has not selected structures with maximum entropy (Schrödinger, 1944), minimum free energy (Nicolis & Prigogine, 1977) or maximum thermodynamic efficiency (Westerhoff & van Dam, 1987), and in fact much of the functioning of biological replication may have been structured so as to prevent relaxation to such a state. Also here simple physical–chemical considerations do not suffice to understand biology. As also proposed by Schrödinger (1944), biology warrants its own explanatory principles.

Of course physics too has undergone a tremendous evolution since the days of Schrödinger and Carnap (Schrödinger, 1944; Carnap, 1966). It has been recognized that far away from equilibrium, physical–chemical systems may relax towards metastable states rather than to equilibrium, and anyway such states are typically well isolated from each other in the form of local minima as in any other search landscape (Bäck et al., 1997; Frauenfelder & McMahon, 2001). The states can be more complex than the equilibrium state, i.e. appear to be more organized than the latter. Such physical self-organizing systems have been proposed to be at the basis of the tremendous organization that is observed in biology. Accordingly, parts of modern physics address the generation and maintenance of complex dynamic structures, and how new properties may emerge from nonlinear dynamic interactions. However the mechanisms that have been proposed such as the Brussellator (Nicolis & Prigogine, 1977) are themselves nonverifiable/nonfalsifiable. This is because they were formulated in much too general terms, causing loss of the specificity of the biological system at hand. Testing of nonlinear phenomena requires precision, hence a precise matching of mathematical model and experimental system. Wolf et al. (Wolf et al., 2000) have recently worked towards such a testing of a proposed self-organization mechanism for synchronization of the glycolytic oscillations in a population of yeast cells, but this may only serve as an incomplete example. This brings us to the second type of understanding, i.e. on the basis not of the principles of underlying sciences but of principles that are discovered in the science at hand, i.e. on the basis of newly discovered principles of biological systems. Here there is the issue whether anything is to be expected from the search for such theories.

Metabolic and hierarchical control analysis are theories that may serve as examples of theories that are custom-made for biological systems (Westerhoff & Hofmeyr, 2005). By making an idealized description of intracellular networks, i.e. metabolic networks for the former theory and gene-expression or signal transduction networks for the latter, a mathematical set of definitions can be made and laws can be deduced from the time-transformation invariance and from stability against fluctuations (cf. Hornberg et al., 2005; Peletier et al., 2003; Westerhoff & van Dam, 1987). These theories are in a sense comparable to theories in physics in that they derive from observations that falsified alternative hypotheses, and led to conjectured new laws, which could then be deduced from postulated fundamental properties (axioms) of the system. Other 'laws' that derive more as a result of induction from experimental observations may also be found for biological systems, such that proteins are encoded by mRNAs which are in turn encoded by pieces of DNA, and the law that for every natural substance on this planet that can be broken down to yield free energy, there is an organism that does precisely that.

On the basis of this experience, we expect that many more theories will be established for living systems. These will differ from those we already know from physical chemistry and then not only in terms of their precise meaning, but perhaps also in terms of their structure. Perhaps such biological theories will be less general, more condition dependent, and much more complex. This remains to be seen. Automated hypothesis generation from experimental data may show new ways in this respect (King et al., 2004).

3.4. Systems biology

Our contention that the molecular biology of living systems is neither physics nor biology, but rather a science in its own right, suggests that it is entitled to a name. Such names already exist, i.e. systems biology and integrative biology. We shall here use the former. We propose that systems biology attempts to establish principles of operation of biological systems such as the living cell. It should thereby find its own concepts rather than reduce them to physical chemistry. It should strive to be quantitative enough to be able to understand the emergence of functional properties from nonlinear interactions between components of biological systems. It should also appreciate that such interactions depend on the precise state that the biological system is in. This has the consequence that laws should address specific conditions rather than be completely general. For instance a law could be that the glycolytic pathway can engage in oscillations provided that the elasticities of the following stated reactions fall within the following range.... The law should not be of the generality of physics i.e., that the glycolytic pathway might engage in oscillations under any, unspecified conditions. Systems biology should synthesize the following features

(1) Information on expression levels is contained in the DNA and is expressed through mRNA into proteins which then catalyse reactions.
(2) The expression levels are not simply determined by transcription activities of the DNA in a dictatorially hierarchical fashion, but controlled by a combination of extracellular signals and intracellular concentrations.
(3) The concentrations of intracellular substances are determined by all the intracellular processes and extracellular signals together.
(4) The intracellular processes are determined by the expression levels of the enzymes, by the kinetic parameters of those enzymes, as well as by extracellular signals and intracellular concentrations.
(5) Much of biological regulation is one of circular or spiraling causality (Rosen, 1991; Westerhoff & Hofmeyr, 2005), i.e. a concentration of a substance may co-determine the concentration of another substance at later times and be co-determined by the concentration of that other substance at earlier times.

(6) Due to nonlinear interactions, qualitatively new properties may emerge; whether this happens depends on the precise magnitudes of the parameter values.
(7) Part of the structure and dynamics of the living cell may be prespecified by evolution, by its mother cell and by the synthetic machinery therein.
(8) Living organisms are the product of dynamic interactions between structures and chemical reactions, where the latter determine the former and the former determine the latter to quite significant extents.
(9) Much of biological mechanism and regulation is not determined by any single factor but by a multitude of factors.
(10) The simplicity of mechanisms that serves as Occam's razor in the decision between competing theories in physics is of comparatively lower real value in biology. Functionality and fitness and empirical facts rule over simplicity. The actual mechanisms in systems biology may be more complex than possible because of coselection for other purposes in evolutionary optimization, because evolution may have led to systems that are optimal locally but not globally, and because simplicity in human eyes may be complex in systems biology terms (and *vice versa*).

Much of life is associated with organizational and intelligence aspects that 'emerge' from molecular behaviour (Kell & Welch, 1991). Although these emergent properties are not in conflict with physics and chemistry, much of physics and chemistry traditionally shies away from complexity, hysteresis and nonlinearity (although other parts such as those dealing with superconductivity, lasers, ferroelectricity and other highly nonlinear phenomena cannot escape it). As we discussed above, their paradigms favour the kind of simplicity and Occam's razor strategy that may not be relevant for biology. We propose that this makes systems biology (the part of biology that focuses on this kind of complexity) its own science with, indeed, its own methodology and its own philosophical foundations. We shall here then seek to contribute to the development of a philosophical basis for this new science by describing some of the modes in which it operates in practice.

4. TOWARDS A SYSTEMATIC METHODOLOGY OF SYSTEMS BIOLOGY

Other chapters in this book describe philosophical aspects that underlie modern systems biology. Here we shall set down some of the methodologies of systems biology as we observe them. As a conceptual context coming from practitioners of systems biology, this may then serve for the further development of the philosophy of this science.

4.1. The goals of systems biology

A discussion of what is or should be the methodology of systems biology requires us to be explicit about our goals in systems biology. The main one, of course, is to understand more general principles underlying the behaviour and mechanistic workings of the complete biological systems that sustain life. After all, and as we discussed above, systems biology should be a science and not just a technology for analysing special cases. Systems biology should discover new scientific laws, which may relate as much to physical–chemical, organizational and fitness aspects as to biochemical principles. With respect to this aim, mathematics should not take the form of modeling but rather constitute a way of codifying proposed or verified laws or principles. A case in point is the connectivity law of metabolic control analysis (see Fell, 1996; Heinrich & Schuster, 1996; Kell & Westerhoff, 1986; Westerhoff & van Dam, 1987), which can be most strictly formulated after defining a new property (i.e. the elasticity, see above) in mathematical terms.

A second aim then is the ability to understand the inner workings of particular living systems. Ultimately this is best done by having a computational or mathematical model of the system in terms of its components and the quantitative nature of the interactions between them. Such a model could be the result of 'simulation' and 'fitting', the model being adjusted in terms of its structure and/or its parameter values, until it describes the observed system behaviour. That description may then constitute understanding. Such a description corresponds to a mechanistic explanation but now in the systems sense.

However, as in other kinds of modelling (Corne et al., 1999; Kell & Knowles, 2006), we want more: A third aim derives from the ability to make predictions about the possible future behaviour of the system on the basis of changes we might make to our models. This creates possibilities of further testing the quality of the model, which is the third aim of modelling. Using a model to make such predictions forbids its further adjustment whilst calculating the prediction; no fitting should be involved at such a stage. The same is true in machine learning (Duda et al., 2001; Hastie et al., 2001; Rowland, 2003). A related, fourth aim of modelling is the use of the model for technological or therapeutic purposes.

The fifth or ultimate aim of systems biology combines the above; it is the aim of accomplishing the mission of the life sciences and understanding living systems in molecular terms, thereby opening such 'applied' avenues as prognosis, diagnosis, preventive medicine and lifestyle adjustment, therapy, drug design and biotechnology.

Here we have addressed the understanding of biological systems more than their explanation in an evolutionary context. Where we addressed explanation this is in terms of the direct causal mechanisms rather than those that derive from divergence and selection for fitness or stability or observability. After all, biological systems live in the absence of evolution. Our discussion has also refrained from discriminating explictly between the two chief strategies for scientific understanding, i.e. by unification through subsumption to laws and understanding in terms of causation through mechanisms.

4.2. Systems biology: What it is

From the above aims and from the background of the limitations of molecular biology and functional genomics, one may surmise which activities are necessary for a successful systems biology. Many of the tools and techniques of functional genomics are in place as are the techniques from molecular biology and biochemistry. In view of the complexity of the subject matter, and because a focus on parts is ultimately not advised, our present strategy is to focus on a single system of life that is relatively autonomous. Ultimately this should result in a complete living organism being the object of study, and as scientific data and knowledge become distributed and available to all via the Internet this is increasingly possible in a coherent manner. At first these are likely to be unicellular microorganisms, or relatively autonomous subsystems thereof. The mathematical tools will be discussed in more detail below.

Figure 2 therefore shows some of the elements of the systems biology agenda (Kell, 2006). It gives a certain primacy to the system of interest as a circle in the centre. However, while specifics of methods will vary between organisms and

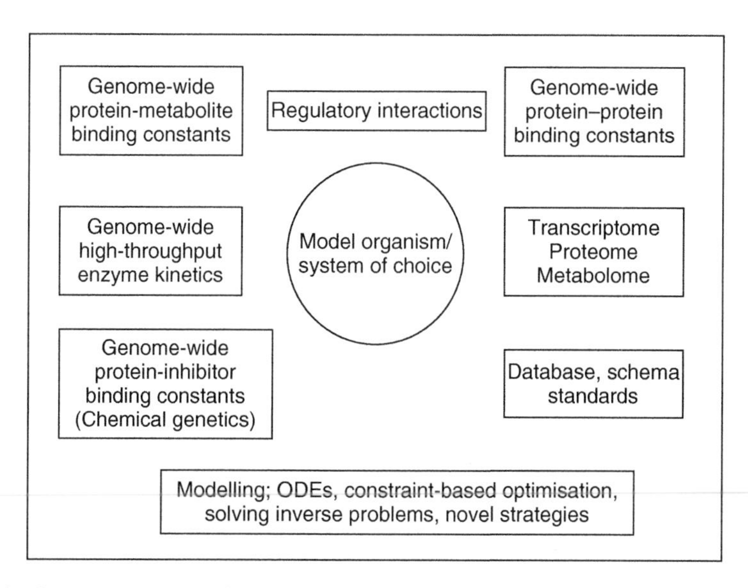

Figure 2 Some elements of the systems biology agenda. These are purposely not interconnected in this figure for reasons of clarity.

systems (e.g. the optimal extraction method for the transcriptome of *Streptomyces coelicolor* – an organism with an unusually high GC content – differs substantially from that for the transcriptome of other organisms), we shall more or less ignore these specifics and here concentrate on generic issues and methodologies.

4.3. The spiral of knowledge

We maintain that for systems biology as well as for science generally, scientific thinking should consist of an interplay between (i) the mental worlds of knowledge and ideas and (ii) the physical world of observations and sense-data. Figure 3 sketches a straightforward view of the relationships between the two worlds, which is usually described as a cyclic interplay between experimental observation and theory, with induction on the basis of experimental observations leading to new, more acute experiments testing the hypotheses. The new experiments should then lead to a further adjustment of the intellectual world view and good hypotheses that derive therefrom. We note then that functional genomics without the systems biology dimension might remain in such a cycle of data collection, pattern recognition and the generation of *ad hoc* empirical 'laws' and hypotheses describing those data phenomenologically. The application of

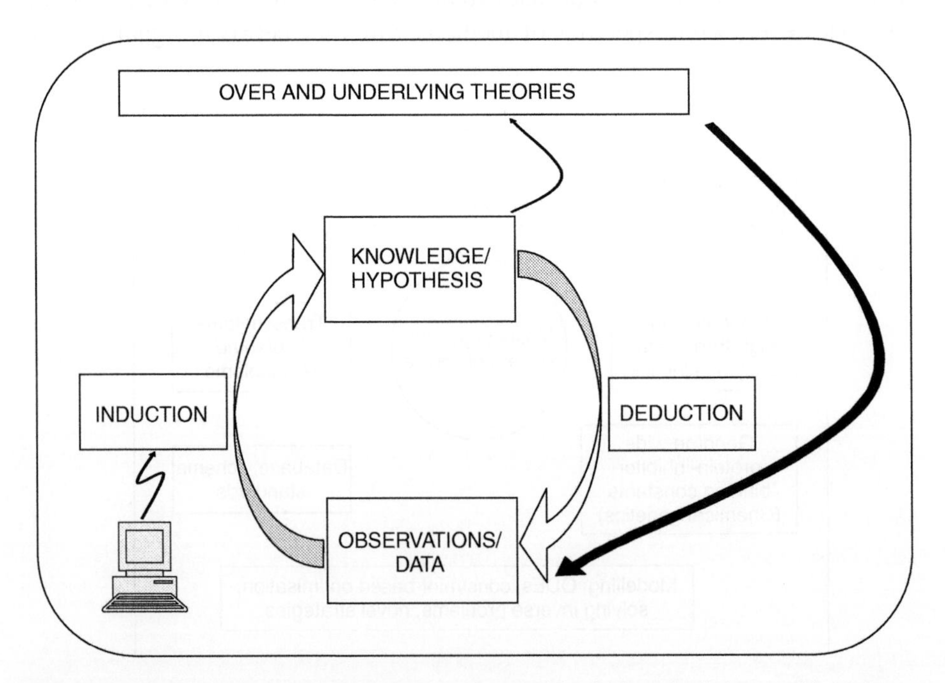

Figure 3 An iterative interplay between the world of ideas and the world of data as the hallmark of both science and systems biology.

systems biology in addition to functional genomics should lead to a progression of insight that is also outside the range covered by the primary dataset. The developing insight is effectively a third dimension, which is one of the aspects that systems biology may help add to functional genomics.

An example would be the observation in a large number of datasets that mRNA for a protein A always goes up or down together with that of protein B. This would lead to the empirical law that proteins A and B always behave similarly. This empirical law would reside on the same conceptual plane as the primary data set and would therefore fit into the cycle picture of Fig. 3. Here the broad aim of functional genomics could be seen to have been satisfied, and experimentation could stop. However, systems biology would search further for the cellular control and regulation hierarchy to find that the two corresponding genes are regulated by the same transcription factor; it would then search for interactions responsible for the correlation. Not only would this explain the observed correlation of mRNA-A and mRNA-B, it would also predict exceptions to these correlations, e.g. when a second transcription-factor footprint would map to gene A but not to gene B. In this way understanding will slowly but steadily grow outside the primary data set and elucidate more and more of cell biology, hence add a dimension of understanding.

We therefore recognize that systems biology may be among the sciences that is better described by a spiral of knowledge rather than a cycle (cf. Fig. 4). A further addition to the traditional vision is that of a box with overlying and underlying theories, with a deductive arrow stemming from that (cf. Figs. 3 and 4). Indeed, any law or hypothesis of systems biology should be consistent with underlying physical–chemical principles and in good systems biology any such hypothesis should therewith also be deduced in part from those underlying principles (this may seem a superfluous remark but we have seen systems biology-type theories that were inconsistent with the second law of thermodynamics and principles of electric fields).

4.3.1. Systems biology: The inductive versus the deductive mode

The recent developments in postgenomics have caused the empirical branch of systems biology, which is closest to functional genomics and stems from the developments in molecular biology (Westerhoff & Palsson, 2004), to develop most strongly. This branch emphasizes the observation component, i.e. the measurement of the dynamic variables. It then establishes patterns in the observed dynamic responses of the system to perturbations, whereby it uses mathematics for the analysis of multidimensional systems. This functional genomics activity tends towards systems biology because it accommodates the feature that the various molecules in the living cell vary coordinately in concentration. Often it is not yet the science of systems biology because it sticks to the observation of the correlations, without necessarily understanding their basis or whether they are

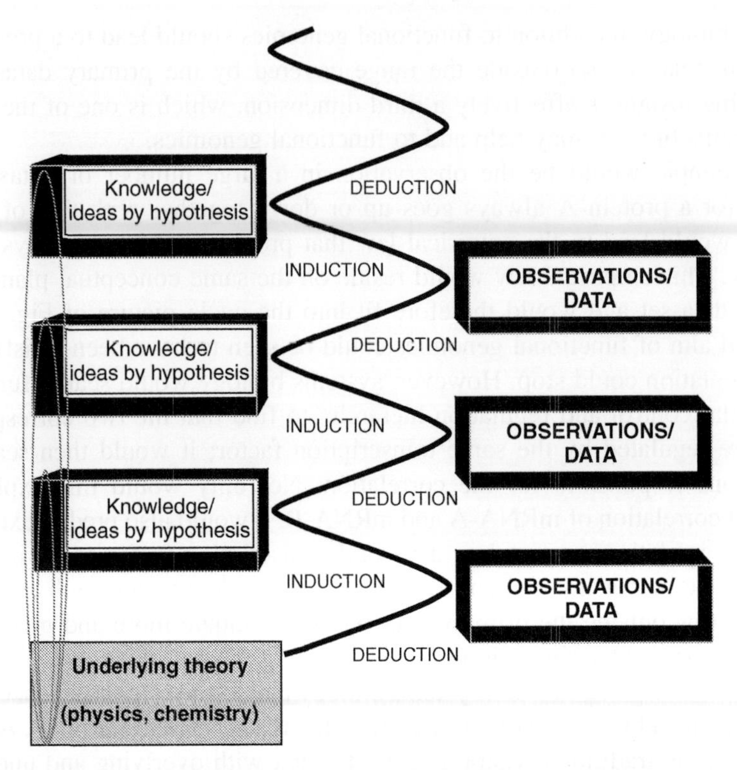

Figure 4 The advancement of Science and of Systems biology as a spiral.

Since the hypotheses are (hopefully) not the same at each turn of the cycle of Fig. 3, one may also or better view the iterative interplay between the elements of Fig. 3 in terms of a spiral.

in an explicit sense causal. This is not to say that this activity is not extremely useful, however, since observations of correlations between the transcriptome of tumours and their response to chemotherapy may help therapy tremendously, long before any mechanistic basis for understanding (and one might comment that this is widely true in medicine).

Functional genomics does become part of the science of systems biology when it makes the step of induction of Fig. 3. In practice, this has not yet happened very often. It seems important to redress the balance by transforming this empiricism into a principled hypothesis-generating arc that leads from data to knowledge. One way in which this can be done is to map the mRNA concentrations that vary coordinately onto the known regulatory maps of cell biology. Perhaps this leads to the recognition of coherent regulation of a pathway, or of a limited number of super-regulators. Either result would lead to a hypothesis which could then be tested further.

The deductive mode of reasoning is a classical obsession of biology, and remains entirely relevant. In the present context, it ranges from branches of mathematical biology and metabolic control analysis which have been deduced from underlying principles, to proposed flux patterns (Reed & Palsson, 2003), or distributions of control (e.g. Hornberg et al., 2005).

By contrast, much of postgenomics and systems biology, in which often we lack reasons or sufficient background knowledge that might lead us to realistically plausible hypotheses, has been data-driven, with a good hypothesis being the result, not the starting point, of the initial investigation. This brings with it a requirement for a different kind of experimental design, in which rather than seeking to hold everything constant except one parameter we seek to vary conditions as much as possible (but in a controlled manner!) to produce a 'training set' of data to establish rules that are likely to generalize well to apply to examples not previously encountered (Kell & King, 2000). This entirely different way of thinking also discriminates the methods of classical statistics (that start with a model and test the goodness of fit of data to that model) from those of machine learning (that start with data and determine the model that best fits those data) (Breiman, 2001).

The chief element of this integrated view of the relation between ideas and data is the recognition that induction is not simply the reverse of deduction (Carnap, 1966; Kell & Welch, 1991). Deductive reasoning starts with an axiom or set of axioms (i.e., a mental construct, the world of ideas, such as 'all swans are white') and a hypothesis such as 'Alice is a swan' that together allow one to deduce with logical certainty that provided Alice is a swan one may make an observation in the expectation that Alice will be found to be white and the data found to be consistent with the hypothesis. Alternatively if Alice is found to be black then either Alice is not a swan or the axiom should be modified (axioms are by definition true). This hypothetico-deductive framework, in which hypotheses can be falsified by data but not proved true, was the focus of Karl Popper's agenda to demarcate 'science' from 'pseudo-science' (Medawar, 1982; Popper, 1992), although one must remark that in the real world some favoured hypotheses can survive in the face of any number of inconvenient facts (Gilbert & Mulkay, 1984; Kell, 1988; Kuhn, 1996).

The inductive mode of reasoning generalizes from patterns observed in a number of actual cases, and thus goes from the world of data to the world of ideas: If Alice is a Swan and is white, Bob is a swan and is white, and George is a swan and is white, an induction might be that 'all swans are white'. Now it has been known since the time of Hume that such induction is logically insecure, in the sense that a single black swan shows it, and that the fact that the sun has risen every morning throughout one's life does not mean it will probably do so tomorrow. However, the existence of black swans is no less harmful to the hypothesis on which the deduction is based that all swans are white than it is to the same view arrived at inductively, and it is not at all clear why induction should in fact be so disfavoured.

The systematic genome sequencing programmes did not set out with any specific hypotheses, save that the provision of such data might be of value (Kell & Oliver, 2004), and Sulston has stressed the importance of hypothesis-free measurements at appropriate stages in the growth of a science (Sulston & Ferry, 2002). Equally, the development of technology is also free of specific hypotheses (again save that their availability would be of scientific value), and it is hard to imagine working in a modern laboratory without techniques (cloning, sequencing, PCR, mass spectrometry, etc.) that have only been available for a comparatively short time (and many of which secured Nobel prizes for their developers). Equally, we see that many measurements, especially in postgenomics (Kell & King, 2000), are designed to be data-driven rather than hypothesis-driven (hypothesis-dependent). Thus in systems biology, science advances by an iterative and spiralling interplay between deductive and inductive reasoning, with a substantial amount of technology development also involved.

Our description of the (preferred) development of systems biology as a spiral, should not be taken to imply that we think of this as unique to systems biology. The development of many other natural sciences may be and have been described in similar terms. They can easily be represented as 'the cycle of knowledge' (Fig. 3).

It should also be mentioned that in many presentations of the novelty of systems biology to audiences of biologists, physicists and chemists, the cycle of knowledge is presented as something that can now finally be brought into effect in biology. This has reasons. First, in biology the experimental activities have become so complex and extensive, and demand such extensive experimental expertise, that the corresponding scientists have had little opportunity to engage in the complete cycle of knowledge. Second, molecular cell biology has long been incomplete in the sense that at any moment an as yet unknown molecule could turn up and explain experimental phenomena without having implications of the theories being tested or examined. For instance, when a hypothetical regulatory effect proposed by a theory is tested by an experiment, an additional, parallel effect would most often turn up, incapacitating the experimental testing of the theory. With functional genomics, it has become possible to have a complete inventory of virtually all relevant molecules, removing this limitation to the testing of theories. Third, in the case of systems biology, the complexity is often so great that the experimental and theoretical parts of the cycle cannot be within the expertise of the same individual. Therewith the cycle of knowledge is also relevant to indicate the roles various individuals in a project have with respect to each other.

4.3.2. *Systems biology: The top-down/analytic versus the bottom-up/synthetic strategies*

Strategies and methodologies for systems biology come in a number of flavours, often discriminated as top-down and bottom-up, but also potentially including middle-out (e.g. Brenner, 2001; Noble, 2003). While the true understanding of complex living systems and/or their subsystems will likely involve the judicious and iterative blending of each, it is convenient to use this distinction as a means of discriminating the necessary methodologies.

Analytical or top-down systems biology tends to start from the system as a whole. In a way it comes from the direction of holism and moves towards molecular mechanism. Either from empirical relations between genome-wide patterns of gene expression, or by calculating properties of genome-wide networks, it induces or proposes the occurrence of more general principles, such as the feature that metabolic networks correspond to small world, scale-free networks (Barabási & Oltvai, 2004; Wagner & Fell, 2001) and that genetic networks abound in certain regulatory motifs (Itzkovitz & Alon, 2005; Milo et al., 2002; Yeger-Lotem et al., 2004). These views may then be tested.

In the leaner, 'Synthetic' or bottom-up branch of systems biology, one typically starts with a qualitative ('structural') and often simple model of molecules interacting with each other in networks, then seeks to determine what system properties might emerge from the nonlinear interactions. By then parameterizing the equations that describe these interactions and inserting parameter values that correspond to actual subsystems, more or less realistic predictions of system properties are achieved. When the predictions are accurate, the proposed mechanisms of emergence of the functional properties are considered to have become more likely. This method is reductionist in that it prefers to deal with simple parts of the true system but not so simple as to lose important aspects of the interactions and the emergence of interesting functional properties. 'Bottom-up' methods start with purified entities (e.g. proteins) that allow the measurement of the parameters, while 'top-down' methods seek to infer their values via 'reverse engineering' of the parameters values through fitting of the calculated system behavior to experimentally observed system behaviour.

4.3.3. The bottom-up approach to systems biology

Our own prejudices – given a historical focus more on metabolic than signalling systems (Kell et al., 1989; Kell & Westerhoff, 1986; Mendes et al., 1996; Pritchard & Kell, 2002; Raamsdonk et al., 2001; Teusink et al., 2000; Westerhoff & Kell, 1987; Westerhoff & Kell, 1988; Westerhoff & Kell, 1996; Westerhoff et al., 1991), and on unicellular organisms rather than the more obviously (cf. Davey & Kell, 1996; Kell et al., 1991) differentiated 'higher' organisms – leads us to concentrate more on the 'bottom-up' approach (Fig. 5), embodied in the 'silicon-cell' concept (Westerhoff, 2001): if we can measure all of the 'local' properties of individual players in a complex system, including their interactions, we can bolt the system together and whatever new properties may emerge will indeed emerge and produce the 'whole system' properties that can indeed be compared with those of the intact system. The apotheosis of this approach to

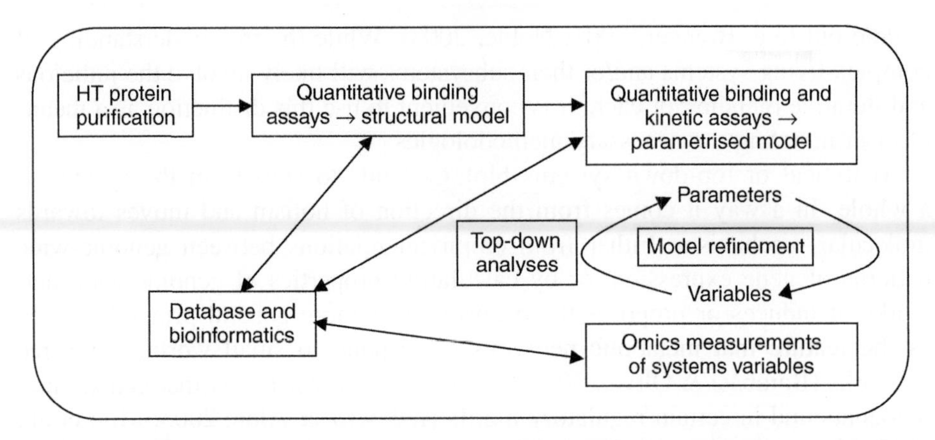

Figure 5 A largely 'bottom-up' strategy for systems biology.

date is the demonstration that the operation of yeast glycolysis under particular conditions can indeed be rather well predicted on the basis of the 'properties' of the isolated enzymes which participate in the overall process (Teusink et al., 2000) (and see (Pritchard & Kell, 2002)). It takes its strongest form when the interactive properties of all the relevant components of the system are put into a precise mathematical model, that is a computer replica ('silicon cell', see below) of the actual system; and if the system behaviour is then calculated successfully.

Occasionally it is argued that such a silicon-cell replica of an actual living cell would be completely reductionistic and therewith incapable to deal with the systems biology of the living cell. This is incorrect. Save for vital force influences, and given an initial physiological condition (cf. below), all there is in the living cell, at least in one way of looking at it, is a large number of molecules and all their interactions. Therewith, all that matters is the components and the relational properties of those molecules. If molecules and interactions (in their spatial context) are precisely reproduced in a computer program, then all system behaviour should emerge. The crux resides in the live interaction between the molecules both in the cell and in the computer program. Here one type of macromolecule carries out a process for a little while, by which it changes its environment in terms of a few, nameable properties such as the concentration of micromolecules like ATP, whilst leaving the rest of its environment unaltered (see below). The change in environment leads to a change in behaviour of other types of macromolecules in the same environment in the same cell (e.g. other enzymes in the same metabolic pathway). The altered behaviour of the latter molecules will again change the environment of the first macromolecule and therewith the behaviour of the former. In this way the activity of the first molecule depends on its own properties through the dynamic activities of the other molecules. Loosely formulated, it is the resonance with other molecules that determines much of the behaviour of each individual molecule. In biology, this part of the molecule's behaviour often leads to important function. An example of the molecular behaviour that only originates in the dynamic interactions with the other molecules, is found with the molecules that are 'responsible for' the cell cycle. None of these would have a cyclic activity in the absence of the others, and this collective cycling is assumed to be the only biological function of these molecules.

The ultimate silicon-cell strategy completely recovers the emergence of functional behaviour of molecules from this resonating with the other molecules. A completely reductionistic approach would look only at the behaviour of the individual molecules, perhaps in an environment that is a frozen representation of the molecules' environment in the living organism. It then sees the behaviour of the living organism as the sum of these molecular behaviours, and thereby misses the extra molecular behaviour that stems from the cycle of interactions running through the other molecules. It would not comprehend the cell cycle, as it would perhaps observe but not explain the cycling.

An important issue is whether the silicon cell requires only molecular knowledge or also systems knowledge to start from. For sure, it does not require systems knowledge of the resonating type (cf. above). On the other hand, the systems of interest are nonlinear and the response of the molecules to the changes in their immediate environment do depend on the average state around which these changes occur, such as intracellular pH and ionic strength. The latter are indeed established by the system as a whole, and in this sense systems properties that correspond to the static physiological state do enter the silicon-cell models. These properties are static in the sense that they could be determined by taking a photograph (Kell and Mendes, 2000), or when they are time dependent, by a movie of the system around the macromolecule of interest. These properties are essentially parameters for the functioning of the interacting macromolecules, whereas the properties that create emergent properties are dynamic variables (cf. below).

As in fundamental physics, there could be cases where it is not really possible to consider macromolecules separately from their molecular environments. In these cases, their complete environment is codetermined by the dynamic behaviour of the macromolecules of interest. Then also, that entire environment consists of variables that are influenced by the macromolecules under study. This might (but would not have to) happen with regards to amino-acid residues in the system of the surrounding amino acids in a protein, or in MAP kinase cascades when all the kinases and phosphatases form a supercomplex, a scaffold.

The silicon-cell approach assumes that there is substantial possibility to consider macromolecules separately from their environments. In cases where parts of that immediate environment is not separable, that part needs to be taken together with the macromolecule. This then still does not incapacitate the silicon-cell approach. If the inseparability is so massive that effectively the entire living cell has to be treated as a single macromolecule, the silicon cell approach does become impractical.

This issue has been alluded to in Boogerd et al. (2005). In the philosophical sense, they have defined the generation of new properties in those systems where macromolecules can be considered as separable from their physical–chemical environment as weak emergence. The cases where macromolecules are not separable from their environment would lead to strong emergence. We would here suggest that it will be possible to make all essential properties of living organisms emerge from silicon-cell-type models. This then implies that all functional properties of living systems come from weak emergence. We base this conjecture on the experience that free-energy transduction, gene expression, cell cycling and developmental biology can be generated by such models (cf. www.siliconcell.net). However, it is a conjecture at present; although these functional properties can be calculated, it has not been verified by experimental testing whether the models generate the functional properties in a quantitatively correct way and from the actual kinetic properties of the constituent macromolecules. And then, there are cases where function arises, where such calculations have not yet been possible, such as in the cases of epigenetic regulation of gene expression.

4.3.4. Parameters and variables and who controls whom

An important distinction to be made in systems biology (and not only there) is between parameters and variables. Parameters are elements set to fixed values by the system itself or controlled externally by the experimenter, while variables are those elements that change during the course of an experiment. (Note that the elapsed time, though in fact a variable, is normally considered an honorary parameter.) In an isolated metabolic system in which protein synthesis and degradation are not occurring, the parameters are then the concentrations, and especially the kinetic and binding constants, of the enzymes involved, as well as the 'fixed' concentration of 'external' substrates. The variables are then the time-dependent concentrations of the intermediary metabolites and the flux(es) through the pathway or network of interest. Two facts are to be noted. First, only parameters can control variables; and variables cannot control other variables. Parameters are controlled neither by other parameters nor by variables. Secondly, normally it is variables that are measured experimentally, as such measurements of changes are easier – and this statement includes all the 'omics' ('expression profiling') methods such as transcriptomics, proteomics and metabolomics. Given these facts, it is seen that there has therefore been a very great dearth of systematic measurements of the properties that we actually wish to measure, viz. the binding and kinetic constants of individual proteins (and other molecules). Such measurements were commonplace in the 1960s and early 1970s (a large number of papers in the journal *Biochemistry* at that time were entitled 'purification and properties of <some enzyme>'), and we need these times to return to biology, with concomitant modernization of the way in which and the scale at which the experiments are done. Indeed, in an account of what needs to be done by bottom up systems biology, one finds many 'old-fashioned' looking terms (cf. Table 1).

4.3.5. Strategies for determining binding and kinetic constants for individual proteins

In the spirit of Mrs Beeton (Beeton, 2000), 'first get your protein'. While these will still require purification, often via dual affinity tags, they will normally

Table 1 Some methodologies of significance for 'bottom-up' systems biology

Stages	Methodologies	Comments	Selected references
'First get your protein'	Cloning, expression and purification	Choice of hosts and vectors, tags, growth media, glycosylation and refolding	
Qualitative binding assays	Mass spectrometry and FTIR	Allows production of a structural model. The binding of some elements may depend on that of others.	(Muckenschnabel et al., 2004; Wharton, 2000; Zehender et al., 2004)
Quantitative binding assays	Mass spectrometry	High-resolution methods such as FTICR are useful	(Last & Robinson, 1999)
High-throughput kinetic methods	Optical, mass spectrometry and calorimetry		(Shen et al., 2004; Ward & Holdgate, 2001)
Omics measurements	Microarrays and mass spectrometry		(Aebersold & Mann, 2003; Goodacre et al., 2004; Schena, 2000)
Bottom-up model	ODE modelling		(Mendes & Kell, 1998)

be prepared by recombinant means. We shall not dwell here in detail on these methods, save to note that the systematic production of nominally *all* the proteins of baker's yeast (*S. cerevisiae*) has been performed by Snyder and colleagues (e.g. Phizicky et al., 2003; Zhu et al., 2001) and in this sense the industrialization of such processes has begun (see also, e.g. for *C. elegans* – http://sgce.cbse.uab.edu/). It is also worth pointing out that even in well-established recombinant hosts there is a nonlinear interplay between the specifics of the recombinant vector, the exact host strain and the growth and production media used to induce the synthesis of the target protein of interest in a form that allows successful purification and refolding.

The next stage is represented by qualitative binding assay, by which we seek the 'structural model' that describes the players including substrates, products and effectors of enzymatic reactions, protein–protein and protein–nucleic interactions and so on (see Fig. 5).

4.4. The special role of mathematics in systems biology: Calculating emergence

As do most commentators (e.g. Hood, 2003; Ideker et al., 2001; Kitano, 2002; Naylor, 2004/2005), we (Kell, 2004; Kell, 2005; Kell, 2006; Kell & Knowles, 2006; Westerhoff & Palsson, 2004) consider systems biology to involve an interplay between theory, computation/modelling and experimental activities. This interplay is strongly catalysed by the development of new technologies, and in fact it is these developments more than anything else that has accelerated the subject (Hood, 2003). It should be noted that Fig. 6 differs rather significantly from Fig. 3, which we presented as our standard paradigm for scientific activity. Indeed, we should like to suggest that in systems biology as in other systems sciences, the role of mathematics is more fundamental than it is in sciences that deal with single entities of much lower inherent complexity.

Of course, mathematics helps the analyses of the rather complex datasets in helping to establish correlations, which then feed into the inductive mode of Fig. 3. It helps ordering the data, then remaining in the empirical box of Fig. 3. It also helps formulate the hypotheses and theories inside the box theory of Fig. 3. And it may help deduce experimental implications from the theories, helping the deductive process depicted in Fig. 3. The reasons for modeling are numerous, and covered elsewhere (Kell & Knowles, 2006; Klipp et al., 2005), and include testing whether the model is accurate, in the sense that it reflects, or can be made to reflect, known experimental facts, analysing the model to understand which parts of the system contribute most to some desired properties of interest, hypothesis testing, allowing one to analyse the effects of manipulating experimental conditions in the model without having to perform complex and costly experiments, and seeing what changes in the model would improve the

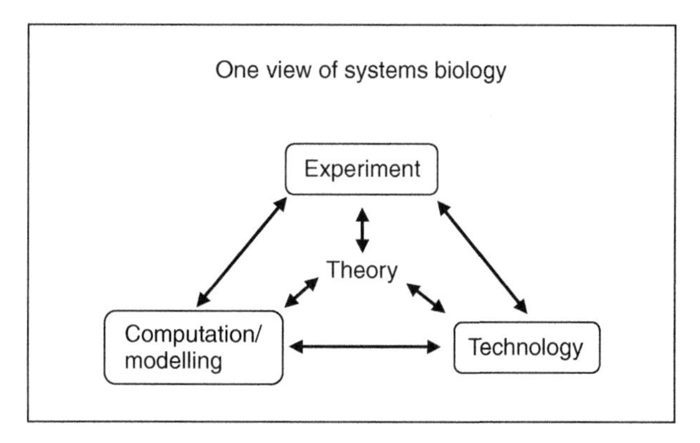

Figure 6 Systems biology as an iterative interplay between theory, experiment and technology development and modelling.

consistency of its behaviour with experimental observations. While these roles of mathematics may be stronger in systems biology than in other sciences, they are not qualitatively different.

The special role of mathematics (which we take to include numerical computation) in systems biology derives from the following. It is an aim of systems biology to understand how properties emerge in the interactions of components of systems. The emergence of these new properties should be completely determined by all those interactive properties. If the interaction properties of the components are correctly known on the basis of experiments with the individual molecule species, then emergence of the new properties in a precise computer model is inescapable. The very emergence is thus not in this direct sense subject to experimental testing. In this aspect systems biology is not subject to experimental testing either. It may be subject to computational testing, however.

In molecular biology similar situations may arise. The properties of a molecule are proposed to have an effect on its behaviour, such as that the adjacency of two glutamate residues in a protein are responsible for the binding of calcium. Usually in molecular biology no time nor effort is wasted in calculating whether indeed in principle the adjacency of the two glutamate residues could enhance calcium binding; this is considered 'obvious' (actually, it may not be quite obvious; protein dynamics calculations should perhaps be carried out; but in view of the many nonlinear interactions involved, this is akin to invoking systems biology). In systems biology it is more often not trivial to see whether a proposed mechanism for emergence could account for the emerging property, even independent of whether the proposed interactive properties are real experimentally. It involves a computational experiment to check if indeed the proposed interactions could generate the emergent behaviour. This is so because the interactions are so complex that an immediate intuitive prediction is impossible, and because the emergence depends on the particular magnitude of the parameter values, i.e. on the particular condition the system is in. (We note, though, that in a sense, such questions about protein engineering are not quantitative, since changing one or both of a pair of adjacent glutamates to alanine may perfectly well change the structure and dynamics of the enzyme irrespective of any effect on their ability to bind calcium.) It is of course well known that even simple systems can exhibit very complex dynamics (Abraham & Shaw, 1992; May, 1976). Accordingly, computation here plays something of the role of experimentation in other sciences. The hypothesis that an experimentally established set of interactions is responsible for certain emergent behaviour in the system needs to be tested by performing calculations.

Although this situation is new to much of the life sciences and was not made very explicit in the original philosophies of physics (Carnap, 1966), it is standard to present-day physics and chemistry. In particle physics and in statistical thermodynamics, certain properties may be known experimentally. The question is then asked whether those properties may be responsible for certain observed behaviour, and the answer is obtained solely by numerical experimentation.

We recently carried out this type of numerical experimental systems biology when proposing that the compound acetaldehyde might be 'responsible' for the synchronization of glycolytic oscillations between individual yeast cells (Reijenga et al., 2005a). Putting in the actual structure of the network in so far as we could, we calculated that the synchronization should indeed occur. More recently, we posed the hypothesis that the glycolytic oscillations in yeast are not controlled at a single step such as the proposed pace-maker enzyme phosphofructokinase, but at many points in the network at the same time. Again numerical experiments based on what was already known experimentally about the interaction and networking in the system, served to verify the hypothesis in the numerical sense (Reijenga et al., 2005b).

We should like to emphasize that in no way do we wish to detract from the importance of experimental work for systems biology. If anything, experimentation is more important to systems biology than to molecular biology, in view of the strong dependence of what actually happens on the precise parameter values. It is just that mathematics is also more important to systems biology than it is to molecular biology.

4.4.1. *Precision, silicon cells and the calculation of emergence*

The calculations we referred to here are often deductive in the sense that they start from a hypothesis and calculate whether indeed the proposed mechanisms of emergence deliver the proposed emergent property. However, calculations in the sense of numerical experiments can also be used to induce general properties.

Indeed, this was involved in the origin of one of the more distinctive laws of systems biology, i.e. the summation theorem as discovered by Jim Burns and the late Henrik Kacser (Kacser, personal communication).

The emergence of properties from nonlinear systems depends on the values of the parameters. The consequence has long been overlooked by theoretical biologists and biologically inspired physicists. The latter supposed that it was good enough to show that some, phenomenological model of the biological system could produce the emergent property of interest. In this manner, Turing modelled developmental biology (in a way that is now known to be wrong, even though parts of the self-organization mechanisms may still act), and Nicolis and Prigogine modelled glycolytic oscillations in yeast. They did find that in such a phenomenological model (with oversimplified and in fact unrealistic rate equations and rather arbitrarily chosen parameter values) the emergent phenomena occurred. For different rate equations or different parameter values, the emergent property did not emerge from the calculations. Hence, to verify whether a proposed systems biology mechanism is indeed responsible for an observed emergent property, the model must be precise in terms of its structure and parameter values. Until recently the handicap was of course that such precise parameter values were not available. (Consequently, the above should not be taken to question the importance of this earlier work in biological physics and theoretical biology.)

With the advance of experimental techniques and thanks to the effort of many scientists, it is now becoming possible to make the required precise models. We refer to these precise models as 'computer replicas' of the real network of interactions or 'silicon cells' (Westerhoff, 2001). In a sense, the silicon cell strategy is entirely reductionist, yet at the same time upwardly compatible with holism (Snoep & Westerhoff, 2005). All the molecules known to act in a network are represented by a computer replica. At present this most often takes the form of a rate equation and a reaction equation for each enzyme. The rate equations, i.e. the reaction equations as well as the values of the parameters therein, should have been established experimentally (here we recognize the irreducibility discussed above) and are all inserted into the computer replica of the network. All the computer then does is let the replica behave through the integration of the equations in time. Emergent properties, if any, should then show up in the computer calculations (*modulo* the statistical error in the measurements).

In this manner, ordinary and partial differential equations may be used to calculate life, i.e. to produce a silicon cell that will display the main properties of the real cell, inclusive of the emergent properties. The implications are unprecedented for the sciences: If there is any place in the natural world where qualitatively new properties emerge, this is life.

In terms of philosophy, we are becoming iconoclastic here however. Emergent properties are sometimes defined as the properties that are irreducible. If properties can be calculated then by some kinds of definition they are not emergent. We consider this definition inappropriate, and it may stem from an oversight of the distinction between linear and nonlinear calculations. Properties that can be calculated from a linear superposition of properties of the components of a system (such as their total mass) should indeed not be called emergent. The important distinction comes when qualitatively new properties can be calculated in systems with essential nonlinear interactions. Only then are the properties new, they were not present in the components, and should indeed be said to 'emerge' (Solé & Goodwin, 2000, *pace* Boogerd et al., 2005).

We here make the challenging statement that life is calculable and can therefore be captured in a computer model. Within 10 or 20 years a silicon cell will have been constructed that accurately describes the main elements and behaviour of a living cell, and therefore can be rightfully considered a replica of the cell. Of course there are some exceptions with respect to a straightforward calculation of all aspects of life. These include deterministic chaos, systems that are extremely heterogeneous, and life beyond its simplest form already present in unicellular microorganisms. This said, a Digital Human, both generated and available *in silico* at a suitably coarse-grained level, will be a fantastic boon for both academic researchers and the Pharmaceutical industry alike; for the latter it may be expected to decrease substantially the present enormous attrition rates of candidate drugs.

The issue of biological evolution too is much more important than suggested by our virtual lack of reference here. However, we have decided here to focus on life as it is at a certain moment in evolutionary history, not on how it came about in the sense of evolution. We think that the explanation of life as such is already a significant and challenging problem that requires systems biology for good answers. Perhaps with this treatise, and certainly with the entire book, we hope to have attracted Philosophers of science to a rapidly developing biology which may well be the place where things are happening in philosophy right now.

ACKNOWLEDGEMENTS

DBK thanks the BBSRC and EPSRC for financial support, and for keeping alive nonhypothesis-dependent science during the dark days and now, and Steve Oliver and Ross King for many interesting discussions. HVW thanks Fred Boogerd and Frank Bruggeman and many participants in the symposium for many thoughts and much food for thought, before, during and hopefully after the event. He is also grateful to the BBSRC, NWO, EU-FP6 and IOP-Genomics for various modes of support.

REFERENCES

Abraham RH & Shaw CD. *Dynamics: The Geometry of Behaviour.* Addison Wesley, Redwood City, CA, 1992.

Aebersold R & Mann M. *Mass spectrometry-based proteomics.* Nature: 422, 198–207, 2003.

Anderson S, Bankier AT, Barrell BG, Debruijn MHL, Coulson AR, Drouin J, Eperon IC, Nierlich DP, Roe BA, Sanger F, Schreier PH, Smith AJH, Staden R & Young IG. *Sequence and organization of the human mitochondrial genome.* Nature: 290, 457–465, 1981.

Bäck T, Fogel DB & Michalewicz Z. *Handbook of evolutionary computation.* IOP Publishing/Oxford University Press, Oxford, 1997.

Barabási A-L & Oltvai ZN. *Network biology: understanding the cell's functional organization.* Nature Reviews Genetics: 5, 101–113, 2004.

Beadle GW & Tatum EL. *Genetic control of biochemical reactions in Neurospora.* Proceedings of the National Academy of Sciences: 17, 499–506, 1941.

Beeton I. *Mrs Beeton's Book of Household Management.* Oxford Paperbacks, Oxford, 2000.

Beynon RJ, Doherty MK, Pratt JM & Gaskell SJ. *Multiplexed absolute quantification in proteomics using artificial QCAT proteins of concatenated signature peptides.* Nature Methods: 2, 587–589, 2005.

Blattner FR, Plunkett G, Bloch CA, Perna NT, Burland V, Riley M, ColladoVides J, Glasner JD, Rode CK, Mayhew GF, Gregor J, Davis NW, Kirkpatrick HA, Goeden MA, Rose DJ, Mau B & Shao Y. *The complete genome sequence of Escherichia coli K-12.* Science: 277, 1453–1462, 1997.

Boogerd FC, Bruggeman FJ, Richardson RC, Stephan A & Westerhoff HV. *Emergence and its place in nature: A case study of biochemical networks.* Synthese 145: 131–164, 2005.

Breiman L. *Statistical modeling: The two cultures.* Statistical Science: 16, 199–215, 2001.

Brenner S. Discussion. In: Complexity in biological information processing (Eds.: Bock G & Goode J), Wiley, Chichester, 150–159, 2001.

Brent R. *Functional genomics: Learning to think about gene expression data.* Current Biology: 9, R338–R341, 1999.

Carnap R. *Philosophical foundations of physics.* Basic Books, New York, 1966.

Corne D, Dorigo M & Glover F. *New ideas in optimization.* McGraw Hill, London, 1999.

Davey HM, Davey CL, Woodward AM, Edmonds AN, Lee AW & Kell DB. *Oscillatory, stochastic and chaotic growth rate fluctuations in permittistatically-controlled yeast cultures.* Biosystems: 39, 43–61, 1996.

Davey HM & Kell DB. *Flow cytometry and cell sorting of heterogeneous microbial populations: the importance of single-cell analysis.* Microbiological Reviews: 60, 641–696, 1996.

Duda RO, Hart PE & Stork DE. *Pattern classification.* 2nd ed. John Wiley, London, 2001.

Dunn WB, Bailey NJC & Johnson HE. *Measuring the metabolome: current analytical technologies.* Analyst: 130, 606–625, 2005.

Dunn WB & Ellis DI. *Metabolomics: current analytical platforms and methodologies.* Trends in Analytical Chemistry: 24, 285–294, 2005.

Fell DA. *Understanding the control of metabolism.* Portland Press, London, 1996.

Fleischmann RD, Adams MD, White O, Clayton RA, Kirkness EF, Kerlavage AR, Bult CJ, Tomb JF, Dougherty BA, Merrick JM & et al. *Whole-genome random sequencing and assembly of Haemophilus influenzae Rd.* Science: 269, 496–512, 1995.

Frauenfelder H & McMahon BH. *Relaxations and fluctuations in myoglobin.* Biosystems: 62, 3–8, 2001.

Fröhlich H & Kremer F. *Coherent Excitations in Biological Systems.* Springer, Berlin, 1983.

Gilbert GN & Mulkay M. *Opening Pandora's box : a sociological analysis of scientists' discourse.* Cambridge University Press, Cambridge, 1984.

Gleick J. *Chaos: making a new science.* Abacus, New York, 1988.

Goffeau A, Barrell BG, Bussey H, Davis RW, Dujon B, Feldmann H, Galibert F, Hoheisel JD, Jacq C, Johnston M, Louis EJ, Mewes HW, Murakami Y, Philippsen P, Tettelin H & Oliver SG. *Life With 6000 Genes.* Science: 274, 546–567, 1996.

Goldbeter A, Gonze D, Houart G, Leloup JC, Halloy J & Dupont G. *From simple to complex oscillatory behaviour in metabolic and genetic control networks.* Chaos: 11, 247–260, 2001.

Goodacre R, Vaidyanathan S, Dunn WB, Harrigan GG & Kell DB. *Metabolomics by numbers: acquiring and understanding global metabolite data.* Trends in Biotechnology: 22, 245–252, 2004.

Hastie T, Tibshirani R & Friedman J. *The elements of statistical learning: data mining, inference and prediction.* Springer-Verlag, Berlin, 2001.

Heinrich R & Schuster S. *The regulation of cellular systems.* Chapman & Hall, New York, 1996.

Hill TL. *Free Energy Transduction in biology: Steady State Kinetic and Thermodynamic Formalism.* Academic Press, New York, 1977.

Hood L. *Systems biology: integrating technology, biology, and computation.* Mechanisms of Ageing and Development: 124, 9–16, 2003.

Hornberg JJ, Bruggeman FJ, Binder B, Geest CR, de Vaate AJ, Lankelma J, Heinrich R & Westerhoff HV. *Principles behind the multifarious control of signal transduction. ERK phosphorylation and kinase/phosphatase control.* FEBS Journal: 272, 244–258, 2005.

Ideker T, Galitski T & Hood L. *A new approach to decoding life: systems biology.* Annual Reviews of Genomics and Human Genetics: 2, 343–372, 2001.

Ihekwaba A, Broomhead DS, Grimley R, Benson N & Kell DB. *Sensitivity analysis of parameters controlling oscillatory signalling in the NF-kB pathway: the roles of IKK and IkBa.* Systems Biology: 1, 93–103, 2004.

Ihekwaba AEC, Broomhead DS, Grimley R, Benson N, White MRH & Kell DB. *Synergistic control of oscillations in the NF-kB signalling pathway.* IEE Proceedings Systems Biology: 152, 153–160, 2005.

Itzkovitz S & Alon U. *Subgraphs and network motifs in geometric networks.* Physical Review E; Statistical, Nonlinear, and Soft Matter Physics 71, 026117, 2005.

Keizer J. *Statistical Thermodynamics of Nonequilibrium Processes.* Springer, Berlin, 1987.

Kell DB. *Protonmotive energy-transducing systems: some physical principles and experimental approaches.* In: Bacterial Energy Transduction (Ed.: Anthony CJ), Academic Press, London, 429–490, 1988.

Kell DB. *Metabolomics and systems biology: making sense of the soup.* Current Opinions in Microbiology: 7, 296–307, 2004.

Kell DB. *Metabolomics, machine learning and modelling: towards an understanding of the language of cells.* Biochemical Society Transactions: 33, 520–524, 2005.

Kell DB. *Metabolomics, modelling and machine learning in systems biology: towards an understanding of the languages of cells.* The 2005 Theodor Bücher lecture. FEBS Journal: 273, 873–894, 2006.

Kell DB, Brown M, Davey HM, Dunn WB, Spasic I. & Oliver SG. *Metabolic footprinting and Systems Biology: the medium is the message.* Nature Reviews Microbiology: 3, 557–565, 2005.

Kell DB & King RD. *On the optimization of classes for the assignment of unidentified reading frames in functional genomics programmes: the need for machine learning.* Trends in Biotechnology: 18, 93–98, 2000.

Kell DB & Knowles JD. *The role of modeling in systems biology.* In: System modeling in cellular biology: from concepts to nuts and bolts (Eds.: Szallasi Z, Stelling J & Periwal V), MIT Press, Cambridge, 3–18, 2006.

Kell DB & Mendes P. *Snapshots of systems: metabolic control analysis and biotechnology in the post-genomic era.* In: Technological and Medical Implications of Metabolic Control Analysis (Eds.: Cornish-Bowden A & Cárdenas ML), (and see http://dbk.ch.umist.ac.uk/WhitePapers/mcabio.htm). Kluwer Academic Publishers, Dordrecht, 3–25, 2000.

Kell DB & Oliver SG. *Here is the evidence, now what is the hypothesis? The complementary roles of inductive and hypothesis-driven science in the post-genomic era.* Bioessays: 26, 99–105, 2004.

Kell DB, Ryder HM, Kaprelyants AS & Westerhoff HV. *Quantifying heterogeneity: Flow cytometry of bacterial cultures.* Antonie van Leeuwenhoek: 60, 145–158, 1991.

Kell DB, van Dam K & Westerhoff HV. *Control analysis of microbial growth and productivity.* Society for General Microbiology Symposium: 44, 61–93, 1989.

Kell DB & Welch GR. *No turning back, Reductonism and Biological Complexity.* Times Higher Educational Supplement 9th August, 15, 1991.

Kell DB & Westerhoff HV. *Metabolic control theory: its role in microbiology and biotechnology.* FEMS Microbiology Reviews: 39, 305–320, 1986.

King RD, Whelan KE, Jones FM, Reiser PGK, Bryant CH, Muggleton SH, Kell DB & Oliver SG. *Functional genomic hypothesis generation and experimentation by a robot scientist.* Nature: 427, 247–252, 2004.

Kitano H. *Computational systems biology.* Nature: 420, 206–210, 2002.

Klipp E, Herwig R, Kowald A, Wierling C & Jehrach H. *Systems biology in Practice: Concepts, Implementation and Clinical Application.* Wiley/VCH, Berlin, 2005.

Kuhn TS. *The structure of scientific revolutions.* Chicago University Press, Chicago, 1996.

Lakatos I. *Philosophical Papers 1: The methodology of scientific research programmes.* Cambridge University Press, Cambridge, 1978.

Last AM & Robinson CV. *Protein folding and interactions revealed by mass spectrometry.* Current Opinion in Chemical biology: 3, 564–570, 1999.

Laughlin RB. *A different Universe: reinventing physics from the bottom down.* Basic Books, New York, 2005.

May RM. *Simple mathematical models with very complicated dynamics.* Nature: 261, 459–467, 1976.

Medawar P. *Pluto's republic.* Oxford University Press, Oxford, 1982.

Mendes P & Kell DB. *Non-linear optimization of biochemical pathways: applications to metabolic engineering and parameter estimation.* Bioinformatics: 14, 869–883, 1998.

Mendes P, Kell DB & Westerhoff HV. *Why and when channeling can decrease pool size at constant net flux in a simple dynamic channel.* Biochimica Biophysica Acta: 1289, 175–186, 1996.

Milo R, Shen-Orr S, Itzkovitz S, Kashtan N, Chklovskii D & Alon U. *Network motifs: simple building blocks of complex networks.* Science: 298, 824–827, 2002.

Muckenschnabel I, Falchetto R, Mayr LM & Filipuzzi I. *SpeedScreen: label-free liquid chromatography-mass spectrometry-based high-throughput screening for the discovery of orphan protein ligands.* Analytical Biochemistry: 324, 241–249, 2004.

Nagel E. *The Structure of Science: Problems in the Logic of Scientific Explanation.* Routledge, London, 1961.

Naylor S. *Systems biology, information, disease and drug discovery.* Drug Discovery World 6, 23–40, 2004/5.

Nelson DE, Ihekwaba AEC, Elliott M, Gibney CA, Foreman BE, Nelson G, See V, Horton CA, Spiller DG, Edwards SW, McDowell HP, Unitt JF, Sullivan E, Grimley R, Benson N, Broomhead DS, Kell DB & White MRH. *Oscillations in NF-kB signalling control the dynamics of target gene expression.* Science: 306, 704–708, 2004.

Nicolis G & Prigogine I. *Self-organization in Nonequilibrium Systems: From Dissipative Structures to Order Through Fluctuations.* Wiley, New York, 1977.

Noble D. *The future: putting Humpty-Dumpty together again.* Biochemical Society Transactions: 31, 156–158, 2003.

Peletier MA, Westerhoff HV & Kholodenko BN. *Control of spatially heterogeneous and time-varying cellular reaction networks: a new summation law.* Journal of Theoretical biology: 225, 477–487, 2003.

Phizicky E, Bastiaens PI, Zhu H, Snyder M & Fields S. *Protein analysis on a proteomic scale.* Nature: 422, 208–215, 2003.

Popelier PL & Joubert L. *The elusive atomic rationale for DNA base pair stability.* Journal of the American Chemical Society: 124, 8725–8729, 2002.

Popper KR. *Conjectures and refutations: the growth of scientific knowledge.* 5th ed. Routledge & Kegan Paul, London, 1992.

Primas H. *Chemistry, Quantum Mechanics and Reductionism.* Springer, Berlin, 1981.

Pritchard L & Kell DB. *Schemes of flux control in a model of Saccharomyces cerevisiae glycolysis.* European Journal of Biochemistry: 269, 3894–3904, 2002.

Raamsdonk LM, Teusink B, Broadhurst D, Zhang N, Hayes A, Walsh M, Berden JA, Brindle KM, Kell DB, Rowland JJ, Westerhoff HV, van Dam K & Oliver SG. *A functional genomics strategy that uses metabolome data to reveal the phenotype of silent mutations.* Nature Biotechnology: 19, 45–50, 2001.

Reed JL & Palsson Bϕ. *Thirteen years of building constraint-based in silico models of Escherichia coli.* Journal of Bacteriology: 185, 2692–2699, 2003.

Reijenga KA, Bakker, BM, van der Weijden CC & Westerhoff HV. *Training of yeast cell dynamics.* FEBS Journal: 272, 1616–1624, 2005a.

Reijenga KA, van Megen YM, Kooi BW, Bakker BM, Snoep JL, van Verseveld HW & Westerhoff HV. *Yeast glycolytic oscillations that are not controlled by a single oscillophore: a new definition of oscillophore strength.* Journal of Theoretical biology: 232, 385–98, 2005b.

Richard P, Teusink B, Westerhoff HV & van Dam K. *Around the growth phase transition S. cerevisiae's make-up favours sustained oscillations of intracellular metabolites.* FEBS Letters: 318, 80–82, 1993.

Rosen R. *Life itself.* Columbia University Press, New York, 1991.

Rowland JJ. *Model selection methodology in supervised learning with evolutionary computation.* Biosystems: 72, 187–196, 2003.

Schena M. *Microarray biochip technology.* Eaton Publishing, Natick, MA, 2000.

Schrödinger E. *What is life?* Cambridge University Press, Cambridge, 1944.

Scrutton NS, Basran J & Sutcliffe MJ. *New insights into enzyme catalysis – Ground state tunnelling driven by protein dynamics.* European Journal of Biochemistry: 264, 666–671, 1999.

Shen Z, Go EP, Gamez A, Apon JV, Fokin V, Greig M, Ventura M, Crowell JE, Blixt O, Paulson JC, Stevens RC, Finn MG & Siuzdak G. *A mass spectrometry plate reader: monitoring enzyme activity and inhibition with a Desorption/Ionization on Silicon (DIOS) platform.* ChemBioChem: 5, 921–927, 2004.

Snoep JL, van der Weijden CC, Andersen HW, Westerhoff HV & Jensen PR. *DNA supercoiling in Escherichia coli is under tight and subtle homeostatic control, involving gene-expression and metabolic regulation of both topoisomerase I and DNA gyrase.* European Journal of Biochemistry: 269, 1662–1669, 2002.

Snoep JL & Westerhoff HV. *Silicon cells.* In: Systems biology (Eds.: Alberghina L & Westerhoff HV), Springer, Berlin, 2005.

Solé R & Goodwin B. *Signs of life: how complexity pervades biology.* Basic Books, New York, 2000.

Sulston J & Ferry G. *The common thread: a story of science, politics, ethics and the human genome.* Bantam Press, London, 2002.

Sutcliffe MJ & Scrutton MS. *Enzymology takes a quantum leap forward.* Philosophical Transactions of the Royal Society A: 358, 367–386, 2000.

Teusink B, Passarge J, Reijenga CA, Esgalhado E, van der Weijden CC, Schepper M, Walsh MC, Bakker BM, van Dam K, Westerhoff HV & Snoep JL. *Can yeast glycolysis be understood in terms of in vitro kinetics of the constituent enzymes? Testing biochemistry.* European Journal of Biochemistry: 267, 5313–5329, 2000.

Vaidyanathan S, Broadhurst DI, Kell DB & Goodacre R. *Explanatory optimisation of protein mass spectrometry via genetic search.* Analytical Chemistry: 75, 6679–6686, 2003.

Wagner A & Fell, DA. *The small world inside large metabolic networks.* Proceedings of the Royal Society B: 268, 1803–1810, 2001.

Walter G, Bussow K, Cahill D, Lueking A & Lehrach H. *Protein arrays for gene expression and molecular interaction screening.* Current Opinion in Microbiology: 3, 298–302, 2000.

Ward WH & Holdgate GA. *Isothermal titration calorimetry in drug discovery.* Progress in Medicinal Chemistry: 38, 309–376, 2001.

Westerhoff HV. *The silicon cell, not dead but live!* Metabolic Engineering: 3, 207–210, 2001.

Westerhoff HV, Hellingwerf KJ & van Dam K. *Thermodynamic efficiency of microbial growth is low but optimal for maximal growth rate.* Proceedings of the National Academy of Sciences USA: 80, 305–309, 1983.

Westerhoff HV & Hofmeyr JH-S. What is Systems Biology? From genes to function and back. In: Systems biology (Eds.: Alberghina L & Westerhoff HV), Springer, Berlin, 2005.

Westerhoff HV & Kell DB. *Matrix method for determining the steps most rate-limiting to metabolic fluxes in biotechnological processes.* Biotechnology Bioengineering: 30, 101–107, 1987.

Westerhoff HV & Kell DB. *A control theoretical analysis of inhibitor titrations of metabolic channelling.* Comments on Molecular and Cellular Biophysics 5, 57–107, 1988.

Westerhoff HV & Kell DB. *What BioTechnologists knew all along . . . ?* Journal of Theoretical Biology: 182, 411–420, 1996.

Westerhoff HV, Koster JG, van Workum M & Rudd KE. *On the control of gene expression.* In: Control of Metabolic Processes (Ed.: Cornish-Bowden A), Plenum Press, New York, 399–412, 1990.

Westerhoff HV & Palsson BO. *The evolution of molecular biology into systems biology.* Nature Biotechnology: 22, 1249–1252, 2004.

Westerhoff HV & van Dam K. *Thermodynamics and control of biological free energy transduction.* Elsevier, Amsterdam, 1987.

Westerhoff HV, van Heeswijk W, Kahn D & Kell DB. *Quantitative approaches to the analysis of the control and regulation of microbial metabolism.* Antonie van Leeuwenhoek 60, 193–207, 1991.

Wharton CW. *Infrared spectroscopy of enzyme reaction intermediates.* Natural Products Reports: 17, 447–453, 2000.

Wolf J, Passarge J, Somsen OJ, Snoep JL, Heinrich R & Westerhoff HV. *Transduction of intracellular and intercellular dynamics in yeast glycolytic oscillations.* Biophysical Journal: J 78, 1145–1153, 2000.

Yeger-Lotem E, Sattath S, Kashtan N, Itzkovitz S, Milo R, Pinter RY, Alon U & Margalit H. *Network motifs in integrated cellular networks of transcription-regulation and protein-protein interaction.* Proceedings of the National Academy of Sciences USA: 101, 5934–5939, 2004.

Zehender H, Le Goff F, Lehmann N, Filipuzzi I & Mayr LM. *SpeedScreen: the "missing link" between genomics and lead discovery.* Journal of Biomolecular Screening: 9, 498–505, 2004.
Zhu H, Bilgin M, Bangham R, Hall D, Casamayor A, Bertone P, Lan N, Jansen R, Bidlingmaier S, Houfek T, Mitchell T, Miller P, Dean RA, Gerstein M & Snyder M. *Global analysis of protein activities using proteome chips.* Science: 293, 2101–2105, 2001.

3

Methodology is Philosophy

Robert G. Shulman

SUMMARY

Systems biology aims to explain functions of the whole organism in terms of component molecules, cells, or structures. Given the varied disciplines that participate in this goal, e.g., genomics, proteomics, and cognitive neuroscience, it is timely to consider if their common strategies can be identified and addressed most generally at their philosophical level. The scope of this book proposes that scientists should respond to the novel questions raised by systems biology with new experimental and theoretical methods. Any methodology both determines and reflects a philosophy so that necessarily methodology is philosophy.

More specifically I propose that systems biology can be studied by a methodology that my laboratory has been developing for more than 30 years, which will be exemplified by one set of experiments. We have conducted magnetic resonance spectroscopic (MRS) measurements of fluxes and metabolites non-invasively in humans, animals, perfused organs, and microorganisms. These experiments measure the parameters that metabolic control analysis (MCA) needs to relate higher level functions to the physical properties of the constituent molecules. The philosophical basis of this study will be presented and contrasted with alternatives.

The studies to be discussed started with ^{13}CMRS measurements of the flux from labeled glucose into muscle glycogen in humans and were subsequently supplemented by *in vivo* concentrations of key metabolites. An MCA analysis of these results showed that the flux was under supply control by plasma glucose concentration and glucose transporters, mediated by insulin. The allosteric, phosphorylated enzyme Glycogen Synthase in the pathway was shown not to

Systems Biology
F.C. Boogerd, F.J. Bruggeman, J.-H.S. Hofmeyr and H.V. Westerhoff (Editors)
Isbn: 978-0-444-52085-2

control flux but to maintain metabolite homeostasis. Hence these results jumped two levels, relating the constitutive biomolecules to homeostasis and control at both the metabolic and physiological levels. These results were connected to the level of systems biology by explaining the metabolic nature of Non Insulin-Dependent Diabetes (NIDD), a disease with both genetic and life style contributions. MRS experiments comparing rates of glycogen synthesis in NIDD patients with healthy controls showed the genetic deficit to lie in a reduced recruitment of glucose transporters to the muscle plasma membrane. Contributions of exercise and obesity upon this recruitment and its mechanism are all being fruitfully studied. Effects of genetics, environment, and the individual's contingent history upon the flux of these pathways are reflected in the biochemical measurements.

Philosophically, these studies are driven by a search for mechanism, where the systemic function is defined in chemical or physical terms. In alternative philosophies, the systemic function is defined by 'clear and distinct' intuitions, such as the explanatory nature of concepts like mind, gene, or structure–functions connections, rather than by mechanisms built upon scientific hypotheses and data. Methods that look for physical correlates of function beg the question of mechanism, rather than answering it. Given the progress made and promised by the studies presented, I see no need to abandon the normal methods and philosophies of physical science.

1. INTRODUCTION

Philosophy had always claimed the right to monitor and judge other fields but in the seventeenth century science declared its freedom from outside values. Descartes conspicuously proclaimed his independence of the existing norms, saying (Descartes, 1996a) of the sciences of his day that 'inasmuch as they derive their principles from Philosophy, I judged that one could have built nothing solid on foundations so far from firm', thereby rejecting a millennium of scholasticism. We need not follow his direction, in no small part this essay intends to reject crucial features of his brilliant, inconsistent philosophy, but the very freedom that scientists have to do so, to hold meetings on the search for a proper Philosophy, derives from his skepticism. He was free, in Holland, to reject the norms, and to create a philosophy built on his scientific studies. And we also are able, once we move away from the prevailing norm with its great worldly rewards, to develop a philosophy that suits our scientific activity. We are in fact even freer because four centuries of scientific progress has made scientists masters of their domain. In contrast to Descartes' time, when Scholastic philosophy made the rules, today traditional philosophy has relinquished its right to be prescriptive, and has yielded to natural science the right to set standards for the methods and goals of understanding nature. A great contemporary philosopher reluctantly acknowledged (Gadamer, 1993) that the ability of philosophy 'to preserve a unity within the totality of what is . . . meets with ever greater mistrust'. The particular mastery of nature offered by science satisfies humanity, Hans-Georg Gadamer acknowledged, and science has assumed the traditional role of philosophy to prescribe normative ways of understanding, at least, he specified, within science's 'particularity'. Scientists' freedom, to create a philosophy that satisfies their needs, is reinforced and accepted by the retreat of traditional philosophy from that prescriptive role.

The efforts that have claimed center stage in the contemporary human endeavors to understand and control the world are the products of scientific methodology. Although many of the questions about the natural world do not have definite answers, still the great authority of science brings humans to science for answers about the unknown. However, there are many mansions in the house of science – there is no one scientific method, and scientific methods of explanation vary. Psychology that considers the brain to function like a Turing computer, sociology that depends upon quantitative statistics, economics that depends upon individuals making rational choices all have developed methods that are properly called scientific, and all these methods, with their implicit goals, have created understandings of their fields that are most properly called philosophy. Hence it is not surprising that the pursuit of a philosophy of systemic biology raises questions about the methods proposed to study the subject, because those methods lead not only to specific explanations but also create a philosophy in which values are defined by the methods proposed and the assumptions made.

Before developing, for the case at hand of systemic biology, the interplay between the philosophies that scientists create and their activities as scientists, let us return to Descartes. With the clarity offered by distance and by Descartes' autobiographical account in his Discourse on Method we can ask how Descartes' personal scientific methods influenced, or even determined, his philosophy.

Imagine that we are in his (enviable) position, in his later formulation of Method, of having discovered analytical geometry in his early twenties. He had started with algebra and geometry, each based upon intuitive, a priori assumptions, whose deductive consequences formed the classical fields of pure mathematics. These fields had neither material content, nor were they dependent upon their applications; they were based on incontrovertible conclusions derived from what appeared to be absolute certainties. Parallel lines never meet, the angles of a triangle add up to 180°. He had combined these fields showing that algebraic functions could be performed by geometric procedures. From this fusion, within a few months he solved problems that had puzzled mathematics since its origin: finding methods for geometrically performing the basic operations of addition, subtraction, division, and multiplication, taking roots, doubling the cube, and providing many solutions of practical questions with great applications to physical reality.

Is it any wonder then that a decade later when presenting a philosophy in his Discourse on Method to replace the Scholastic norms, he should require a starting point with the certitude of mathematics, whose assumptions were intuitive and whose conclusions could be deduced from the assumptions? In his words he planned to accept 'nothing more than what was presented to my mind so clearly and distinctly that I could have no occasion to doubt it.' Questions raised by this 'whole' were to be divided up 'into as many parts as possible' and by starting this way with 'the most simple and easy to understand, to rise little by little, or by degrees, to knowledge of the most complex'(Descartes, 1996b). The Discourse on Method described the methods used in his discovery of analytical geometry and his resolve that 'not having restricted this Method to any particular matter, I promised myself to apply it as usefully to the difficulties of other sciences as I had done to those of Algebra'.

The Discourse on Method is often considered the origin of modern science and modern philosophy. Its clear formulation and subsequent methodological modifications provide guidance for understanding the roles of a priori assumptions and empiricism in both science and philosophy. In broader terms, Descartes' method has developed into the normative method of science. The clear and distinct assumptions of Descartes' scientific method changed in his lifetime in response to empirical evidence, and his method moved significantly towards the modern empirical-hypothetic model of physical science. As a philosopher and a geometrician, Descartes demanded certainty of knowledge but as historians have shown 'He settled for empirically confirmed hypothesis when the subject was nature' (Massa, 1996). The need for a clear and distinct beginning that was incontrovertible wavered in the face of Descartes' wide-ranging experimental studies, and the growing explorations by his contemporaries. For modern physical science, the a priori nature of its starting assumptions were soon replaced by hypotheses that were continually modified in response to experiment and observation. Newton's laws were innovative hypotheses formulated in response to and in explanation of experimental results. As hypotheses they were civilization's glory but in time, as Einstein showed, they were, like all our understanding, contingent – not absolute.

Until recently modern physical science was confined to questions about the non-living world. Hypotheses had replaced a priori certainty in the formulation of enquiries, and Cartesian a priori certainties had been banished. However, the recent advances of physical science in understanding the molecules and processes of living entities have encouraged physical scientists to address the larger questions of living organisms. That is the very subject of systems biology and of this book. Genes, minds, languages, and social structures all are being explored in molecular terms. As these phenomena come from other realms of human experience, from continuing philosophical concerns embodied as common sense, their unexamined authority often seems intuitively clear and distinct to the physical scientist. However, their molecular explanations have been elusive. Proposed methods for discovering these mechanisms are the subject of this chapter. In these new directions important questions arise as to the choice of methodology to guide research. Normative methods, responsible for the progress of physical science, are being challenged by questions raised by these larger, less readily defined subjects. To which I have added, as evidenced in my title, that the methodology used in physical studies of life processes, if different from the well-established method of physics, will necessarily interplay with the changing philosophical goals, each forming and being formed in turn.

On the other hand, outside of physical science, in the realms of thought centered on more traditional philosophy, and most markedly in philosophy itself, the acceptance of intuitive, a priori concepts has remained the norm. The concept, in the tradition reaching back to Plato, that any intuitively acceptable phenomena was a suitable question for scientific investigation was formulated by Descartes and remains active today. The confidence that we know clearly and distinctly what the 'whole' is, in my opinion, has been responsible for the confusions that have been developing in modern biology. For example, Descartes' concept of mind identified areas of certainty so that, in contrast to the uncertainty he felt was introduced into observation by dreams and illusion, Mind allowed him to claim with confidence that 'I think, therefore I am'. Even though Cartesian certainties remain active in most philosophies today, unlike physical science where they have been replaced by hypotheses, some modern philosophers are arguing that the pragmatic value of a concept should prevail over its claims to absolute certainty. Richard Rorty (Rorty, 1989) and others have shown that philosophers after Descartes have struggled repeatedly, and inconclusively, to reconcile the immaterial but absolute certainty attributed to mind with the empiricism of observation and self. I see value for scientists in Richard Rorty's re-evaluation of philosophy, which denies the validity of absolutes. He argues that concepts are valuable to the extent they are useful, and not because they are related to standard philosophical questions.

In agreement with Rorty, I regard all understanding of the larger organismic properties, like mind or biological systems, not to be absolute descriptions, found in nature, but contingent descriptions invented by humans. Any absolute definition of the nature of a subject arises from a philosophical understanding and is based on an implicit methodology which is nonscientific. From this position, how can we begin to explain qualities of biological systems such as mind in terms that share the robust, albeit still contingent, understanding derived from the laws of molecular physics? How can we bootstrap ourselves into a physical understanding of higher order functions? How can we take advantage of the validity of physical science to study properties we cannot describe? And finally, how can we distinguish those assertions about higher order functions that can be explained in molecular terms from those that cannot?

The answer is near at hand if we examine what biological scientists actually do. We are, in fact, quite opportunistic using whatever starting points or methods available. We adopt neither a top-down approach starting from a fixed view of the properties of biological system, nor a bottom-up view in which molecular features would be studied regardless of their possible relevance to function. Robert Brandon (Brandon, 1996) analyzed scientific methodology by rejecting both holism in which the higher order function is clear and distinct, as well as the reductionist assumption that enough knowledge about biomolecules will somehow explain the function. Brandon claimed that biology follows neither approach. Nor, he argues, should it. Instead, he says, biologists are indifferent to this distinction and move freely in both directions. Their goal is to find a causal mechanism, and such an understanding can only be achieved, he suggests, by considering parts and whole together. The piston can only be understood as part of an automobile engine. The engine and the car itself must be regarded in these explanations as physical entities. Many disciplines have developed functions for the automobile – for the psychologists it may represent power and freedom, for the oil industry it is the customer with wants and needs, for the sociologist and the economist it plays important roles in technical societies – and although these concepts have great insights in their particular fields they would not help the automotive engineer to design or repair an automobile. We seek to explain the function of living systems in terms of its constituent molecules and their behavior as described in physical and chemical laws. This is possible provided that the function itself can be described in molecular terms. To define the function in nonphysical terms, for example as correlates for psychological or evolutionary concepts is not wrong, but it does beg the question and leave for future study the more difficult problem of molecular mechanism that connects the neural correlate with a psychological concept or the genetic correlate with an inheritable trait.

The methodology which forms my approach to these problems builds upon the experimental program my laboratory has been following for the past three decades (for a review see Shulman & Rothman, 2001 and 2005). I propose to build upon this methodology and, with relevant results achieved by others, to offer a philosophy that can satisfy the multilevel explanations demanded by systems biology. Systems biology, as defined in the introductory chapter of this book, is based on recent research directions. Its first requirement was for novel 'quantitative experimentation under physiological conditions'. The new methodology that I propose can support a practical philosophy of systems biology is an NMR method for following the concentrations and reaction rates of metabolites noninvasively, *in vivo*. This method has been applied to the study of metabolism in humans, animal models, and microorganisms. It is called magnetic resonance spectroscopy (MRS). It resolves and measures NMR spectroscopic signals from *in vivo* metabolites, e.g., glucose, ATP, glutamate, and glutamine that are usually, but not always, small molecules. Measurements of the proton or natural abundance ^{13}C signals of such metabolites can be used to determine their concentrations. These concentrations can be localized using the NMR methods that support the parallel method of magnetic resonance imaging (MRI) which images the internal body. After having identified the metabolites by their characteristic NMR spectrum their concentrations are localized to particular body regions by MRI methods.

The theoretical component of the methodology based on the MRS experimental results is supplied by metabolic control analysis (MCA). An excellent review of the concepts of MCA can be found in Fell (1997). MCA, also described elsewhere in this book, provides definitions of experimental parameters and a logical analysis of the results so as to evaluate these parameters. The parameters describe the control of *in vivo* fluxes in terms of the *in vitro* properties of the constituent enzymes. They also describe the effects of these enzymes and the *in vivo* fluxes on concentrations of metabolites. Therefore, MCA provides a framework for relating the measured properties of the constituent enzymes to the *in vivo* properties of metabolic pathways. It allows one to go up one level of complexity, from molecules to metabolism.

Metabolic fluxes are measured by labeling a molecule, usually at the beginning of a pathway, and following the label by MRS. The most useful label has been ^{13}C, which is only 1.1% naturally abundant. The weak natural abundance allows signals to be obtained from unlabeled metabolites only when they are more concentrated. One particularly useful signal, even in the absence of labeling, has been from the large glycogen molecule, which was early shown to give a well-resolved MRS signal, acting in the NMR spectrum like a small molecule because of its mobile molecular structure. Infusing 99% labeled ^{13}C glucose into the human or animal blood stream allows the flow of glucose into glycogen to be measured even more quantitatively by ^{13}C MRS in the human skeletal muscle, heart, or liver (Fig. 1). This particularly quantitative measurement of fluxes has allowed our experiments to illustrate the inseparability of methodology and philosophy.

Our experiments have explored the pathway of glycogen synthesis, figuratively shown in Fig. 2. (For a review see Shulman & Rothman, 2001.) Plasma glucose is taken up by the muscle via passive glucose transporters. For the ranges of plasma glucose concentrations studied, the transporters operate in a linear range. They can be recruited by insulin, thereby increasing their velocity, but in the experiments described insulin levels were maintained constant. The flux from glucose to glycogen was measured by clamping plasma glucose at two different levels and measuring the rate of incorporation of the glucose into muscle glycogen by ^{13}C MRS. To improve the signal strength, plasma glucose was enriched by infusing ^{13}C-enriched glucose. The plasma glucose concentrations

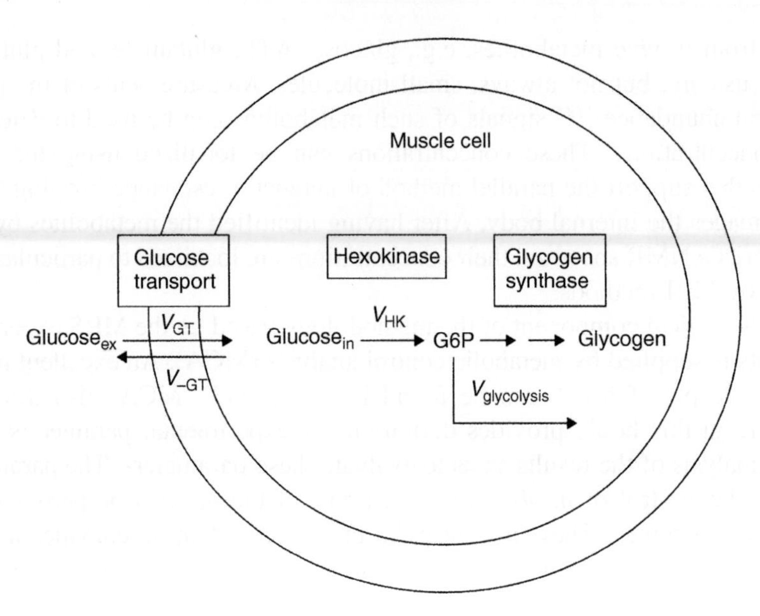

Figure 1 Diagram of potential impaired steps in insulin-stimulated muscle glycogen synthesis in type 2 diabetes.

G6P, glucose-6-phosphate; glucose$_{ex}$, extracellular glucose; glucose$_{in}$, intracellular glucose; $V_{glycolysis}$, net velocity of the glycolytic flux of glucose-6-phosphate; V_{GT}, velocity of glucose transport into the muscle cell; V_{-GT}, velocity of glucose transport out of the muscle cell; V_{HK}, velocity of glucose phosphorylation of hexokinase. (Reprinted from *Metabolomics by* In Vivo *NMR* (eds Shulman RG and Rothman DL) copyright Wiley, England, 2005.)

and ^{13}C enrichments were monitored every few minutes in blood samples. The flux of glycogen synthesis was calculated by measuring the rate of ^{13}C glucose accumulating in a selected volume of the gastrocnemius muscle, and the label flow was converted to mass flow from the substrate enrichment. The validity of the measurement had been established by previous *in vivo* and *in vitro* ^{13}C MRS studies. The MRS accuracy was higher than in the previous method of muscle biopsies and was therefore able to provide meaningful values of the fluxes.

The concentrations of two metabolites in the pathway were also measured in MRS experiments. The intramuscular concentration of glucose was measured by ^{13}C MRS showing that it was very low. Hence, flow through the glucose transporters was essentially unidirectional and linear with glucose concentration.

From the experimental measurements of flux the elasticity (as defined in MCA) of this step with respect to glucose was unity. The responsivity of the flux to glucose was measured, under insulin clamp, by increasing glucose concentrations and measuring flux changes. The responsivity of flux changes to glucose changes was also unity. As responsivity equals the product of elasticity and flux

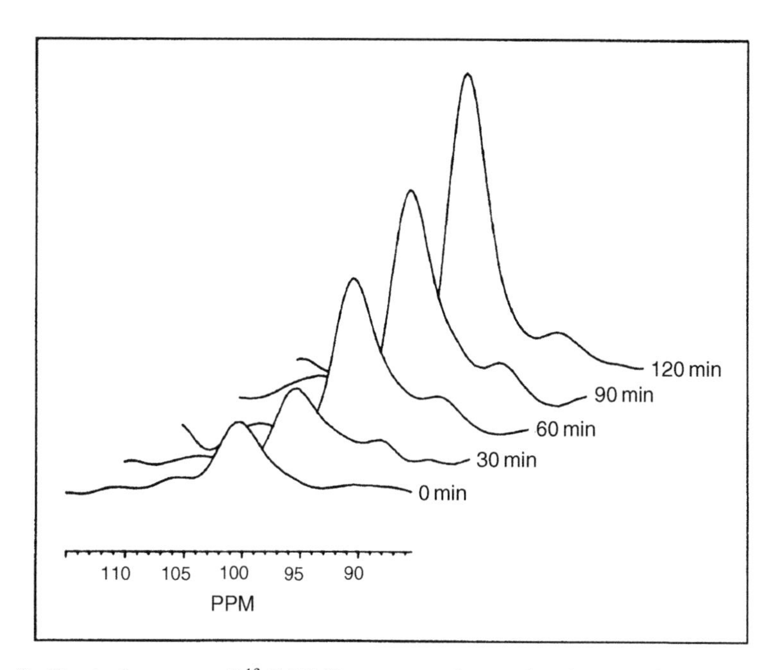

Figure 2 Typical course of ^{13}C NMR spectra of muscle glycogen in a representative normal subject during the hyperglycemic-hyperinsulinemic clamp study.

The C-1 peak of glycogen appears at 100.4 ppm. (Reproduced from Shulman GI, Rothman DL, Jue T, Stein P, DeFronzo RA, Shulman RG, *New Engl J Med* 322:223–228, 1990 by permission of Massachusetts Medical Society).

control coefficient these results showed that the flux was completely controlled by glucose transporters and plasma glucose concentrations, the first step in the pathway. The pathway of glycogen synthesis, in which the flux was controlled by the concentration of the substrate, glucose, is an experimentally determined example of supply control (Fell, 1997).

At this point the summation theory of flux control coefficients showed that all the enzymes in the pathway following the first step had negligible flux control. This means that glycogen synthase (GSase), an allosteric enzyme under phosphorylation control, is not contributing significantly to the control of the flux of glycogen synthesis. As the velocity of GSase is elaborately controlled by numerous second messengers, hormones, activating control pathways that in turn activate kinases and/or phosphatases which regulate the enzyme's activity, this result raises the question as to the function of these regulating pathways. MCA answers this question theoretically, because a top-down analysis of the pathway would show that GSase is expected to be exercising metabolite control. Experimental support came from the third *in vivo* measurement made possible by MRS. The ^{31}P spectrum, of a nucleus that is 100% abundant, was measured by MRS non-invasively in the human muscle. It had been long established by MRS studies that the ^{31}P spectrum gave very well resolved, strong signals from ATP and Phosphocreatine in the muscle, as well as from inorganic phosphate. Guided by the knowledge of the pathway and by the expectation from MCA that the metabolite concentrations were being regulated during changes of flux, the ^{31}P MRS spectrum enabled us to resolve, identify, and quantitate the glucose-6-phosphate (G-6-P) concentrations under different conditions. The results basically showed that G-6-P concentrations changed negligibly when the flux of glycogen synthesis doubled under *in vivo* conditions – a classic case of Metabolite Control or metabolic Homeostasis (Schafer et al., 2004).

Implications of the methodologies used here for the multilevel concerns of systemic biology can be derived from experimental results. In general terms the MRS data showed how the human organism is maintaining constancy of metabolite concentrations during changes in fluxes. On the physiological level, the rate of glycogen synthesis serves to maintain constant concentrations of blood glucose in the face of increased rates of glucose delivery to the blood, generally by ingestion. (During decreased delivery rates glucose levels are maintained by hepatic production.) This is an autoregulatory process in which glucose concentrations drive their uptake and also enhance this flux by stimulating insulin levels that recruit more transporters. The physiological need for such homeostasis is known from the pathological consequences of chronic high glucose levels seen in diabetes.

The biochemical pathway of glycogen synthesis serves this physiological need by varying its flux in response to the blood glucose levels. Furthermore biochemical homeostasis, in which pathway intermediates are constant during changes in flux, is maintained by the complex signaling pathways that regulate GSase activity. Therefore, the MRS experiments have explained flux and metabolic control at both the biochemical level and the higher level of systemic physiology. Most importantly these explanations have been in terms of the measured, *in vitro* properties of the constituent enzymes, structures, and metabolites – all of which have been thoroughly explored at the molecular level. By the methodology of *in vivo* MRS coupled with the explanatory concepts of MCA, we have jumped two levels of explanation, deriving a quantitative understanding of a major process in systemic physiology from molecular data. The methodology of these experiments satisfies the criteria established for systemic biology and has provided prototypical results of the sort expected from this new field.

When one questions the immediate future of these methods it is obvious that several-fold improvements in the sensitivity and resolution of MRS will soon be realized, enabling many more pathways to be followed, although there is no scarcity of possible experiments with present capabilities. Much grander improvements are being developed in which sensitivity will be increased by orders of magnitude. These include gains introduced by polarized nuclei or by taking advantage of the high signal strength of water resonances, commonly used in MRI experiments but which are being increasingly applied to measuring metabolic parameters, such as oxygen consumption and energy production.

The future will also require advances in the theories of metabolic control, which presently guide and interpret experimentation. A most promising study of 'Control in Multi-level Reaction Networks' by Hofmeyr and Westerhoff (Hofmeyr and Westerhoff, 2001) is relevant for extending the experimental results discussed above. They have developed a formal framework for expressing the control of the whole system in terms of its component modules. In our experiments, the pathway of glycogenesis would be one such module and the area of systemic physiology of glucose would be identified as another module. The theory also allowed for the modular properties to change as a result of interactions with other elements of the whole system. The modules are not necessarily at the same level of organization so that the physiological and biochemical levels of our experiment would be suitable inputs for this framework. Interactions between the modules are 'only by means of regulatory or catalytic effects – a chemical species in one module may affect the rate of a reaction in another module by binding to an enzyme or a transport system or by acting as a catalyst'. In our experiments interactions between physiological and biochemical modules are effected by glucose.

2. FROM MOLECULES TO DIABETES VIA METABOLISM AND SYSTEMIC PHYSIOLOGY

Experiments using these methods have extended the physiological understanding to explain the control of glucose metabolism in Non-Insulin-Dependent Diabetes (NIDD). In NIDD, the pancreas secretes insulin but the body does not use it effectively to remove glucose from the blood. The disease has a genetic component, as evidenced by family histories and by its high concordance in identical twins. We know that it also has life style contributions, as a controlled diet and active exercise can contribute to delaying or ultimately avoiding the high blood glucose and its harmful consequences. In early life, the pancreas overproduces insulin which compensates for its ineffectiveness, so that blood glucose concentrations are maintained in the normal range. However, in later life, the overproduction may cease, creating high concentrations of blood glucose that will subsequently damage eyes, muscle, or other organs. On the basis of these properties and the definition of the disease by centuries of medical science, and armed by earlier biochemical studies, we studied this disease in humans by ^{13}C MRS experiments. (for review see Shulman & Rothman, 2001 and 2005, Shulman & Schafer, 2005). We located the particular chemical step in the pathway of glycogen synthesis in NIDD patients that is responsible for the slower clearance of glucose under insulin stimulation (Fig. 3). The MRS experiments

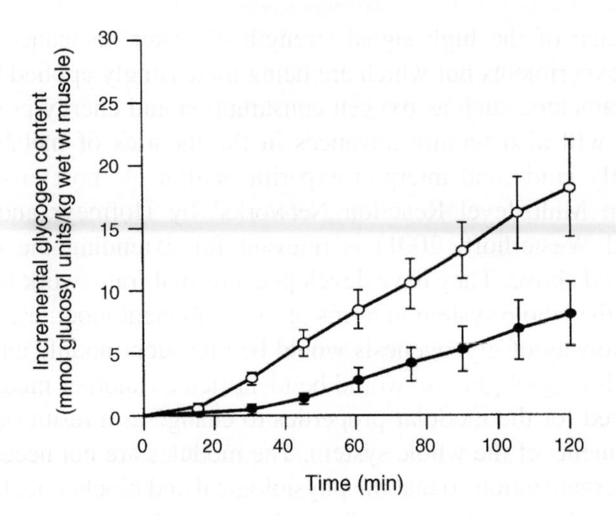

Figure 3 Time course of muscle glycogen concentration calculated from the ^{13}C MRS spectra during an insulin and glucose infusion for noninsulin-dependent diabetes mellitus subjects and healthy controls.

The subjects with diabetes (solid symbols) synthesize glycogen slower than the control subjects (open symbols). Quantitation of the rates showed that insulin-stimulated glycogen synthesis is the major metabolic pathway of glucose disposal in both groups and that the impairment in muscle glycogen synthesis accounts for the reduced glucose disposal in the subjects with type 2 diabetes. (Reprinted from *Metabolomics by* In Vivo *NMR* (eds Shulman RG and Rothman DL) copyright Wiley, England, 2005.)

showed that the flux was controlled by recruitment of glucose transporters to the plasma membrane, thereby controlling the flux of glycogenesis. This recruitment was less responsive to insulin in the diabetic subjects and was responsible for the slower clearance of glucose, which created the disease. The disease did not appear until later life because in the early years the pancreas overproduced insulin to compensate for its reduced effectiveness. Furthermore, we showed how dietary protocols and exercise could restore normal glucose clearance rates by normalizing this particular step. In this disease, we had found the particular pathway whose controlling step was genetically deficient in the patients and had shown that the deficiency worsened in response to the environmental factors of inactivity and obesity. The large questions of nature and nurture were playing out in a chemical pathway. We could also measure and begin to study how the individual's contingent history affected the onset of the disease caused by pancreatic failure. Despite the well-established environmental influence of diet and exercise some slim, athletic subjects became ill while other overweight, couch potatoes did not.

The research described has built upon the cumulative advances in medicine, chemistry, and physics through the centuries. Diabetes was recognized in the

Hellenic Age, and the anomalously high blood glucose had been identified as a pathology whose solution would be beneficial. Reliable advances in the understanding and control of the causes of this disease have been made by our magnetic resonance methods. This progress has assuaged the need for questioning the methods or philosophy of science. Some uncertainties such as pancreatic failure continue as challenges, practical contingencies, which further study will probably explain. Disagreements about how to proceed abound, but presumably many such questions can be resolved by experiment and theory. It is an optimistic, positivistic experimental direction – there is a well-identified question about the ability to store glucose and we are learning more and more about it. There being no reason why we will not continue to close in on the reality – and if there presently are limits to our understanding of pancreatic failure, they are not disquieting because the methods in place can be expected to improve our understanding and control.

3. MRS AND MCA FORM A SUCCESSFUL METHODOLOGY FOR SYSTEMS BIOLOGY

This series of MRS experiments, guided and interpreted by MCA, has created a reliable molecular explanation of a systems-wide human disease condition. The understanding developed by these methods shows the powers implicit in a philosophy built upon the criteria of physics and chemistry and studying a system that can be defined in physical and chemical terms, with inputs from biochemistry and genetics.

Here we return to the initial formulation that as scientists we are free to choose our methods. The method I have been advocating, in the glucose experiments reaching from molecules to diabetes, is that normative physical explanations of mechanism, based on experimental and theoretical physical certainties can, in many important cases, span the gap. However, the systemic property itself must be described in molecular terms that are compatible with the methodology that will be used. In the diabetes/glycogen study, ^{13}C MRS methods for glycogen detection were turned to the question of diabetes because it had been established that diabetes was a disease of glucose storage and in mammals glycogen was the glucose storage compound.

Similarly, present studies of brain energetics, which encompass the rates of glucose oxidation and their relation to neurotransmitter release, are making rapid advances in understanding neuronal activities (Shulman & Hyder, 2004). On the other hand functional imaging studies of psychological concepts, like memory or consciousness, are becoming mired in scientific uncertainties. The conceptual assumptions, e.g., memory or consciousness, that are assumed to be brain activity in an experimental task, are overwhelmed in imaging results by additional brain activities, called 'context' that are not controlled in the experiments, showing the inadequacies of the concepts (Shulman & Hyder, 2004; Fodor, 2000). In these inconclusive investigations, modern functional imaging methods are being selectively interpreted to subserve and explain the 'clear and distinct' assumptions made by contemporary psychology. These concepts are being studied because philosophical and psychological reasons have claimed them to be clear and distinct and therefore, following Descartes, to be worthy of study. However, concepts of Mind from psychology or everyday experience, which incorporate Cartesian assumptions about mind, are not providing a sound scientific explanation of brain activity.

Of course, advocates of large-scale revolutionary changes, such as are envisaged for systems biology, could counter that this is not a time for normal science. I have argued, with diabetes as an example, that questions being asked about systems can be answered by novel but normal science. I cannot be sure that the great data banks becoming available would not be more productive if we pursued revolutionary methodologies than if we persisted in treating them as the data of normal scientific enquiry. However, I can see that, at least in the systemic field of cognitive neuroscience, the new synthesis is not a heroic, revolutionary hypothesis integrating physical data, as Karl Popper proposed, but rather the imposition of Cartesian-like certainties from psychology upon physical methods.

The scientific methodology of cognitive neuroscience resonates with philosophies proposed by other modern scientists. We are familiar with the claims – of particle theorists that universal truths of 'what is' will be explained by a final theory or of particle experimentalists proposing that everything will be explained by a God particle, or of computer scientists building computational theories of mind or, closer to our subject, of Francis Crick proposing that consciousness, like DNA, will be explained by its constituent atoms and molecules – to realize that philosophy is often an extension of scientific method. These examples illustrate the temptations felt by scientists to generalize the results of their scientific methodology and to use them to define the questions posed by larger systems, e.g., cosmology or mental activities of brain. However, the ironic nature of these modern claims, simultaneously assured and exaggerated, suggests that generalizations of scientific methods into philosophical principals for understanding 'what is' must not remain unexamined. These proposals are similar to cognitive neuroscience in that they have offered intuitive certainties as guidelines for research. They represent philosophical principals of Cartesian certainties rather than of modern scientific methodology. The philosophy responsible for this kind of proposal for new scientific synthesis must be examined closely in the formulation of systems biology.

My preference for the advantages of normal rather than revolutionary science echoes an exchange between Karl Popper and Thomas Kuhn in their two essays on 'Science and Pseudoscience' in a collection of papers on philosophy of science. For Popper (Popper, 1998), the value of science depended upon the formulation of revolutionary hypotheses which could be falsified by experiment. However, my views are in sympathy with Kuhn's response (Kuhn, 1998) whose main point was 'a careful look at the scientific enterprise suggests that it is normal science, where Sir Karl's sort of testing does not occur, rather than extraordinary science which most nearly distinguishes science from other enterprises'.

The high value placed upon normal science in the study of systems biology requires that the conditions allowing normal science are satisfied in the particular system considered. Kuhn's central description of normal science was that the research could be done within a paradigm of theories, experiments, and beliefs guiding contemporary science. Diabetes satisfied these conditions, when we include MRS and MCA, because it was identified by a chemical property – chronic hyperglycemia. Accordingly it was understood by identifying the errant steps in glucose metabolism. This rather direct identification was readily made when the flux and control of the glucose storage pathway were measured by MRS and analyzed by MCA. However, the identification of hyperglycemia as the pathological condition had not served to allow the gene(s) responsible for NIDDM to be identified despite years of intensive study, presumably because the causal relationships between genes and the disease were not nearly as direct as they were for the metabolic pathways. In this case, and serving as a guide for the examples of system biology that we propose can be further studied by these methods, the chemical or physical nature of the disease must first be in hand and then the molecular mechanisms can be understood, particularly by the methods we have discussed. But to start from an a priori assumption about the nature of the systemic function, which is necessarily contingent, and then to seek connections with the ultimate molecular properties of a gene or a particular protein, while it might succeed in the rare case, does not offer the general approach to systemic function that systems biology considers its goal.

No one would deny the value of great revolutionary scientific hypotheses that might relate a specific molecule or section of DNA to a systemic function, but while waiting I suggest it is more profitable to continue employing MCA and *in vivo* MRS to study systemic questions by normal science. These methods reflect a reductionist philosophy in which physical understanding is the least contingent goal and where other fascinating but contingent insights about the world are only valued insofar as they usefully lead to physical understanding.

4. CONCLUSION

I have proposed a methodology built upon experimental *in vivo* MRS experiments whose results are analyzed by MCA. To fulfill the goals of systems biology, i.e., to explain systemic properties in molecular properties, this methodology requires that the systemic property be described in physical terms, with chemical components or processes identified. In this event, causal mechanisms can be traced from the molecular level to the systems level. This method does not promise to explain systemic properties that are defined in non-physical terms, as molecular correlates of such properties only beg the question of causality. The choice of our method defines what can be known with the relative certainty offered by physics and that is the philosophy usually known as physical reductionism.

REFERENCES

Brandon RN. *Reductionism versus Holism versus Mechanism.* In: Concepts and Methods in Evolutionary Biology. (Series: Cambridge Studies in Philosophy & Biology) Cambridge University Press, Cambridge UK, 179, 1996.

Descartes R. (1637) *Discourse on Method and Meditation on First Philosophy* (Ed.: Weissman D), Yale University Press, New Haven, 11, 1996a.

Descartes R. (1637) *Discourse on Method and Meditation on First Philosophy.* (Ed.: Weissman D), Yale University Press, New Haven, 131, 1996b.

Fell D. *Understanding the Control of Metabolism*, Portland Press, London/Miami, 1997.

Fodor J. *The Mind Doesn't Work That Way.* MIT Press, Cambridge, MA, 2000.

Gadamer H-G. *Reason in the Age of Science,* 8th Ed, MIT Press, Cambridge, MA, 2, 1993.

Hofmeyr J-HS & Westerhoff HV. *Building the Cellular Puzzle: Control in the Multi-Level Reaction Networks.* Journal of Theoretical Biology: 208, 261–285, 2001.

Kuhn T. *Logic of Discovery or Psychology of Research.* In: Philosophy of Science: The Central Issues (Eds.: Curd M & Cover JA) W.W. Norton & Company, New York/London, 11–26, 1998.

Massa L. *Physics and Mathematics.* In: Discourse on Method and Meditation on First Philosophy. (Ed.: Weissman D), Yale University Press, New Haven, 273, 1996.

Popper K. *Science: Conjectures and Refutations.* In: Philosophy of Science: The Central Issues. (Eds.: Curd M & Cover JA) W.W. Norton & Company, New York/London, 3–10, 1998.

Rorty R. *Contingency, Irony, and Solidarity.* Cambridge University Press, Cambridge, 11, 1989.

Schafer JRA, Fell DA, Rothman DL & Shulman RG. *Protein phosphorylation can regulate metabolite concentrations rather than control flux: The example of Glycogen Synthase.* Proceedings of the National Academy of Sciences, USA: 101(6), 1485–149, 2004.

Shulman GI & Rothman DL. *MRS Studies of the Role of the Muscle glycogen Synthesis Pathway in the Pathophysiology of Type 2 Diabetes.* In: Metabolomics by In Vivo NMR. (Eds.: Shulman RG & Rothman DL), John Wiley and Sons, Ltd, England, 45–57, 2005.

Shulman RG & Hyder F. *Brain and Mind: an NMR Perspective.* In: Brain Energetics & Neuronal Activity: Applications to fMRI and Medicine. (Eds.: Shulman RG & Rothman DL) John Wiley and Sons, Ltd, England, 295–309, 2004.

Shulman RG & Rothman DL. *^{13}C NMR of Intermediary Metabolism: Implications for Systemic Physiology.* Annual Reviews of Physiology: 63, 15–48, 2001.

Shulman RG & Schafer JRA. *Summarized Reflections on Metabolism.* In: Metabolomics by In Vivo NMR (Eds.: Shulman RG & Rothman DL) John Wiley and Sons, Ltd, England, 175–184, 2005.

4

How can we understand metabolism?

David A. Fell

SUMMARY

Metabolism is used as an example of a complex biological system where the explanations that are given for how the properties and activities of the components account for function at the system level have been changing. It is argued that the difficulties in arriving at satisfactory explanations that contribute to understanding arise because biological systems are the result of evolution. One issue is that the system functions attributed to the components are not known *a priori* but are biological hypotheses, and errors in explanations in metabolism have arisen through incorrect assumptions about function. Another issue is that the study of knock-out mutants with no overt phenotypic differences from the wild-type shows that some gene products may confer a significant selective advantage, but their contribution to functionally relevant properties may be too small to be discernable given the usual degree of variability in biological experimentation. Lastly, the concept that each component makes a discrete and identifiable contribution to system function may be true of human artefacts constructed from discrete components, usually with intermediate levels of organisation such as modules, but need not be true of evolved systems. If the functional characteristics of biological systems generally arise from the interactions of many components, with each component making small, but not critical, contributions to various aspects of the overall function, then such systems may be amenable to simulation, given the properties of the components and their interactions, but the models may not be reducible to simpler levels of explanation.

Systems Biology
F.C. Boogerd, F.J. Bruggeman, J.-H.S. Hofmeyr and H.V. Westerhoff (Editors)
Isbn: 978-0-444-52085-2

1. INTRODUCTION

The implied promise that drove the rise of genome sequencing was that, once we had the genetic blueprint of an organism, we would understand the basis of its attributes and its vulnerabilities, first for an average member of the species, but ultimately for specific individuals. If this is realisable at all, then it should surely be possible for an organism's metabolism, because this is an aspect of the phenotype that we know is coded in the genome in a very simple way: genes specify enzymes and transporters, these proteins catalyse chemical and transport reactions, and the resulting network of transformations constitutes the metabolism. Indeed, by the time genome sequencing got under way, metabolism was widely regarded as a problem that had been solved. If that were really true, then there should be little difficulty in designing the changes in metabolism necessary to make cells produce more of a desired product (metabolic engineering) or in selecting a step in metabolism to inhibit so that a pathogenic organism dies whilst the host is unaffected (target selection in drug development). But in neither field is it the norm to be able to accurately predict beforehand the consequences of an intervention. In this chapter, I will consider how our modes of explanation of metabolic phenomena are changing with the rise of systems biology, and whether we are becoming better placed to claim that we understand metabolism. Although I will draw examples from metabolism, other cellular processes such as signal transduction and the cell cycle show essentially similar attributes.

2. TRADITIONAL PRINCIPLES OF METABOLISM

Biochemistry textbooks have almost always contained a section called 'principles of metabolism' that was until recently largely based on the ideas prevalent in the 1960s and 1970s. The distinction between reversible and irreversible reactions and the need to bypass irreversible reactions by one or more different reactions to reverse a pathway (as in the case of glycolysis and gluconeogenesis) were underpinned by arguments from classical thermodynamics. Other concepts seemed to have no stronger foundations other than being claims about recurrent features in metabolism or statements of opinion. The concept of the control of a metabolic pathway by a single rate-limiting step was strongly promoted by Krebs (1957), having originated with Blackman (1905) in the context of the response of plant photosynthesis to environmental factors. There was no real theoretical underpinning of the concept, to the extent that Denton & Pogson (1976) justified it as a truism.

However, the rate-limiting step concept was strengthened by the discovery of feedback inhibition in metabolic pathways (Umbarger, 1956; Yates & Pardee,

1956), that is, the frequent occurrence of the end metabolite of a pathway acting as inhibitor of one of the earlier enzymes in its production. Great significance was attached to the discovery of a control mechanism that was a common motif in man-made devices. Again, however, there was no detailed theory of the operation of feedback inhibition; it was assumed to control the rate of a metabolic pathway by analogy with known control mechanisms. These different strands came together into the generalisation that metabolic pathways began with a committed step (an irreversible reaction) that was often controlled by feedback inhibition and which was the rate-limiting step of the pathway. The best attempts to systematise these beliefs were by Rolleston (1972) and Newsholme and colleagues (Crabtree & Newsholme, 1985; Newsholme and Start, 1973).

These ideas were also presented as criteria for the identification of the rate-limiting step in a pathway. As so often in biology, teleological explanation had come to the fore: the features of the pathways were explained in terms of the function of controlling metabolic rates.

Needless to say, a number of shortcomings with these ideas began to emerge. First, on occasions different groups would examine the evidence and nominate different steps as the rate-limiting step of a pathway. For example, there were conflicting claims for the site of control of mitochondrial respiration, as summarized by Groen et al. (1982). The traditional axioms of metabolic control were not so specific as to be able to decide between competing interpretations. Secondly, the representation of pathways was sometimes made to fit the generalisations rather than the other way round. For example, it is rare to see the pathway for synthesis of the amino acid threonine drawn correctly in any book. The first step, aspartate kinase, uses ATP and is represented as an irreversible committed step, further underlined by the fact that the enzyme is feedback inhibited (by threonine, and usually lysine, depending on the organism). As it happens, the equilibrium constant of the aspartate kinase reaction does not favour the direction of threonine synthesis (Black & Wright, 1955; Chassagnole et al., 2001); the reaction is a close analogue of that of phosphoglycerate kinase from glycolysis when working in the gluconeogenic direction. Hence we have a first pathway step that counters the principles because, although it converts ATP to ADP, it is reversible and is catalysed by a feedback-inhibited enzyme; recent evidence suggests it is far from being a rate-limiting step, which probably does not exist in the threonine synthesis pathway (Chassagnole et al., 2001). Another counter-example to the generalisations is mammalian serine synthesis; this three-enzyme pathway has two reversible steps at the beginning, followed by an irreversible step that is product-inhibited (through non-cooperative, uncompetitive inhibition) by serine. Most control of the pathway is in the last enzyme (Fell & Snell, 1988). This different control pattern is not the result of some chemical or thermodynamic imperative, as the bacterial pathway begins more conventionally with a serine-inhibited enzyme (Pizer, 1963).

3. THE RISE OF SYSTEMS ANALYSIS OF METABOLISM

The occasional criticisms of the rate-limiting step concept made little headway until various individuals and groups began to explore the properties that would be expected of metabolic pathways given that they were composed of enzymes (and transporters) which exhibited nonlinear kinetics with respect to substrates, products, and effectors. Some approached this through computer simulation of actual metabolic systems (Chance et al. 1958; Garfinkel & Hess, 1964, Park & Wright, 1973), though the modest computer power available at the time and lack of good numerical software for integration of systems of nonlinear differential equations made this a difficult undertaking. The other approach was algebraic analysis, often of model metabolic systems, to deduce general properties that might be expected of connected groups of enzymes, usually under steady-state conditions. A key concept introduced by Higgins (1963) was to replace qualitative descriptors such as 'rate-limiting' by sensitivity coefficients that quantitatively measure the response of the rate of a pathway to a change in a parameter such as the activity of an enzyme. Sensitivity analysis itself was not, of course, the innovative step, as it was already used in engineering and economics, but its application in metabolic biochemistry was. Higgins' ideas were extended independently by Kacser & Burns (1973) and Heinrich & Rapoport (1974). Although these two groups initially went about their analyses in somewhat different ways, they eventually agreed that their ideas were compatible and unified them under the title of *Metabolic Control Analysis* (Burns et al., 1985; Cornish-Bowden & Cárdenas, 1990). A key sensitivity coefficient in their analyses was the flux control coefficient, which measures the relative response of the metabolic pathway rate (or flux) to a relative change in an enzyme activity.

At about the same time, Savageau (1969, 1976) was developing *Biochemical Systems Theory*, which included approximations to dynamic behaviour as well as analysis of metabolic steady states. Crabtree & Newsholme (1985) also adopted algebraic analysis, by a methodology widely regarded as intermediate between metabolic control analysis and biochemical systems theory. Both these approaches also implemented sensitivity analysis.

It can be argued that these developments did not introduce any new theory into metabolism, but allowed the formulation of a mathematical description; indeed, Henrik Kacser always insisted on referring to metabolic control analysis and not metabolic control theory. However, with the new analysis tools, old principles could be tested and new generalisations about the characteristics of metabolic systems proposed.

An example of a new generalisation was the flux summation theorem of Kacser & Burns (1973). They showed that the sum of all flux control coefficients of the enzymes in a system on a specified metabolic flux would be one. This surprisingly suggests that there is a limit to the total amount of control that can be exerted on a metabolic flux. Further interpretation depends on ancillary assumptions that cannot be rigorously claimed to be necessary and invariant properties of all metabolic pathways. For example, if all flux control coefficients are constrained to be zero or positive (as would be the case in a linear sequence of enzymes all of which had rates accelerated by their substrates and inhibited by their products), then no flux control coefficient could be greater than one, and could be equal to one only if all the other flux control coefficients were zero. A flux control coefficient of one would mean that the enzyme is a linear controller of the metabolic flux, in other words, a rate-limiting step. However, Kacser & Burns (1973) showed that there was no particular reason to expect such a situation, because their algebraic analysis made it apparent that the flux control coefficient of any enzyme contained terms relating to the kinetic properties of all the other enzymes. Furthermore, they showed that many of the traditional identifying characteristics of the supposed rate-limiting enzymes could not in fact reliably indicate the value of the enzyme's flux control coefficient. Rather, they claimed it was more likely that the total available flux control of one would be shared, though not equally, between all the enzymes of the system. In large metabolic systems with many enzymes, this implied that the average flux control coefficient would be small, and there was no *a priori* reason to expect any enzyme to have a large flux control coefficient. This conclusion cannot be claimed to be an invariant principle, though, and has often been contested. In branched metabolic pathways (which are usual in metabolic networks), some of the flux control coefficients are naturally negative, which relaxes the limitation on the magnitude of the positive flux control coefficients. Indeed, it is possible to think of scenarios where the summation total is different from one, though many of these discrepancies can be avoided by ensuring consistency between the precise definition of the flux control coefficient and the experimental system and its mathematical representation (e.g. Kholodenko et al., 1995). However, 30 years of measuring flux control coefficients have turned up relatively few instances of exceptions to the generalisations of a summation total of one and have been consistent with a tendency for control to be shared between enzymes (e.g. Fell, 1997). Furthermore, the generalisations can be used as a basis for explaining how the genetic phenomenon of dominance of wild type over mutant phenotypes arises (Kacser & Burns, 1981; Porteous, 2004). Another phenomenon that can be explained in similar terms is that of the threshold for the onset of disease symptoms in mitochondrial cytopathies (Letellier et al., 1994). Therefore an argument in favour of metabolic control analysis indeed being a theory is that it not only furnishes explanations in its own domain of metabolism, but it is compatible with observations in other fields of biology such as genetics.

Views of the role of feedback inhibition in metabolic pathways have also been affected by metabolic control analysis and the other forms of sensitivity analysis of metabolic systems. Kacser & Burns (1973) analysed the role of feedback inhibition in a linear pathway on an enzyme at the beginning of the pathway. They showed that the effect of feedback inhibition was to lower the flux control coefficient of the inhibited enzyme and to increase the flux control coefficients of the steps consuming the feedback metabolite. This was an unpopular conclusion, as it contradicted the assertion that feedback-inhibited enzymes were prime candidates for rate-limiting steps. However, the prediction has been subsequently borne out because over-expression of a number of feedback-inhibited enzymes (notably phosphofructokinase; Schaff et al., 1989; Thomas et al., 1997) has resulted in little impact on metabolic flux. Savageau (1974) analysed the role of feedback inhibition in branched metabolic pathways, where there can be different configurations for the feedback loops from the end-product metabolites on enzymes in their own branch and the common branch. The two main types are the nested and sequential configurations (see Fig. 1), which he was able to show would have distinct performance characteristics in terms of metabolite homeostasis, cross-talk between the branches, and the range of conditions under which they would perform appropriately. It is interesting to note that the same sub-networks in different organisms can have different patterns of feedback control (for example, the pathways for synthesis of aromatic amino acids), showing that the regulatory interactions are more plastic than the enzymic reactions themselves.

Hofmeyr & Cornish-Bowden (1991) also examined the properties of feedback inhibition loops. Their approach used a combination of the algebraic relationships of metabolic control analysis and computer simulation of hypothetical pathways. Amongst the issues they addressed was the function of the cooperativity that

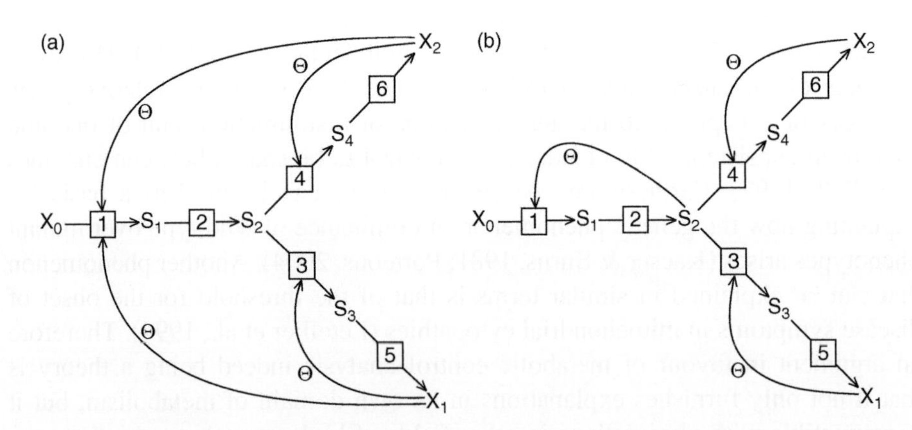

Figure 1 Feedback inhibition patterns in branched pathways.

(a) Nested inhibition; (b) sequential inhibition. Any step shown above can be regarded as being composed of an arbitrary number of enzyme reactions.

is frequently observed in the kinetics of feedback-inhibited enzymes, and that causes the rate of the isolated enzyme reaction to show much greater sensitivity to changes in the concentration of the inhibitor. As feedback-inhibited enzymes had in the past been regarded as candidates for rate-limiting steps, their cooperativity was interpreted as a mechanism that improved the control of metabolic rate. In fact, the analysis by Hofmeyr & Cornish-Bowden showed that varying the degree of cooperativity of the inhibition of an enzyme in a simulated system made little difference to the control of the metabolic rate, but resulted in better homeostasis of metabolite concentrations in response to perturbations of the system as cooperativity increased. They were also able to show that uncompetitive inhibition is necessary to make the control pattern of mammalian serine synthesis (see above) work effectively (Hofmeyr & Cornish-Bowden, 1996).

These are merely selected examples of how the application of new modes of analytical thinking can increase our understanding of the way metabolism functions and can lead to predictions that are different from those derived using axioms such as the existence of rate-limiting steps. However, they are sufficient to illustrate a number of issues:

(1) As already stated, these methods of systems analysis contain no theory of how the regulatory and control mechanisms of metabolism are or should be constructed; they are frameworks for deducing the consequences of particular configurations of enzymes and mechanisms.
(2) The analyses point at features that might be generally expected in metabolic pathways, whilst at the same time they allow the deduction of feasible circumstances under which such generalisations would not be true. In other words, there is always demonstrable scope for exceptions that preclude any of the results, such as the flux summation theorem, being presented as invariant laws of metabolism. However, there are aspects that can be seen to be consistent with other concepts in biology: the tendency for flux control to be shared between enzymes so that the values of flux control coefficients are small can be seen as a likely result of maximising the fitness of metabolism through natural selection subject to limitations on the amount of protein that can be committed to enzymes (Kacser & Beeby, 1984; Heinrich & Schuster, 1998; Klipp & Heinrich, 1999). The link with the concept of dominance in genetics has been mentioned above.
(3) A fundamental weakness of teleological explanations is evident: if they are to contribute to understanding, the system-level function fulfilled by a component must be correctly identified. In biology, however, the functions of a component at this higher level (as distinct from the activities of

the component itself)[1] are not necessarily certain; attributed functions are themselves hypotheses that need to be tested before they are used as axioms for the development of further theories. In the case of feedback inhibition, the initial assumption was made too readily that the control of metabolic rate (an observable property of the system formed by an enzyme network) was the function that the inhibitory activity served. Designation of enzymes as the rate-limiting steps of pathways was another such error, attributing a function to such enzymes that in fact they could not perform. This issue of identification of function will be considered further below. In attempting to show how the generic properties of metabolic systems could have arisen through evolution by natural selection (as described in the previous point), it is important not to commit the same mistake: any claim for a potential contribution of some aspect of the activities of metabolic networks to the fitness of an organism (the system of which metabolism is a component part), such that it has therefore been subject to favourable selection, is also a hypothesis that needs to be tested.

I have stated that I believe systems analysis of metabolism has improved our understanding, but could this be over-confidence? What are the limitations to our analyses? One limitation is that the models we analyse, whether theoretically or experimentally, are very much more simple than the metabolism of a cell, in many respects. Even the structure of the pathways we consider is simplified; much of the time we consider a few steps, most often in a linear sequence, and occasionally with a single branch-point or in a cycle. Further, most of the steps are represented as conversions of a single substrate to a single product, whereas enzymes that work in this way are a minority; three quarters or more use more than one substrate or yield more than one product, or both. The cosubstrates and coproducts in such reactions create links between different metabolic pathways and couple the fluxes of carbon to fluxes of nitrogen, sulfur, phosphate and electrons. We often assume that we can isolate purely metabolic levels of control from the control and regulation of metabolism by signal transduction, gene transcription and translation, post-translational modification, and enzyme degradation. It is true that it is possible in principle to extend our analyses to include such phenomena (e.g. Kahn & Westerhoff, 1991; Small & Fell, 1990; ter Kuile & Westerhoff, 2001), but experiments to test the conclusions become more difficult. On top of this, it is impossible to adequately represent the full

[1] 'Function' can be used in different senses in biology; at one level it can be used to refer to activities performed by components of a biological system, but it also extends to properties the system gains from having the component. I am not including the former at all in my use of the term function, only the latter. Wouters (2003) divides these higher level aspects of function into three further categories, but I do not wish to be that specific here given that we are not being prescriptive about what level of organisation constitutes a system in systems biology.

complexity of the molecular environment in cells in our theories. How much all these details matter is controversial. In complex multilevel systems, it is inconceivable that all the properties at one level have equally strong influences over the properties and behaviour at higher levels, but even this presupposes that levels or modules can be identified as discrete entities on the basis of time or distance scales or of some other measure of the clustering of the components involved. The paradox is that the only way to be certain that a simplified model is an adequate representation of the properties of a more complex system is to have a more detailed, complex (theoretical or experimental) model whose properties can be compared. But usually, the reason for making a simpler model is that the more complex model is not tractable in the first place. This may offer a role for simulation, which will be considered at the end of this chapter, but first there is a more fundamental question to be addressed.

4. SHOULD WE EXPECT METABOLISM TO BE UNDERSTANDABLE?

As mentioned above, function plays a key role in biological explanations. To some extent this is a legacy of religious and anthropocentric belief systems, but the theory of natural selection implies that unless a component of an organism is truly vestigial, it makes some contribution to the fitness of an organism and it is this that constitutes its function. At a high enough level of organisation in the organism, this is unlikely to be difficult to discern or controversial. No one is likely to argue against locomotion as a function of legs, or gas exchange as a function of lungs. It is how these higher level functions arise from the activities of lower level components that is more problematic, and this has been increasingly evident as techniques have been developed for targeted gene knockouts. In every genome that has been sequenced to date, there is a substantial proportion of putative genes for which no biological activity is known. This may start at over 60% of the genes for a newly sequenced organism, and falls with time, but it is difficult to get this down even to 50%. One technique for identifying possible functions of gene products is to selectively knock out each gene in the organism in turn and determine the effect on the phenotype. This has been done systematically for the yeast *Saccharomyces cerevisiae* in the Eurofan project (Dujon, 1998), but a large fraction of the knockouts show no overt phenotype. There are two main explanations for the roles of these genes: either they only prove necessary under environmental circumstances that have not been examined in the laboratory or the difference that they make is too small to be detected by normal laboratory experiments. This was tested by Thatcher et al. (1998), who took a random selection of knockouts with no overt phenotype and grew them in competition with the parent wild-type strain in a chemostat. In the majority of cases, the wild type eventually out-grew the mutants, though it took many generations before this happened, implying that the selective disadvantage suffered by the mutants was very small, such that their effects on the phenotype, though unmeasurable in any short-term comparison, are sufficient to account for the maintenance of the genes in the wild type. Even if this is only true for a subset of the genes with no overt phenotype, its profound implications do not seem to have been absorbed. How can a hypothesis for the link between the activity of a gene product and the selective advantage it offers the organism be framed and tested when there is such a small difference in the properties and behaviour of an organism whether it is present or absent?

There is no reason to believe that such results are unique for yeast. Mice with the myoglobin gene knocked out (Garry et al., 1998) had an unchanged exercise capacity and a normal ventilatory response to hypoxia; their muscles appeared to function normally. It turned out that there were multiple differences in the anatomy and biochemistry of the mutant mice that compensated for the lack of myoglobin (Grange et al., 2001), but that does not change the fact that it is now much more difficult to make a simple statement about the function of myoglobin in mammalian muscle.

It is possible to take this argument further and ask why there should be any expectation that it ought to be possible to link the activities of any and every component of a biological system with a discrete function. The fact that biologists think that this is a reasonable approach to understanding function was satirised by Lazebnik (2002) in his article 'Can a biologist fix a radio?'. His point was that a radio can only be understood in terms of its characteristic functional modules, but within the modules, the functions were implemented by the interactions between the components, and most of the characteristics of the components had no relevance to their contribution to the module. His criticism of biologists for their obsession with minute description of the characteristics (colour, morphology, behaviour of the system when the component is removed) was apt, but does it follow that living cells ought to be analysed by delineating the functional modules within? The function served by a radio, as for any other artefact, is known from the start; its modular structure results from both the designer and the constructor consciously choosing to break the problem of making the radio into a set of discrete sub-problems. This is not the route taken by biological evolution. In fact, when biologically inspired genetic algorithms are used to computationally evolve electronic circuits to fulfil a specified function (Bennett et al., 2000), the performance of the sub-tasks becomes a single unified process, and although known circuit designs can be 'rediscovered' in this way, entirely novel designs of a dense, non modular nature can be produced (Koza et al., 1999). If this is a fair model of the evolution of biological networks to implement functions, it illustrates there is no reason why we should expect to recognise any similarity between the arrangement of components in the network and the way we would have thought of constructing a module to carry out the function.

Sauro (2004) and Deckard & Sauro (2004) have therefore attempted to invert the problem of recognising the functions implemented by networks of enzymes by first defining output functions and then, again with genetic algorithms, evolving enzyme networks in the computer that compute the function. They hope that this will allow the inference of a similar function for a module in a biological network if it is observed to contain a similar configuration of components. This is an ingenious approach, but requires for its success that we can hypothesise appropriate functions and set up the genetic algorithm to evolve appropriate circuits.

In this section, I have attempted to show that there could well be limitations to our ability to understand completely how a cell works, partly because we may have difficulties in defining the functions being implemented and partly because we may not recognise the collection of components that implement the function. This does not necessarily mean that we cannot predict the responses of a biological system, as we have another type of modelling available that I have not yet discussed: computer simulation.

5. IS SIMULATING CELL METABOLISM THE SAME AS UNDERSTANDING IT?

Although I have presented theoretical analysis of models as a separate issue from computer simulation, most practitioners of the former also build simulations of the networks and phenomena they are analysing. There are a number of reasons for this. First, simulations of a model can replace experimental observations as source material for developing hypotheses about behaviours of the system. Next, once a theoretical analysis has been formulated, the simulation can serve as an illustrative case, especially for the benefit of those biologists who would rather look at a graph than read an equation. In both these instances, the lack of experimental noise in the simulated system (at least, unless the simulation is stochastic) and the freedom to accurately make any desired alterations to the properties or quantity of a component, irrespective of their practical feasibility, are distinct advantages. Furthermore, it is usually easy to simulate larger systems with additional processes beyond those represented in the theoretical analysis, where they would make the analysis less tractable. In this way, it may be possible to test whether factors that were omitted from the theoretical analysis are truly of lesser importance, or, looking at it the other way round, whether the processes and components included in the theoretical model are sufficient to generate the main aspects of the behaviour of a more realistic model of the biological system.

Through these approaches, simulation of metabolism and other cellular networks has proved an important tool in developing our understanding, especially as biochemical simulations can be developed and solved relatively rapidly with modern software. However, simulation alone without an analytical model has some weaknesses. The main one is that even a small model can have a large number of parameters, for example, in metabolism, these may be K_ms, Hill coefficients, equilibrium constants, and limiting rate values for the component enzymes. Hence it soon becomes impractical to determine whether a particular behaviour is robust over the whole feasible range of every parameter, or whether different behaviours emerge in different regions of parameter space. This contrasts with an analytical model, where it should usually prove possible to specify boundary constraints for the observation of a particular behaviour. Thus simulating a model will generally not contribute as much to the understanding of a biological process as does an appropriate analytical model. On the other hand, even in the simplest cell there are thousands of different interacting components, and it is inconceivable that an analytical model can be made of the whole of such a complex system. Currently, it is not possible to simulate a whole cell either, but it is not inconceivable that this could be done. The advent of high-throughput analytical technologies (the '-omics') is accelerating the rate of data capture, and consortia are already forming to build simulations of *E. coli* (http://www.ecolicommunity.org/), *S. cerevisiae* (http://www.siliconcell.net/ysic/), and the hepatocyte (http://www.bmbf.de/en/1140.php). What understanding would we have obtained if we were to succeed in producing a silicon replica of a living cell? In one respect, we would have substituted a highly complex experimental object with a highly complex simulation, which might not be any easier to understand. It would still be necessary to develop hypotheses about the factors and processes that determine the model's behaviour, and then to test those hypotheses by appropriate interventions. Some of the advantages of simulation of small models would still be present. The implementation of the interventions would be much easier in the case of the simulation, and the results might be more clear-cut, even if they involved such small changes in a concentration or rate that they would not be detectable experimentally. We would still be able to test whether the components and processes described in the simulation are sufficient to account for known behaviour, though tracking down the reasons for any discrepancies is likely to prove very difficult on account of the size of the model and its parameter space.

My view is that models of this size will be useful for integrating information and concepts about the cell's components in order to predict the behaviour, and that as a result we will be more successful in predicting the impact of mutations, environmental alterations and drug treatments. Whether we will really understand why the model has made the predictions it does is a different matter.

Understanding will more likely come from analysing subsystems of such a model, and less likely from the full model itself, but as discussed earlier, complex evolved systems may be simulable without being readily comprehensible.

REFERENCES

Bennett FH, Koza JR & Yu J, Mydlowec W. *Automatic synthesis, placement and routing of an amplifier circuit by means of genetic programming*. Lecture Notes in Computer Science: 1801, 1–10, 2000.

Black S & Wright NG. *β-aspartokinase and β-aspartyl phosphate*. Journal of Biological Chemistry: 213, 27–38, 1955.

Blackman FF. *Optima and limiting factors*. Annals of Botany: 19, 281, 1905.

Burns JA, Cornish-Bowden A, Groen AK, Heinrich R, Kacser H, Porteous JW, Rapoport SM, Rapoport TA, Stucki JW, Tager JM, Wanders RJA & Westerhoff HV. *Control analysis of metabolic systems*. Trends in Biochemical Sciences: 10, 16, 1985.

Chance B, Holmes W, Higgins JJ & Connelly CM. *Localization of interaction sites in multi-component transfer systems: Theorems derived from analogues*. Nature (London): 182, 1190–1193, 1958.

Chassagnole C, Fell DA, Rais B, Kudla B & Mazat JP. *Control of the threonine-synthesis pathway in* Escherichia coli: *A theoretical and experimental approach*. Biochemical Journal: 356, 433–444, 2001.

Cornish-Bowden A & Cárdenas ML (Eds.). *Control of Metabolic Processes*. NATO ASI Series, Plenum Press, New York, 1990.

Crabtree B & Newsholme EA. *A quantitative approach to metabolic control*. Current Topics in Cellular Regulation: 25, 21–76, 1985.

Deckard A & Sauro HM. *Preliminary studies on the* in silico *evolution of biochemical networks*. ChemBioChem: 5, 1423–1431, 2004.

Denton RM & Pogson CI. *Metabolic Regulation*. Chapman Hall, London, 1976.

Dujon, B. *European Functional Analysis Network (EUROFAN) and the functional analysis of the* Saccharomyces cerevisiae *genome*. Electrophoresis: 19, 617–624, 1998.

Fell DA. *Understanding the Control of Metabolism*. Portland Press, London, 1997.

Fell DA & Snell K. *Control analysis of mammalian serine biosynthesis*. Biochemical Journal: 256, 97–101, 1988.

Garfinkel D & Hess B. *Metabolic control mechanisms VII. a detailed computer model of the glycolytic pathway in ascites cells*. Journal of Biological Chemistry: 239, 971–983, 1964.

Garry DJ, Ordway GA, Lorenz JN, Radford NB, Chin ER, Grange RW, Bassel-Duby R & Williams RS. *Mice without myoglobin*. Nature: 395, 905–8, 1998.

Grange RW, Meeson A, Chin E, Lau KS, Stull JT, Shelton JM, Williams RS & Garry DJ. *Functional and molecular adaptations in skeletal muscle of myoglobin-mutant mice*. American Journal of Physiological – Cell Physiology: 281, C1487–94, 2001.

Groen AK, Wanders RJA, Westerhoff HV, van der Meer R & Tager JM. *Quantification of the contribution of various steps to the control of mitochondrial respiration*. Journal of Biological Chemistry: 257, 2754–2757, 1982.

Heinrich R & Rapoport TA. *A linear steady-state treatment of enzymatic chains; general properties, control and effector strength*. European Journal of Biochemistry: 42, 89–95, 1974.

Heinrich R & Schuster S. *The modelling of metabolic systems: Structure, control and optimality*. Biosystems: 47, 61–77, 1998.

Higgins J. *Analysis of sequential reactions.* Annals of the New York Academy of Sciences: 108, 305–321, 1963.

Hofmeyr JHS & Cornish-Bowden A. *Quantitative assessment of regulation in metabolic systems.* European Journal of Biochemistry: 200, 223–236, 1991.

Hofmeyr JHS & Cornish-Bowden A. *Predicting metabolic pathway kinetics with control analysis.* In: BioThermoKinetics of the Living Cell. (Eds. Westerhoff HV, Snoep JL, Wijker JE, Sluse FE & Kholodenko BN), BioThermoKinetics Press, Amsterdam, 155–158, 1996.

Kacser H & Beeby R. *Evolution of catalytic proteins or on the origin of enzyme species by means of natural selection.* Journal of Molecular Evolution: 20, 38–51, 1984.

Kacser H & Burns JA. *The control of flux.* Symposium Society Experimental Biology: 27, 65–104, 1973, reprinted in Biochemical Society Transactions: 23, 341–366, 1995.

Kacser H & Burns JA. *The molecular basis of dominance.* Genetics: 97, 639–666, 1981.

Kahn D & Westerhoff HV. *Control theory of regulatory cascades.* Journal of Theoretical Biology: 153, 255–285, 1991.

Kholodenko BN, Molenaar D, Schuster S, Heinrich R & Westerhoff HV. *Defining control coefficients in non-ideal metabolic pathways.* Biophysical Chemistry: 56, 215–226, 1995.

Klipp E & Heinrich R. *Competition for enzymes in metabolic pathways: Implications for optimal distributions of enzyme concentrations and for the distribution of flux control.* Biosystems: 54, 1–14, 1999.

Koza JR, Andre D, Bennett FH & Keene MA. *Genetic Programming III: Darwinian Invention & Problem Solving.* Morgan Kauffmann, San Francisco, 1999.

Krebs HA. *Control of metabolic processes.* Endeavour: 16, 125–132, 1957.

Lazebnik Y. *Can a biologist fix a radio?* Cancer Cell: 2, 179–182, 2002.

Letellier T, Heinrich R, Malgat M & Mazat JP. *The kinetic basis of threshold effects observed in mitochondrial diseases - a systemic approach.* Biochemical Journal: 302, 171–174, 1994.

Newsholme EA & Start C. *Regulation in Metabolism.* Wiley and Sons, London, 1973.

Park DJM & Wright BE. *Metasim, a general purpose metabolic simulator for studying cellular transformations.* Computer Programs in Biomedicine: 3, 10–26, 1973.

Pizer LI. *The pathway and control of serine biosynthesis in* Escherischia coli. Journal of Biological Chemistry: 238, 3934–3944, 1963.

Porteous JW. *A rational treatment of Mendelian genetics.* Theoretical Biology & Medical Modelling: 1:6, 2004.

Rolleston FS. *A theoretical background to the use of measured concentrations of intermediates in study of the control of intermediary metabolism.* Current Topics in Cell Regulation: 5, 47–75, 1972.

Sauro HM. *The computational versatility of proteomic signalling networks.* Current Proteomics: 1, 67–81, 2004.

Savageau MA. *Biochemical systems analysis: I. some mathematical properties of the rate law for the component enzyme reactions.* Journal of Theoretical Biology: 25, 365–369, 1969.

Savageau MA. *Optimal design of feedback control by inhibition: Steady state considerations.* Journal of Molecular Evolution: 4, 139–156, 1974.

Savageau MA. *Biochemical Systems Analysis: a Study of Function and Design in Molecular Biology.* Addison-Wesley, Reading, Mass., 1976.

Schaaff I, Heinisch J & Zimmerman FK. *Overproduction of Glycolytic Enzymes in Yeast.* Yeast: 5, 285–290, 1989.

Small JR & Fell DA, *Covalent modification and Metabolic Control Analysis: Modification to the theorems and their application to metabolic systems containing covalently-modified enzymes.* European Journal of Biochemistry: 191, 405–411, 1990.

ter Kuile BH & Westerhoff HV. *Transcriptome meets metabolome: Hierarchical and metabolic regulation of the glycolytic pathway.* FEBS Letters: 500, 169–171, 2001.

Thatcher JW, Shaw JM & Dickinson WJ. *Marginal fitness contributions of nonessential genes in yeast.* Proceedings of the National Academy of Sciences USA: 95, 253–257, 1998.

Thomas S, Mooney PJF, Burrell, MM & Fell DA. *Metabolic Control Analysis of Glycolysis in Tuber Tissue of Potato* Solanum tuberosum. *Explanation for the Low Control Coefficient of Phosphofructokinase over Respiratory Flux.* The Biochemical Journal, 322, 119–127, 1997.

Umbarger HE. *Evidence for a negative-feedback mechanism in the biosynthesis of isoleucine.* Science: 123, 848, 1956.

Wouters AG. *Four notions of biological function.* Studies in History and Philosophy of Science Part C: 34, 633–668, 2003

Yates RA & Pardee AB. *Control of pyrimidine biosynthesis in* Escherichia coli *by a feed-back mechanism.* Journal of Biological Chemistry: 221, 757–770, 1956.

5

On building reliable pictures with unreliable data:[1] An evolutionary and developmental coda for the new systems biology

William C. Wimsatt

SUMMARY

The new systems biology (NSB) is a cluster of methodological approaches for the analysis of dynamical behavior in networks that sits at the confluence of a number of disciplines. They are all now experiencing massive increases in qualitatively new actual and potential interactions driven by the data explosions in genomics and proteomics. I urge that developmental and evolutionary perspectives provide particularly useful tools for the analysis of life systems in the NSB. Their character as systems that must develop and evolve means that they possess certain properties, in particular evolvability, genetic and environmental robustness, and differential generative entrenchment for their parts. These properties are themselves very general and robust. They also possess virtues that help to ameliorate a problem of rapidly growing magnitude in the analysis of complex living systems: that the data produced by high-throughput methods ('gene chips') have very high error rates. Some of these errors are undoubtedly products of a technology that is new, noisy, and needs further tuning. Some are almost certainly systematic and, by recognizing this, remediable. Knowledge of the general kinds of systems we are dealing with can help with both.

---

[1] This title is an homage to the famous paper of a similar name by John von Neumann, published posthumously 50 years ago, in 1956. Von Neumann's (1956) paper pioneered the consideration of reliability, in both organic and computer design.

Systems Biology
F.C. Boogerd, F.J. Bruggeman, J.-H.S. Hofmeyr and H.V. Westerhoff (Editors)
Isbn: 978-0-444-52085-2

1. INTRODUCTION

These are exciting times for biology, on multiple fronts. A number of disciplines intersect and are enlightened by the explosive growth of new knowledge in detail and at the system level about genetic and biochemical interactions. New perspectives on the evolution, phylogeny, development, and organization of complex adaptive systems emerge as we learn more about these interacting systems in development at the biochemical, cellular, and multicellular levels affecting differentiation and compartmentalization. The new systems biology (NSB) is riding an expanding wave as we found and rename departments in its name.

We also seem to be reverting spontaneously to talk that was more common in the heyday of 'systems theory' and cybernetics in biology, from the late 1950s to the early 1970s. This reversion is a product of the kinds of knowledge we are gaining. It was more schematic and promissory then; now, while still schematic in many places, it is increasingly richly empirically based and detailed. The timeliness of much of this history is explored elsewhere in this volume by Evelyn Fox Keller.[2] And increasing use of the cybernetic vocabulary comes from both macro- and microdirections: Thus Wallace Arthur (1997) and Eric Davidson (2001) both make rich use of such language and 'wiring diagrams' of interactions between and among genes and their products. Arthur, a self-retooled population biologist, became intrigued by the rich complexities of development. His interest is morphological but reaches down to detailed gene-control interactions relevant to morphological expression.[3] Davidson, a pioneer in the study of gene control from the late 1960s and fairly speaking, a new systems biologist before there was a NSB, has a focus that is more 'bottom-up', analyzing, and articulating gene-control networks and cascades to extrapolate to an overall, developmental architecture (Davidson & Erwin, 2006). The return to cybernetic language is not surprising. In the last decade, we have analyzed 'genetic wiring diagrams' of increasing complexity and scope (Davidson, 2006).

---

[2] Evelyn and I are both historically well placed to remember it though as a biophysicist working on development she was a participant, while I looked on enviously, by then as a philosopher. My connections came through Frank Rosenblatt's broad ranging course in the Fall of 1964 (titled 'Brain Models and mechanisms', but it was really on adaptive systems more generally and many of the readings showed the influence of cybernetics and systems theory). Of all that I read, probably Kacser (1957) came closer to representing the spirit of the NSB. It was not well known, but in many ways the spirit of the NSB was paradigmatically anticipated.

[3] Another example with another approach is provided by Stuart Newman, a theoretical chemist by training, and an evolutionary morphologist Gerd Müller. They undertake systematic exploration of morphological possibilities for cellular constructions and their connections with the underlying chemistry (Newman and Müller 2000; Müller and Newman 2003). Their approach seeks generic constraints on possible modes for assembly of cells into larger morphological structures. In the generality it seeks, it is a methodology more reminiscent of bottom-up approached from physics, but practiced on top-down objects and phenomena.

2. THE NEW SYSTEMS BIOLOGY AND EVO-DEVO

Is evo-devo irrelevant? To some in the NSB, it simply indicates another domain that NSB can do without. I disagree. So would Eric Davidson and Evelyn Fox Keller. So, I expect would Carl Woese, a pioneer in the elucidation of the genetic code and in the phylogeny of very early life. Those who believe they have no need of evolutionary and developmental perspectives have perhaps been misled by a common kind of stereotyping of these disciplines.

Many practitioners of evolutionary developmental biology themselves feel that 'evo-devo' lumps diverse practices and perspectives, making them appear more monolithic than they feel. To some, it seems too skewed towards developmental genetics as opposed to higher levels of organization such as either morphology (Love, 2003) or ecology (Gilbert, 2001). Others would worry about the emphasis on long time scales and correlative emphasis on typological conceptions of species away from populational variability (Raff, 1996, p. 21). In this stereotypic image, population genetics, while dynamic, is limited to models of genetic change in terms of selection coefficients and gene frequencies that abstract away from physiology and phenotypic organization completely, and the evolutionary biology of macroevolution is just descriptive and not predictive. No wonder it seems irrelevant. This stereotyping is inevitable for a new discipline that articulates so many prior separate areas, and in which most of the practitioners of one subarea are amateur consumers of most of the others. But unfortunately it masks a great deal of relevant work.

In fact the confluence of developmental genetics with systems approaches and a more macroscopic developmental biology, systematics, and comparative studies has created a new hybrid discipline in which population genetics can move towards more detailed dynamical models of the phenotype. It is doing so both synchronically and through its developmental history, and in which systematics and phylogeny are again used to provide important clues to the organization of development and the course of evolution, predictively as well as descriptively, as they did in the nineteenth century. At the hands of researchers like systematist–geneticist–systems biologist Carl Woese, they are redoing the history of the early origins of life and uncovering surprising things about its nature (Woese, 2004). These include the initially shocking claims of the endosymbiotic origins of eucaryotes (Margulis, 1971) and its subsequent elaboration to discover widespread, now entrenched, symbioses. Morowitz's (1992) claim that large chunks of metabolism represent preserved chunks of earlier biotic environments does have things to say about the origins and nature of life, as well as about its evolution. More recently, Woese's (1998) hypotheses that there was an early stage in which the ancestors of already distinct lineages interchanged genes far more readily than later after the development of mitosis solved a remaining puzzle for systematists with a set of physiological and evolutionary proposals.

These suggest investigations that could be done by systems biologists concerning the origins and plausible evolutionary order of the different complex elements of mitosis. Hypotheses about the origins of life have always had to pass muster on biochemical grounds and plausibility. All these have implications for the broader architecture of living systems within which NSB works. Not everything in either evolutionary biology or developmental biology is relevant to NSB, but some of it is already part of NSB, and other parts will become increasingly relevant over time as we learn to better relate processes, acting on different time and size scales.

We should not fear that evolutionary and developmental biology will simply swallow systems biology, because systems biology is characterized by its approach as much as by its subject matter. Nor must the aim of the NSB be to serve developmental and evolutionary biology, any more than it might be to serve, e.g., oncology or epidemiology. Evolutionary biology, developmental biology, genetics, cell physiology, and biochemistry are using converging methodologies on the common stage of the cell, and recognizing that they must share common assumptions and knowledge to do their respective jobs adequately.[4] Progress in NSB surely will serve all these and just as surely will be served by them. Moreover, the modeling aims and techniques will surely be in at least some part different, and NSB will have a lot to contribute to these areas as well as to derive from them. In large fractions of their domains, evolutionary, developmental, and genetic investigators are being forced to take a systems biology perspective, and so systems biology should grow as their methodologies spread among related disciplines. But it cannot avoid them.

3. THE PROBLEM OF DATA RELIABILITY IN THE ANALYSIS OF LARGE SYSTEMS

To illustrate how and why evolutionary and developmental concerns are central to core issues in systems biology, I want to start with a seemingly unrelated puzzle: How do we get a reliable account of the cell when we do not have totally reliable data about it?[5] This applies both to the analysis of gene-control networks and to biochemical pathways. Although the data I draw upon comes from the former context, it obviously must influence the latter. There are uncertainties

[4] Community ecology and traditional systems ecology also share many of the same methodological approaches, tools, and problems, but on more macroscopic objects. Though I do not discuss them here, they too should be a part of the broadly conceived systems biology.

[5] I was first made aware of the magnitude of the problem of data unreliability by Beckett Sterner, who also provided me with the key reference (Deane et al., 2002). Sterner's input was crucial and my debt is substantial, because this is the key organizing insight of the paper.

about the magnitude of parameter values, but even more about whether components interact at all (both under the conditions studied and in living organisms). To sketch what I will argue, fundamental features of living systems are crucial to how we can deal with these data errors, and these features require recognizing the developmental and evolutionary natures of these systems.

The new use of 'gene chips' presents an array of new possibilities in the massive volume of data produced. Unfortunately, just when we would seem to need more accurate data rather than less, they also appear to present us with much higher error rates. These DNA microarrays use miniscule amounts of different DNA sequences ('probes') to detect RNAs that may be involved in producing active proteins. Chips with tens of thousands of distinct probes are common. The latest and largest number is nearly 400 000 on a single chip. They are used not only for broad censuses of activity, but also for more targeted ones such as identifying interactions in specific metabolic pathways or disease states. The targeting is as simple as the choice of what to spot in the array.[6] They can either detect or compare activity patterns using several different protocols, but so far in a boolean ('yes/no') rather than quantitative manner.

This new technology allows an enormous reduction of labor and coordinated detection of simultaneous activity patterns involving multiple genes or proteins in a cell, something that would have been impossible two decades ago. Gene chips also increase the uniformity of assay procedures. The changes they have provided are not unlike the move from 'single-unit' recording in the neurophysiology of the late 1950s–1970s to the localization of massive changes in activity patterns by brain regions possible with functional nuclear magnetic resonance (fNMR) beginning in the 1980s. This transition produced not only new kinds of data, but also inevitably new orientations in theory and is probably responsible for the emergence of the new discipline of neuropsychology, which draws heavily on the sort of molar data produced by fNMR. The change in orientation and questions that could be addressed was enormous in both cases (though for gene chips, still early in its course), but so also in each, the increase in breadth of information was accompanied by a loss in its specific local quality. As a result, in both cases, we have the continuation of two technologies rather than the replacement of one by another.

Because the reduction in data quality with gene chips is substantial, its accuracy is a key issue. Deane et al. (2002) conducted an evaluation by comparing the results of these 'high-throughput' methods with a database of already known interacting and noninteracting proteins, producing the expression profile reliability (EPR) index. A second method (PVM) involved determining how likely the individual interactions were by asking whether they had paralogs that interacted. The second method picked up only 40% of known interactions,

[6] Wikipedia entry on 'DNA arrays', accessed 23 June 2006.

but had a false positive rate of only 1%. So it was a conservative method for inclusion of an interaction. With the two together, they estimated that of a list of 8000 protein interactions from the Database of Interacting Proteins,[7] about 50% are reliable, and using the latter test, identified 3000 of these as likely true interactions. These are chastening error rates. While-high throughput methods can generate candidates at an enormous rate, validating that they are indeed interactions requires more intensive analysis on an interaction-by-interaction basis. Moreover, Deane et al. note but do not deal with the issue of false negatives – how many interactions may be missed by the high-throughput screen.

Can we do better? One obvious move is to try to improve the quality of the data. There are many possible sources of error. We should be interested particularly in systematic ones because this fraction usually indicates problems we can do something about, by supplementing or recalibrating our methods. It may also indicate sources of systematic bias in methodological or theoretical approach (Wimsatt, 1980, 2007). Thus the high number of false positives noted by Deane likely arises at least in part for a systematic reason: Testing for the possibility of chemical interaction among all possible reactants does not allow the presumably substantial number of interactions that do not occur in vivo because the reactants are spatially or temporally segregated in the organism under natural conditions – sequestered by design. This points to the need to move beyond the current focus of NSB on intracellular dynamics. We can hope to correct these kinds of error, but only by investigating intracellular and intercellular structure and morphology, how they change through development, and how they may act to catalyze and compartmentalize reaction dynamics.

4. DATA ERRORS AND MOLAR SYSTEM PROPERTIES

Do we need to know everything? At this level of analysis, we must treat all sorts of errors as the same: reactions left out or reactions erroneously included, and assume that whether errors of either sort make a difference depends upon the context – upon the network structure of the system, which of its products are crucial, and how sensitive system behavior is to the levels of the products.

At first this suggests detailed local analyses. But at a more molar level, how sensitive system performance is to errors depends upon what kind of system we are analyzing. Software is very breakable, but consequences can be large or small. Y2K turned out to be less of a problem in part not only due to massive

[7] Some care was taken in assembling this database. About 2000 of the 8000 were identified in small-scale experiments (from 800 research articles), and the rest came from four high-throughput screens. Nonetheless, the overlap between these sources was described as 'petite' – the factor originally motivating the calibrations they performed.

preventative changes in software and hardware, but also because many errors were not fatal, and computer users were not unaccustomed to rebooting frozen machines already. (Taking the larger system into account thus reduced the real threat.) How about organic systems? For decades of progress in genetics, using reductionistic analyses (in which more details more accurately known were always better), we were presented with stories about how small changes in a system may change its behavior radically – and indeed that was how genetic changes were supposed to affect molar systems. The paradigm case for this view was the now classic story in every textbook of how a single base substitution in the gene coding for the beta-chain of the hemoglobin molecule could lead on the one hand (in the heterozygote) to greater resistance to malaria, and on the other to severe anemia and 'sickle-cell crisis' and an early painful death among HbS homozygotes. And that was just the beginning: There was a long list of 'single gene' genetic diseases.

This presents a picture in which organisms are like computer programs, so we need to know the genetic constitution and the biochemistry of the system in great detail because a single change could wreak major havoc. Yet software has an organization: We often do not need to know the details of a procedure to write other parts of a program, and object-oriented programming has increased the modularity of code substantially. While there are serious problems caused by inaccurate data, the picture of organic systems that would require complete knowledge to analyze any aspect of system behavior is not accurate: if it were we would be unsurvivable, unevolvable, and unstudyable. In some ways, at a very simple level we are like a house lighting system: There are things, like a failure at the main junction box, that can shut down the whole thing, and blown fuses can temporarily take out subsystems of varying sizes. But most failures in the system are, and are designed to be, strictly local. Thus individual bulbs and appliances can fail without requiring anything more than their replacement or repair because of the parallel (redundant) organization of the house wiring at the lowest level. Unreliable data are critical problems for the NSB, as they must be for the analysis of any complex system,[8] but there are various mitigating conditions.

Faced with the problem of unreliable data, we must find workarounds. Some must come through improvements in technology and the development of better means of testing the accuracy of our data. Some come through choice of questions that are less affected by this problem, though this may skew research and theory construction, as we saw earlier with the differences engendered by 'single' vs. 'multichannel' approaches, both for neurophysiology and for cell biology.

---

[8] Taylor's (1985) study in ecological communities in nature and with simulations showed that interactions left out could make studied components appear causally connected when they were not, or make them apparently independent when they were not. The conclusions would apply for networks more generally.

But significant relief can also come from the nature of the systems we study. Living systems are robust. And they are evolvable. If they were not both of these, living systems could not survive environmental fluctuations and would not have evolved in the cumulative and diversifying manner we see. Analysis of system robustness is a topic of importance in NSB (Bruggeman, 2005, Ch. 5).[9] One particularly robust (and important) feature is that organic system architectures show significant differential generative entrenchment. It is inevitable that some things have more consequences than others in the operation of a system.[10] Elements that play large roles in generating or maintaining the behavior of the system, and for which there are no alternatives, will have high evolutionary stability because their wide usefulness has rendered them irreplaceable and they are increasingly constrained in the ways or degrees to which they can change. The pervasiveness and importance of robustness and entrenchment should make them proper topics for investigation in the NSB in their own right.

They are discussed here for another reason also: They can also be used as tools for identifying and getting relevant knowledge about these systems, and in part for ameliorating effects of unreliable data. There has been an explosion of interest in the last 5–7 years in the role of robustness in the design and evolution of complex systems. This has recently been nicely reviewed and synthesized for organic systems by Andreas Wagner (2005). Robustness can also be a (usually selective) help when reliable information is hard to come by. You may not require much information about variation or exact values of variables in dimensions for which a system is robust if outcomes are relatively insensitive to the details (Levins, 1968). We can tolerate a higher rate of errors in the specification of a system in the analysis of those properties. And system analysis can tell how robust a system is and in what dimensions.

We can also get help at the other extreme: For entrenched things, by contrast, outcomes may be strongly dependent on details but that very necessity has anchored the architecture against change in those respects. Systems that do not maintain them do not survive, so these elements are relatively constant and often more readily determinable (Wimsatt, 2001). So both robustness and entrenchment are important for the analysis of system behavior, and an understanding

[9] Robustness can mean just relative stability of a system property (e.g., rate of production of a metabolic product) across different parameter values (concentrations and reaction rates) in the system, or it can mean stability across addition and subtraction of interactions or relative invariance across an ensemble of systems of a given type. To be evolutionarily stable would often require the latter and stronger conditions. Differential entrenchment, robustness, and evolvability are arguably more robust still, because they are characteristic of evolving and living systems of all kinds.

[10] To preserve deeply entrenched elements in the face of dissipative forces (mutation, etc.), it is important not only that some things be deeply entrenched, but also that some things be lightly entrenched, such that their loss is not costly. These together set up a dynamic that is crucial not only for preserving the most important elements, but also for allowing an explosion of variation that may allow exploring other optima when either internal or external conditions relax (Wimsatt and Schank, 2004).

of their implications are deeply rooted in evolutionary and developmental perspectives. I will explore the role they can have in ameliorating the problem of incomplete and unreliable data in the next two sections.

5. ROBUSTNESS AND THE MANAGEMENT OF UNCERTAINTY

The following strategies and constraints seem reasonable ways of dealing with uncertainty in data:

(1) Particularly if data is important, try to determine it in more than one way. That is, incorporate robust designs to increase reliability into your experimental methodology. This not only reduces errors through cross-checking, but can also be used to detect systematic differences that may lead to technological improvements to reduce errors and to have better knowledge about when data can and cannot be trusted (Wimsatt, 1981; Levins, 1968). The late Sylvia Culp (1995) provided powerful and revealing examples in her analyses of diverse methods in molecular genetics.
(2) Models of smaller circuits or systems require less data to keep the included errors to reasonable levels. Learn how these circuits behave, and their sensitivity and robustness to changed structure and parameter values. If they are robust, use them as 'seeds', taking their outputs as given, and investigate the circuits including and intersecting them.
(3) When modeling larger circuits, look particularly for their robust properties.
(4) Take the values and behaviors emerging from such simulations with a grain of salt. Regard the simulations as exploratory rather than definitive.
(5) After finding a behavior that is somewhat robust, try specifically to 'break' it, determining the conditions under which it fails. These might be informative: it might be a tunable switch or threshold device, breakdown conditions may indicate other variables that must be maintained or other dimensions in which it is designed to be robust.
(6) If you find a property that appears to be biologically important, and it is not robust, be suspicious of your model or the assumed parameter values. This is the complement to the old maxim of adaptive design that 'Nature does nothing in vain'.[11] The more important something is, the more important it is to guarantee its presence. Nature does not guarantee anything, but it is a good working hypothesis. So if the property is fragile in your model, explore the possibility that it is not important, that the model is wrong, or that you have misidentified its function and what it is doing.

---

[11] The primary application of that principle in this context is the reverse engineering one: the more complex is a mechanism, the more important is its function or functions. This may fail to be true if the mechanisms and function have been recently co-opted from another functional system, or 'kluged'.

At multiple levels, Wagner (2005) found recurrent patterns, whether it be for the conformation of an RNA or a protein, the generation of a crucial product, or the production or maintenance of a required morphology confirming this assumption of robustness: The natural system was more robust under neighboring perturbations, whether genetic, structural, or dynamic, than for values of that system drawn at random from possible systems like it. This could be for different reasons, and Wagner investigates their plausibility and scope extensively at various levels of organization.[12] One of his conclusions was that robustness to environmental fluctuations probably was the source of selection that conferred robustness to the effect of mutations as a secondary effect.[13] This is interesting: The kind of results systems biologists can get directly is relevant to evolutionary questions. More generally, one must consider that:

(1) A system property might be robust because state-space neighborhoods where a property is robust are easier to find in an evolutionary search.
(2) Once found, if the property is selectively advantageous, it is easier to maintain in the face of mutation and environmental perturbation if small induced changes in state leave the property relatively unchanged.[14]
(3) Structural changes in the system may change the character of neighboring state-spaces and, in particular, may act to increase the size of what Wagner calls the 'neutral space' in which the relevant property remains unchanged. Something like this must be going on in what Waddington (1957) called 'genetic assimilation', in which selection changes the expression of a property manifested only in the presence of an environmental shock so that it is manifested under a much wider range of conditions.
(4) The probability that something will be entrenched, that other parts of an accumulating adaptive structure should come to depend upon it, should be a monotonically increasing function of its stability and persistence. The nonlinear amplifications of selection found by Wimsatt and Schank (2004)

[12] In his discussion, Wagner focuses on the first two, though I believe that all of them come up in passing elsewhere in his discussion.

[13] My one reservation about this claim is that it turns on the fact that environmental perturbations are much more common than mutations. But if one counts recombination in a system with lots of epistatic effects as producing new mutations (as they likely will at the phenotypic level), the number of mutations goes up enormously – by orders of magnitude.

[14] Kauffman (1969, 1985, 1993) commonly assumed that the given circuits would be realized in only one specific state – that any deviations led to reduced fitness. This made them highly sensitive to mutation and was the major reason why his results seemed to establish that selection could not maintain large complex systems. The larger is the 'neutral neighborhood' for a property, the more easily it is maintained by selection. We called this 'degree of genericity' when we argued that selection and self-organization would work most effectively in concert, when the selected state was multiply realizable, and thus not radically improbable (Wimsatt, 1986, Schank and Wimsatt, 1988). Given the sizeable neutral neighborhoods for features found at many levels of organization in evolved systems (Wagner, 2005), Kauffman's claims were too pessimistic.

suggest that this probability increases nonlinearly as well. Selection should tend to entrench robust or generic features, though one would also expect a steady accumulation of contingent, arbitrary, improbable features that become sufficiently entrenched to persist and become phylogenetically distinguishing features.

A concept used several places above, which should become very useful to NSB and is elaborated to great advantage by Wagner, is that of a 'neutral space'.[15] This is a further abstraction and generalization of the idea of 'neutral percolation surface' introduced by Huynen, Stadler, and Fontana in their important 1996 paper on evolution in RNA configuration space. Huynen et al. looked at the major forms of folded configurations of RNA molecules of length 100 nucleic acid bases and found that a few major forms dominated. They also found that the regions in which specific major forms appeared tended to be connected in mutation space, such that one could often move, one mutation at a time, throughout a connected neighborhood without changing the folded configuration, and thus, to a first approximation, remaining 'neutral', preserving the function or the fitness of the molecule. This meant that a population could 'percolate', one mutation at a time, to distant parts of the space. They also found that most major forms were reachable within a small number of mutations from one another. These 'neutral spaces' thus provided 'don't care' conditions for the composition and behavior of systems using these molecules, but the diversity of 'neutral' positions they could occupy could lead to rapid divergence if external conditions changed and selection for different configurations became advantageous. This idea itself is clearly related to the concepts of a fitness topography and to an energy surface, but is usefully exploited here as applied to discrete state systems.

Wagner's extensive discussions show that this situation or others analogous to it (as he points out, one cannot always formulate problems such that a well-organized discrete space can be defined for them) is characteristic not only of RNA space, but also at other levels of organization as well, in particular, to protein space and to spaces characterizing the dynamics of metabolic systems. This generally means that organic systems are designed so that they have many 'don't care' conditions – that behavior may often not need to be fully specified to be able to predict the dynamics with reasonable confidence. This is not

[15] Wagner's book is also distinguished by the robustness of his review of the evidence. Most claims are addressed using two or more distinct modeling strategies, often several, with the limitations of the different strategies compared. Some of the modeling concepts, for example, the idea of a 'lattice model' of a protein, in which all amino acids are of the same size and separation and their bonds can only take up angles at 90° intervals (0, 180, 270 in two dimensions, and the analogous lattice angles in three dimensions), are lovely and revealing. This one, for example, explores the consequences of topology (one-dimensional connectivity) and the distribution of hydrophobic and hydrophilic sites for the folding configurations.

something we would have a right to expect were we not talking about evolved systems, but it is bound to help with the analysis of complex systems in the presence of unreliable data.

6. GENERATIVE ENTRENCHMENT

Robustness is interesting not just because it makes organisms survivable and evolvable, but because robustness itself seems to bc so pervasive among organisms – in a word, so robust.[16] This is surely why there is so much interest in studying the formal properties of networks, and also why the NSB should not see itself in contrast with evolutionary biology, particularly the new evolutionary developmental biology. Are there other features of the internal complexity of the organism which have this kind of generality? There is one that has significant implications for the behavior of networks: The architecture of development, prima facie, can be used to predict differential rates of evolutionary change in different factors, and identify constraints on how and how much they can change (though generally not the details of how they will change). This is generative entrenchment, and in particular, the differential generative entrenchment of different elements in a causal network (Wimsatt, 1986, 2001; Schank & Wimsatt, 1988, 2000; Wimsatt & Schank, 1988, 2004).

Differential entrenchment is not an accidental feature of evolutionary systems. It is generic. Nor is it avoidable in any of our engineered systems. The importance of generative entrenchment points naturally to a number of architectural and dynamical network properties, particularly redundancy and canalization (ways of getting robustness) and modularity. Each of these act to modulate and commonly to reduce its effects and magnitude. These should all qualify as general properties of interest to the NSB, but they are also of central interest to developmental genetics and evolutionary developmental biology. By collaborating in their analysis, the NSB extends its central importance to these other disciplines.

If we consider a network, a pathway, or a cascade whether of gene activity or of biochemical metabolism, different nodes are differently connected. If we draw a directed graph for the propagation of causal effects in one of these or in any mechanism – including any of the engineered products of our modern technology – we will find that different numbers of nodes are reachable from different starting points in the network. Figure 1 is a randomly constructed directed graph of 20 nodes with 20 edges, generated by computer for our first

[16] This is at least partially due to reasons suggested by Aldana and Cluzel (2003), but Wagner (2005) also provides multiple arguments to this cumulative conclusion.

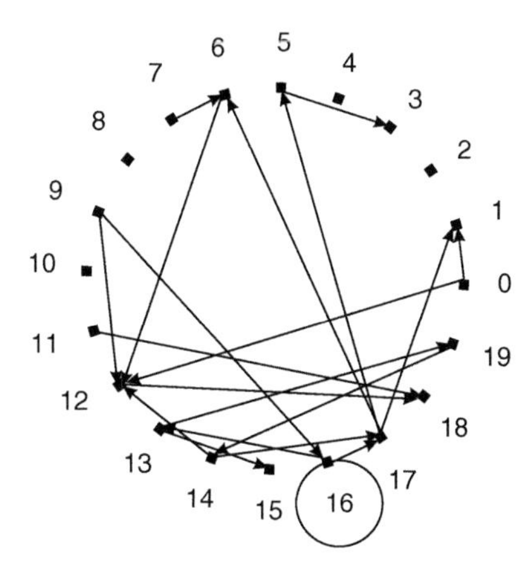

Figure 1 A directed graph representation of a gene-control network with 20 genes and 20 connections output by the program.

Nodes are genes, and a directed arrow indicates action of the gene at the tail and the expression of the gene at the head of the arrow. Mutations act on connections, and may randomly reassign the gene at the head or tail of the arrow. With 20 connections, it thus has 40 mutable sites. This particular graph was produced by 100 mutation events acting on a closed-loop model gene system of 20 genes and 20 connections. It is indistinguishable in generic properties from the one constructed at random. (From Schank and Wimsatt, 1988, p. 51. Figure copyright retained by the author.)

simulations of generative entrenchment in Stuart Kauffman's (1985) model of the evolution of gene-control networks (Schank & Wimsatt, 1988; Wimsatt & Schank, 2004). I co-opt it here to illustrate differential entrenchment.[17]

In Fig. 1 (Schank & Wimsatt, 1988) the connection from 5 to 3 has no further consequences (no arrows leave 3), but the connection from 16 to 13 has many. From node 16, we can travel: 16 → 13 → 19 → 14 → 17 → 5 → 3, with other divergent paths along the way. It is rare – essentially impossible for robust reasons – to find interesting networks in which all nodes have equal influence. Differential generative entrenchment of different nodes in a network is a generic property in Kauffman's sense – virtually all networks will have it. But it is even more powerfully anchored, for it is doubly robust – generic

[17] Directed graphs provide one kind of measure of generative entrenchment, but not the only one and not always the best. Thus Wagner (2005) presents cases where volume throughput of a product at a node correlates better with evolutionary stability than the number of nodes which are reachable (downstream) in a directed graph. He also argues that the necessity of a particular product is not well represented by its topological connectivity. Many cases of the first type (where production capacity matters and is accomplished by duplicating like parts, as with multiple copies of the DNA to make ribosomal RNA, or multiple liver cells) may represent k-out-of-m structures like those discussed in Wimsatt and Schank (1988). In such cases the whole k-out-of-m structure should be treated as a single unit for evaluating entrenchment.

again under different selection regimes: Networks in which all nodes start with equal generative entrenchment will spontaneously break symmetry, generating differential entrenchment under random mutation (another consequence of genericity in Kauffman's sense, also illustrated in Fig. 1). Differential generative entrenchment will also arise spontaneously with the random addition of modifier loci or with environmental fluctuations differentially affecting different genotypes (Wimsatt & Schank, 1988; Wimsatt, 2001). But entrenchment and its conservation under selection make increasing deviations from perfect equality or symmetry (with no differential entrenchment) inevitable, and therefore self-amplifying. Loci persisting longer for any of these reasons have a greater chance of acquiring additional modifier loci, leading to their further entrenchment, and increasing disparities in entrenchment. Cellular differentiation in metazoan evolution presumably inevitably does the same thing and is crucial to the evolution of increasing size, as environmental heterogeneities for cells located in different places in the cell mass become inevitable and specialized transport and coordination mechanisms become essential.

So why should this matter? Loss of a node in a network through which many nodes are reached should cause more disruption than those leading to only a few. (The same goes for changes in its properties, which are more highly constrained if those connections are going to remain unchanged.) So, prima facie, more negative[18] selection coefficients should be assigned to changes in nodes with more nodes and connections downstream. This property is plausibly generic for causal mechanisms of all types. It may be realized differently in mechanisms of different types and appear differently in different representations of their static and dynamic structure, but it seems unavoidable.

Notice also that selective consequences and intensities emerge directly from the structural properties of the systems under consideration. This is important: when this is true, selection coefficients are not external 'add ons' to black-box models of the phenotype, as was true for population genetic models.[19] So the complaint that distances selection models from system structure is not valid for evolutionary models based in differential generative entrenchment. They are properly part of the subject matter of the NSB.

For deeply entrenched traits, the negative consequences of changing them in uncontrolled ways are virtually unconditional. The chances of making a change

[18] This simple way of putting it makes it look like a monotonic relation between variables, but actually we are talking about changes in the means and higher moments of distributions.

[19] In artificial life simulations a systematic distinction is made between simulations in which fitness measures are 'intrinsic' to the artificial organisms and those given externally, where the designer of the simulation specifies the choice rules for what is to be optimized. In the former case, the only way to tell which morph is fittest is to see if any succeeds systematically. While in principle this could be inferred from 'engineering models' of fitness, the interactions 'in practice' inevitably contain unanticipated dimensions and new unexpected consequences.

and getting away with it without changing the entrenchment are virtually zero, unless there is an array of changes that are already neutral (or nearly so) at that level of organization (Huynen et al., 1996; Wagner, 2005). To be neutral any change will usually have to meet an increasing number of constraints that generate the upper-level property that must be preserved in the face of microlevel variations in composition and process. Though the selective consequences of making changes thus are not quite intrinsic properties of the network (fitness still is a relation between system and environment), essentially no changes external to the system in the environment or selection regime will change the outcome: deep generative entrenchment is thus an 'effectively' intrinsic property of that node in the network. But as changes in the network can cause major changes in entrenchment we should be especially interested in structural or dynamical changes in the network which change the entrenchment of the network element in question. And we are interested in the connectivity patterns in networks both in general and in detail. Significant changes in generative entrenchment can emerge from addition or subtraction of a single connection that moves that part of the circuit in the direction of local integration or parcellation (Schank & Wimsatt, 2000).

How general is this? We saw that differential dependencies of components in structures – causal or inferential – are inevitable in nature. And in the symmetry-breaking, we saw that those that have some tend to get more. Their natural elaboration generates foundational relationships. New systems in which some elements play a generative or foundational role relative to others are always pivotal innovations in the history of evolution, as well as – much more recently – in the history of ideas. Mathematics, foundational theories, generative grammars, and computer programs attract attention as particularly powerful ways of organizing complex knowledge structures and systems of behavior. This is a principle of great generality, going well beyond biology to evolved systems generally. Generative systems would occur and be pivotal in any world – biological, psychological, scientific, technological, or cultural – where evolution is possible. Generative systems came to dominate in evolution as soon as they were invented for their greater replication rate, fidelity, and efficiency. We must suppose that even modest improvements in them spread like wildfire. Combinatorial generative power like that found in the genetic code, the immune system, and languages of all sorts (spoken, visual, and written) add another important dimension of amplification best treated more fully on another occasion. Information (contrary to the reductionistic talk of replicator and meme theorists) is a system property, and thus properly leads back to a properly formulated systems biology.

But no runaway processes are unbounded for which we should be thankful, else we would not be here, buried under a heap of Darwin's elephants or some much more phylogenetically primitive sludge.[20] Generative entrenchment also cannot grow beyond limit. At some point the mutation rate (and 'genetic load') gets too great to preserve the structure, and we should expect an equilibrium between entrenchment-building and entrenchment-breaking processes. Michael Lynch et al. (1993) have analyzed this from one perspective (still within traditional population genetics) and described the behavior above the equilibrium as 'mutational meltdown'. Reduced absolute fitness from accumulating mutations decreases population size, leaving fewer possibilities to find ameliorative mutations, and the population goes extinct. His original application was to explain why asexual reproduction was not more common, but similar problems can arise for populations with mutation-sweeping sexual recombination if the overall genetic load is too great. Selection cannot maintain indefinitely large genome sizes, though various kinds of adaptations can enormously increase the size that can be maintained. (Wimsatt & Schank, (1988, 2004) consider different kinds of systemic adaptations involving generative entrenchment and genetic load that can do so.) At some point the design architecture cannot grow more: It faces a complexity catastrophe. The only escape from this is to start over again with these systems as units to build larger differentiated structures. This is the route to new levels of complexity, as Maynard Smith and Szathmár (1997) have argued, and also the route to a new hierarchical systems biology.

REFERENCES

Aldana M & Cluzel P. *A natural class of robust networks*. Proceedings of the National Academy of Sciences: 100 (15): 8710–8714, 2003.

Arthur W. *The Origin of Animal Body Plans*, New York: Cambridge University Press, 1997.

Bruggeman F. *Of Molecules and Cells: Emergent Mechanisms*. Academisch Proefschrift, Vrije Universiteit Amsterdam, 2005.

Culp S. *Objectivity in Experimental Inquiry: Breaking Data-Technique Circles*. Philosophy of Science: 62, 438–458, 1995.

Darwin C. 1859. *The Origin of Species*. Facsimile reprint, Harvard University Press, 1964.

Davidson E. *Genomic Regulatory Systems: Development and Evolution*. New York: Academic Press, 2001.

Davidson E. *The Regulatory Genome: Gene Regulatory Networks in Development and Evolution*. New York: Academic Press, 2006.

Davidson E & Erwin D. *Gene Regulatory Networks and the Evolution of Animal Body Plans*. Science: 311, 796–800, 2006.

[20] A simulation at http://www.athro.com/evo/elephs.html exhibits Darwin's illustration of the power of geometric growth with the reproduction of elephants (1859, p. 64). With a rate of growth you specify, you are invited to estimate how many years it will take until there is a sphere of elephants expanding out beyond the orbit of Pluto at the speed of light. (Not so long.) Long before that, of course, the center would have undergone gravitational collapse and sucked it all in! (Any guesses on the Schwartzchild radius for elephants?)

Deane CM, Salwinski L, Xenarios I & Eisenberg D. *Protein Interactions: two methods for assessment of the reliability of high-throughput observations.* Molecular and Cellular Proteomics: 1, 349–356, 2002.

Gilbert SF. *Ecological developmental biology: Developmental biology meets the real world.* Developmental Biology: 233, 1–12, 2001.

Huynen M, Stadler PF & Fontana W. *Smoothness within ruggedness: the role of neutrality within adaptation.* Procceedings of the National Academy of Science USA: 93, 397–401, 1996.

Kacser H. *Some Physico-chemical Aspects to Biological Organization.* Published as an appendix to Waddington, 1957, pp. 191–249, 1957.

Kauffman SA. *Metabolic Stability and Epigenesis in Randomly Constructed Genetic Networks.* Journal for Theoretical Biology: 22, 437–467, 1969.

Kauffman SA. *Self-Organization, Selective Adaptation and Its Limits: A New Pattern of Inference in Evolution and Development.* In: Evolution at a Crossroads: The New Biology and the New Philosophy of Science, edited by Depew DJ & Weber BH, MIT press, Cambridge, MA, 1985.

Kauffman SA. *The Origins of Order.* New York: Oxford U. P., 1993.

Keller EF. *Making Sense of Life: Explaining Development with Models, Metaphors, and Machines.* Cambridge: Harvard, 2002.

Levins R. *Evolution in Changing Environments.* Princeton: Princeton University Press, 1968.

Love AC. *Evolutionary Morphology, Innovation, and the Synthesis of Evolutionary and Developmental Biology.* Biology and Philosophy: 18, 309–345, 2003.

Lynch M, Bürger R, Butcher D & Gabriel W. *The Mutational Meltdown in Asexual Populations.* Journal of Heredity: 84, 339–344, 1993.

Maynard Smith J & Szathmár E. *The Major Transitions in Evolution.* New York: Oxford University Press, 1997.

Margulis L. *The Origin of Plant and Animal Cells.* American Scientist: 59, 230–235, 1971.

Morowitz HJ. *Beginnings of Cellular Life.* New Haven: Yale U. P., 1992.

Müller GB & Newman SA, eds. *Origination of Organismal Form. Beyond the Gene in Developmental and Evolutionary Biology.* Cambridge: MIT Press, 2003.

Newman SA & Müller GB. *Epigenetic mechanisms of character origination.* Journal of Experimental Zoology (Molecular and Developmental Evolution): 288, 304–317, 2000.

Oyama S, Griffiths P & Gray R, eds. *Cycles of Contingency: Developmental Systems and Evolution.* Cambridge: MIT Press, 2001.

Raff R. *The Shape of Life: Genes, Development, and the Evolution of Animal Form.* Chicago: U. Chicago Press, 1996.

Schlosser G & Wagner G, eds. *Modularity in Evolution and Development.* Chicago: U. Chicago Press, 2004.

Schank JC & Wimsatt W. *Generative Entrenchment and Evolution.* In: Fine A & Machamer PK, eds., PSA-1986: volume 2, East Lansing: The Philosophy of Science Association, pp. 33–60, 1988.

Schank JC & Wimsatt W. *Modularity and Generative Entrenchment.* In: Thinking about evolution: Historical, philosophical, and political perspectives, (eds. Singh RS, Krimbas CB, Paul DB & Beatty J) Cambridge: Cambridge University Press, 2000.

Taylor PJ. *Construction and Turnover of Multi-species Communities: a Critique of Approaches to Ecological Complexity.* Ph.D. dissertation, Department of Evolutionary and Organismal Biology, Harvard University, 1985.

Von Neumann J. *Probabilistic Logic and the Synthesis of Reliable Organisms from Unreliable Components.* In: Automata Studies, Eds. Shannon CE & McCarthy J. Princeton, N.J.: Princeton University Press: 43-98 Princeton University Press: 43–98, 1956.

Waddington CH. *The Strategy of the Genes.* London: Allen and Unwin, 1957.

Wagner A. *Robustness and Evolvability in Living Systems.* Princeton: Princeton University Press, 2005.

Wimsatt W. *Reductionistic research strategies and their biases in the units of selection controversy.* In: Nickles T (ed.) Scientific Discovery-vol. II: Case Studies. Dordrecht: Reidel. pp. 213–259, 1980.

Wimsatt W. *Developmental constraints, generative entrenchment, and the innate-acquired distinction.* In: Bechtel W (ed.) *Integrating Scientific Disciplines.* Dordrecht: Martinus-Nijhoff. pp. 185–208, 1986.

Wimsatt W. *Generative Entrenchment and the Developmental Systems Approach to Evolutionary Processes.* In: Oyama et al., 219–237, 2001.

Wimsatt W. *Re-Engineering Philosophy for Limited Beings: Piecewise Approximations to Reality.* Cambridge: Harvard University Press, 2007.

Wimsatt W & Schank J. *Two Constraints on the Evolution of Complex Adaptations and the Means for their Avoidance.* In: Nitecki M (ed.), Evolutionary Progress, Chicago: The University of Chicago Press, pp. 231–273, 1988.

Wimsatt W & Schank J. *Generative Entrenchment, Modularity and Evolvability: When Genic Selection meets the Whole Organism.* In: Schlosser & Wagner G (eds.), pp. 359–394, 2004.

Woese C. *The Universal Ancestor.* Proceedings of the National Academy of Science USA: 95, 6854–6859, 1998.

Woese C. *A New Biology for a New Century.* Microbiology and Molecular Biology Reviews: 68 (2), 173–186, 2004.

SECTION III

Theory and models

6

Mechanism and mechanical explanation in systems biology[1]

Robert C. Richardson and Achim Stephan

SUMMARY

Mechanistic explanations in cell biology provide a strong case for systems biology; for systems biology involves explaining the properties of cells in terms of the properties and interactions of their molecular constituents. Mechanistic explanations are, essentially, more detailed redescriptions and dynamic explanations of system behavior, which treat cells (or organisms) as complex biochemical systems. As an illustration, we apply this approach to the regulation of diauxic growth of *Escherichia coli* and the involvement of the *lac* operon. We discuss when properties manifested by mechanisms can be considered emergent properties, and contrast mechanistic explanations with properly reductionistic explanations.

---

[1]This work began in March 2005, at the Hanse Institute for Advanced Study (HWK) at Delmenhorst (Germany), in collaboration with Fred Boogerd and Frank Bruggeman. The HWK was a remarkable environment, and we acknowledge their support as the prime movers of this paper. The work continued at the University of Osnabrück in the summer of 2005. In many ways, Boogerd and Bruggeman deserve to be coauthors of this paper. We certainly would not, and could not, have done this work without our four-way collaboration. We do intend to develop another paper with them in which these themes are worked out in a more satisfactory way. RCR also would like to thank the Taft Faculty Committee at the University of Cincinnati for supporting the project.

Systems Biology
F.C. Boogerd, F.J. Bruggeman, J.-H.S. Hofmeyr and H.V. Westerhoff (Editors)
Isbn: 978-0-444-52085-2

1. INTRODUCTION: MECHANISTIC EXPLANATION AND REDUCTION

Recent philosophical discussions often tend to emphasize mechanistic explanation as an alternative to reductive explanation, especially insofar as reduction is construed in terms of theory reduction (cf. Richardson, 2002; Richardson & Stephan, 2007). The latter claims to render higher level explanations either redundant or eliminable, depending on whether, on the one hand, the higher level explanations are retained as at least approximate (as is the case when we compare Newtonian and Relativistic mechanics), or, on the other hand, the higher level explanations are falsified and replaced (as is the case with Aristotelian and Galilean physics). Mechanistic explanation offers a different perspective on the understanding of the relationship of explanations at different levels or, perhaps better, the relationship of explanations incorporating different levels of organization.[2] In mechanistic models, higher level explanations are neither redundant nor eliminable. There is genuine explanatory work done at the higher level, reflecting systemic properties. When this is so, and when that explanatory work is not captured at a lower level, the higher levels of explanation are not eliminable. As Kitano says,

> A system level understanding of a biological system . . . includes the network of gene interactions and biochemical pathways as well as the mechanisms by which such interactions modulate the physical properties of intracellular and multicellular structures.
>
> (2002, p. 1662)

The recent reinvigoration of mechanism within philosophy derives from the idea that the Positivistic approaches to reduction and explanation fail, for one reason or another, to capture scientific practice, especially in the biological sciences (e.g., see Wimsatt, 1976; Bechtel & Richardson, 1993; and Cartwright, 1995). The recent reinvigoration of systems ideas within biology derives from the parallel idea that reductionist ideals are inadequate for understanding some of the most fundamental biological phenomena. We want to join accounts of mechanistic explanation with those of systems biology. Within more Positivistic accounts of reduction, and theoretical integration, the point is the elimination of higher order theories or models. As we understand mechanistic modeling, this is

[2] The appeal to mechanistic explanation has a deep history, dating at least to the seventeenth century. Sometimes, mechanistic explanation means little more than causal explanation, and sometimes simply explanation in terms of the laws of mechanics. Among Logical Positivists, at the beginning of the twentieth century, the thought was that formal, or structuralist, accounts of explanation would capture mechanistic and reductionistic models; these were, explicitly, an attempt to capture the idea of causal/mechanistic explanation in a formal mode. We will focus on mechanistic explanation only conceived as an alternative to theory reduction.

not at all the issue. The point is, rather, to explain some particular phenomenon or to provide a mechanism which underlies some phenomenon. From a scientific perspective, the point is to provide an explanation of some robust phenomenon. From a philosophical perspective, we aim to explain how such modeling differs from eliminative reduction.

Let us begin with mechanism and its rebirth as a philosophical ideal. Most generally, as Carl Craver (in press) says, mechanistic explanations are 'constitutive explanations.' They are not narrowly 'causal' explanations, moving from cause to effect, but explain how some phenomenon or behavior is 'constituted'. Usually, we begin with some set of phenomena, or some behavior, to be explained. The explanation involves the articulation of parts, or components, typically in functional terms. Having detailed the capacities of the parts, and the character of their interactions, mechanistic models are sufficient to predict systemic behavior. The most recent articulations of mechanistic explanation within philosophy have taken two rather different forms. One, initially due to Peter Machamer, Lindley Darden, and Carl Craver (2000), describes the development of interlevel explanatory models. A second, due largely to Stuart Glennan (1996, 2002), fits a more conventional emphasis on theories and laws. Both approaches assume that mechanistic explanation has a particular structure. Let us begin with the structure from Machamer, Darden, and Craver, often referred to as the 'MDC' model. We assume we have, say, an organism with a characteristic behavior we want to explain. These behaviors, on the MDC model, can differ depending on the entities, characteristic activities, and the spatio-temporal organization of the containing system. Spatial and temporal structure on this picture are crucial. The result is a schema for understanding mechanistic explanations. Craver (2002, § 5.3) says, at one point, 'In such schemata, higher level activities (ψ) of mechanisms as a whole (S) are realized by the organized activities (ϕ) of lower level components (Xs), and these are, in turn, realized by the activities (σ) of still lower level components (Ps)'. On Glennan's approach, a mechanism is a complex system, whose parts are so organized that they interact to yield a characteristic behavior (see Glennan, 1996). In a recent version of his view, 'A mechanism for a behavior is a complex system that produces that behavior by the interaction of a number of parts, where the interactions between parts can be characterized by direct, invariant, change-relating generalizations' (2000, p. S344). We do not intend to explore the differences in emphasis between the views, so much as the alternative they offer.

So we begin with a behavior. We might want to explain, for example, the behavior of the bacterium, *E. coli*, and its culinary 'preferences' on various media. For its maintenance *E. coli* needs a reliable source of energy. Sugar solutions are natural choices for the experimentalist and favorites for the bacterium. To explain how *E. coli* uses sugars, we need to see what happens as the various sugars are consumed and processed. This is the study of metabolism. We look toward the multitude of enzymes which break down the sugars and their products; in other words, we look at the metabolic pathways within the cell. However, we look at these in toto. These are pathways for the processing of sugars within a certain sort of cell with characteristic capacities. We might want to explain, to take a very different behavior, a pain withdrawal reflex. To explain it, we characterize the lower level components and their characteristic properties. One set of nerves transmits a signal to the spinal chord from the periphery, which in turn initiates a signal that causes muscular contraction. We can continue the process, with an interest in, say signal transmission. At that point, we shift our focus to the structure of nerves, the molecular mechanisms responsible for the pulses down the axon, and the release of neurotransmitters. At each stage, we focus on a characteristic set of components, and their characteristic behaviors, with the goal of explaining the behavior of entities at a higher level.

We intend to explore the adequacy of these sorts of models of mechanistic explanation, in light of the undeniably mechanistic accounts of the behavior of unicellular organisms. There are a number of questions we will address, though not in sequence. First, we ask whether the current models of mechanistic explanation within philosophy are adequate for the broader range of scientific cases they are intended to cover. This issue has already been raised in an evolutionary context (Skipper & Millstein, 2005); we want to assess the question within the context of molecular cell biology and systems biology. The specific models offered by Glennan and by Machamer, Darden and Craver are not designed for these sorts of cases but should cover them if they are adequate. So, one concern is how generally these philosophical models apply. Second, and more fundamentally, we ask whether mechanistic explanation actually fits a reductive mode of explanation, or whether some properly mechanistic models are not reductionist. We have argued (see Boogerd et al., 2005) that there are emergent phenomena in cell biology that are also mechanistically explainable. The key thought is that cell biology exhibits manifestly mechanistic explanations of systemic behaviors; and these are properly emergent phenomena. We have shown there is at least one scientifically and philosophically interesting sense in which this is so. Here, we propose to further explore the thought that cell biology exhibits emergent phenomena that are mechanistically explicable. If this is so, then the context in which models of mechanistic explanation have been developed may not exhibit the richness they deserve. The goal would be to develop a more enhanced understanding of both mechanism and emergentism, one which moves beyond the more standard mechanism/reductionism and mechanism/eliminativism dichotomies.

Some theories appear to cross levels of organization or involve multiple levels of explanation. These are the sorts of explanatory theories that are most inviting to analysis in terms of mechanisms. Morton Beckner (1959) once called these 'interlevel' theories. In more recent discussions, the point has been repeated by, among others, William Wimsatt (1974, 1976), Lindley Darden & Nancy Maull (1977), and William Bechtel & Robert Richardson (1993). Robert McCauley recognizes this sort of multilevel explanation requires what he calls 'explanatory pluralism' (1996). The pluralism is crucial. Insofar as pluralism has any substance, it must at least acknowledge that higher levels of explanation are not eliminable. Systems level explanations matter. Systems matter. Carl Craver accurately characterizes mechanistic explanations as multilevel causal explanations that 'explain by showing how an event fits into a causal nexus' (2001, p. 68). There is some ambiguity in Craver's endorsement, but it leaves open the interesting thought that multilevel explanations are constitutive and need not refer exclusively to proximate causes.

For the introduction of our own model of mechanistic explanation, it will be useful to draw attention also to a more conventional account of reductionist ambitions. In Kenneth Schaffner's *Discovery and Explanation in Biology and Medicine* (1993), we are offered a comprehensive account of theory development in the biological sciences. He does not take himself to be developing an account of mechanistic explanation as we understand it, though all his models are certainly mechanistic. Most important for our purposes, Schaffner offers a detailed and comprehensive account of what he calls the 'general reduction-replacement model', aimed at understanding the relations between theories or models pitched at different levels of organization.[3] It was originally inspired by the classic model of Ernest Nagel (1961). Nagel originally distinguished the reduced and reducing theories on a number of grounds, including the observation that a reducing theory will typically have greater scope and precision than the reduced theory. Often the thought was that the reducing theory could displace the reduced theory, at least for explanatory purposes. Unlike some other models, such as that from Kemeny & Oppenheim (1956), the reducing theory was supposed to capture the explanatory principles of the reduced theory (see Richardson, 2006). Schaffner recognized that there could be a range of different cases. At one extreme, the phenomena explained by the reduced theory are captured by the reducing theory, though without the explanatory principles characteristic of the reduced theory. These seem paradigm cases of explanatory replacement. At the other extreme, the reduced theory captures the explanatory principles of the reducing theory, though perhaps with some modification. These seem paradigm cases of explanatory reduction. Schaffner's model is premised

[3] We do not want to suggest that theories and models are to be identified. One may describe a phenomenon without explaining it; a model is at most a partial explanation of a phenomenon; and theories are at best collaborative models. Within systems biology, the ambition is sometimes to elaborate models in such a way that they become comprehensive theories. We are skeptical that the ideal can be realized. In any case, a mechanistic model is a model geared toward explaining some behavior (or phenomenon) in terms of the parts and processes within the system, given specified (and realistic) boundary conditions.

on the observation that many cases – perhaps most in the biomedical sciences – lie somewhere in the middle between these two poles.

To make this picture more precise, he explicitly recognizes a temporal component. We begin with the original reduced and reducing theory, or model, T_R and T_B. These are associated with their respective, 'corrected' versions, T_{R*} and T_{B*}. For reduction to occur, four general conditions must then be in place:

(1) Connectability. The primitive terms of the reduced theory T_{R*} must be 'associated with' terms within the reducing theory T_{B*}.
(2) Derivability. The principles of T_{R*} must be derivable from T_{B*}, perhaps supplemented with new correspondence rules.
(3) Enhanced precision. T_{R*} should make more precise predictions than did T_R; and, perhaps, T_{B*} makes more accurate predictions in the domain of T_R than did T_R itself.
(4) T_R and T_{R*} are 'strongly analogous', and so even when T_R is replaced by T_{R*}, given derivability, there is a clear sense in which T_R is explained by T_{B*}.

Models of reduction with a similar import, though differing significantly in details, have been elaborated by Clifford Hooker (1981) and John Bickle (1998).

It is often worthwhile to make general models concrete. When we further focus on *E. coli* in Section 3, the case of the *lac* operon will be a case in point as a splendid exemplar of a mechanistic model within the biological sciences. And, among others, Schaffner treats it that way. It is what he calls a 'theory of the middle range' (1993, p. 98). Theories of the middle range are really paradigms for what has since been called 'mechanistic' models. These are typically models which describe a process. They explain the process in terms of the behavior and organization of components, and they do so in dynamic terms. These models are, moreover, idealized, and implicate entities at more than one level.

Having elaborated his model for the reduction of theories of the middle range, here is what Schaffner says concerning the *lac* operon model:

> Though ... reduction was not the aim of Jacob and Monod's work, significant progress toward a complete chemical explanation of enzyme induction has been a consequence of their research.
>
> (1993, pp. 481–482)

Indeed, Schaffner is not always so modest, claiming later that 'the kind of reduction exemplified by the operon theory' is a 'paradigm of the reduction of biology to physics and chemistry' (1993, p. 487; cf. Schaffner, 1993, pp. 76–82, 160–165, 481–487). Similarly, Sarkar, in an insightful series of essays concerned with reduction and reductionism in biology, describes the *lac* operon as a 'superb example of strong reduction' (1998, p. 140; cf. pp. 55ff.). In cases of strong reduction, explanations of systemic behavior can be constructed entirely on the basis of a more fundamental domain; there is hierarchical organization, and that hierarchy has a spatial organization (see 1998, pp. 43–45; for his reflections on reduction, see especially Chapters 2 and 3). This sort of view of the *lac* operon is typical, though in the sections to follow we will offer reasons to doubt that it is adequate. In the next section, we offer a characterization of mechanistic models; and in the section following it, we illustrate our treatment in terms of the *lac* operon.

2. LEVELS OF ORGANIZATION AND DEGREES OF RESOLUTION

Nature appears to be organized into levels. This is acknowledged by many authors. In his *Space, Time, and Deity*, Samuel Alexander, for example, distinguishes different orders (or levels) of 'empirical existence' (1920, Vol. II, p. 3). Among the more obviously distinguishable levels he mentions 'motions, matter as physical (or mechanical), matter with secondary qualities, life, mind' (ibid., p. 52). A more fine-grained (robust) picture of organization where nature is hierarchically organized stems from Oppenheim & Putnam (1958) who distinguish between elementary particles, atoms, molecules, macromolecules, cells, organisms, and eventually social groups. Another type of hierarchy is referred to by Herbert Simon (1973), who distinguishes different forces: Gravitational forces operate at one scale, electromagnetic forces at another, molecular bonding forces at still another, atomic forces at another, and nuclear forces at yet smaller scales. He observes that they not only operate at different scales but have very different strengths. Indeed, just as they operate at distances differing by orders of magnitude, their relative strengths differ by orders of magnitude as well. The result is a hierarchical ordering of causal forces in the universe, differing in temporal and spatial scale. So we have entities participating in characteristic levels of organization, and the hierarchical ordering is typically in terms of part/whole relationships. This much leaves a number of interesting questions unresolved. For example, even if levels or organization are ordered in terms of compositional relationships that tells us little about how we define levels. Wimsatt suggests they should be identified as maxima of causal regularity. We will tend to identify levels in terms of the characteristic sorts of entities they present, though we see that as wholly consonant with Wimsatt's suggestion. If our science is to respect the structure of nature, then a hierarchy in levels of organization should be reflected in a hierarchy for levels of analysis or levels of explanation. These are difficult and controversial issues, currently unresolved.

Often, philosophers of science also approach complex systems with an ontology of levels. Explanations seem to be given on various levels. Explanations at higher levels seem to reduce to explanations pitched at lower levels; even causal powers seem to reduce to causal powers located at lower levels. For some they even seem to 'drain away'. We think that this picture about both the organization and the explanation of systemic properties and behaviors of complex systems is not well taken. Indeed, it is important to avoid a simple transition from thinking of levels of organization to levels of explanation, though the two are doubtlessly connected. We will carefully distinguish these sorts of questions in what follows.

We begin to draw a different sort of picture that seems to be more adequate, or so we think: at the center of our two-dimensional approach is the notion of different degrees of resolution. Instead of shifting levels, we first of all allow to change the resolution (of description) at which complex systems are described that exhibit systemic properties and behaviors. By doing so, we always keep the focus on the whole system, we do not shift to lower levels. The characteristic entities remain the same. Only the descriptions change. As a consequence, we stay at the same level of organization.

Let us go into more detail. Mechanistic models aim at explaining the behaviors or capacities of complex systems. For example, we may be interested in the behavior or capacities of a person, a trade union, an organism, or a cell. These are clearly different levels of organization and require very different explanations. At whatever level, we start with a description of the behavior to be explained; for example, the growth rate a bacterium exhibits in liquid culture, the ability of a bacterium to digest lactose, or the behavioral symptoms characteristic of schizophrenia. The description of the behavior should be precise enough to identify the explanatory target within a certain degree of tolerance. Exactness is not the issue, as it is not available. Thus

(1) We specify the behavior to be explained within a given degree of tolerance.
(2) We determine the grade of resolution for the description of the behavior to be explained.

The questions are not independent. As we specify a behavior more precisely, we will doubtless need a more exact description of the mechanism involved; conversely, as we have more exact descriptions of alternative mechanisms, we will need more precise descriptions of the behavior to be explained. The idea is this: Explaining the behavior of a system will naturally involve some degree of tolerance. So we might want to explain the rate at which an organism acquires an adaptive response to an environmental stimulus. We will not require that the explanation be exact. The degree to which we allow it to be inexact is the degree of tolerance. The behavior to be explained can be described with varying degrees of 'resolution'. So we may describe an adaptive response at the organismic level, or in terms, say, of organ systems or of cells. At a resolution focused on organisms, the adaptive response is a matter of learning. At a resolution focused on organ systems, it might involve the hippocampus. At a resolution focused on cells, it might focus on cells within the hippocampus. The key thought is that, even with a tolerance level assumed, we may seek explanations with different explanatory depths. These are what we think of as different degrees of resolution. Let us try this with a model from cell biology.

In Fig. 1, we imagine some cellular behavior to be mechanically explained belongs to some level of organization L (e.g., the cellular level, or a molecular level), and it is described by a certain grade of resolution G (e.g., as behavior of a cell or as the behavior of a molecular system).

To construct a mechanistic model, we need to describe the system in terms of parts and processes that are expected to be sufficient to explain some behavior (with some degree of precision). There are typically various grades of resolution possible in the way we describe the system. These can be thought of as focusing on more or less detail, as if we could zoom in or out on the system revealing more or less about what is happening within the system. So, for example, a bacterial culture can be thought of as a culture, or as a collection of cells, which in turn can be thought of as a collection of biochemical networks or as a collection of macromolecular systems. This is a matter of redescription rather than reduction. We redescribe the culture as a system of cells or as a set of macromolecular systems. We are not reducing the system behavior to cellular

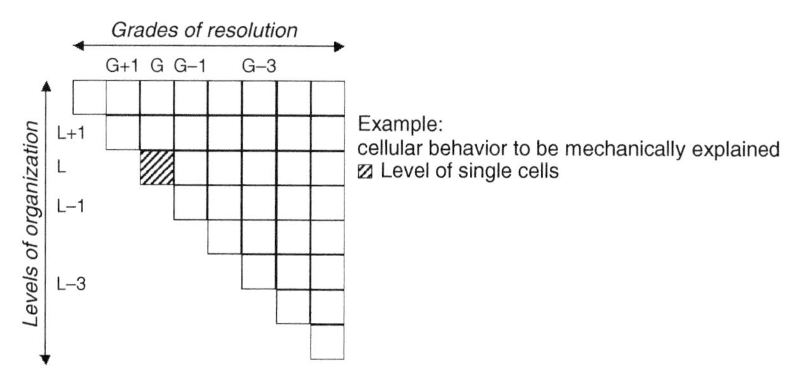

Figure 1 Grades of resolution and levels of organization.

There are two independent dimensions: one concerns levels of organization, and is characterized by a change in the characteristic entities (e.g., organisms, single cells, molecules); the other concerns the resolution with which they are described (e.g., a cell may be described relationally as a component within a system of cells, or as a network of metabolic pathways, or as a system of molecules). One can change the grades of resolution without changing levels of organization. All of these are descriptions of the same system, and hence at the same level of organization, though the system is described differently.

or macromolecular behavior. We are, instead, elaborating on the structure and dynamics of the system. Thus,

(3) We have to find the set of processes and parts to be referred to in the mechanistic explanation and to move toward the corresponding grade of resolution to redescribe the cellular behavior.

Of course, with a reduced tolerance, we may not be able to produce any explanation without proceeding to a greater degree of resolution. The tolerance is a matter of the empirical flexibility allowed in describing a behavior for a system. The resolution is a matter of the descriptive flexibility in describing the mechanisms. As we increase the resolution, we may predict the behavior more exactly, but that also means we lose generality. As we decrease the resolution, we may predict the behavior less exactly but increase generality. The scope or generality is what is captured in the idea of tolerance.

(4) We articulate the mechanism corresponding to the behavior.

In Fig. 2, the behavior of cells is represented as being redescribed as behavior of molecular systems, whereby we use a higher grade of resolution, say G−3. This is not a shift in the level of organization or explanation. We are still dealing with cells, not with isolated pathways or isolated enzymes. We just describe the cells in greater detail, i.e., with a higher degree of resolution.

For example, for its maintenance *E. coli* needs a source of energy – sugars are good carbon sources; to explain how it uses sugars we need to see what happens

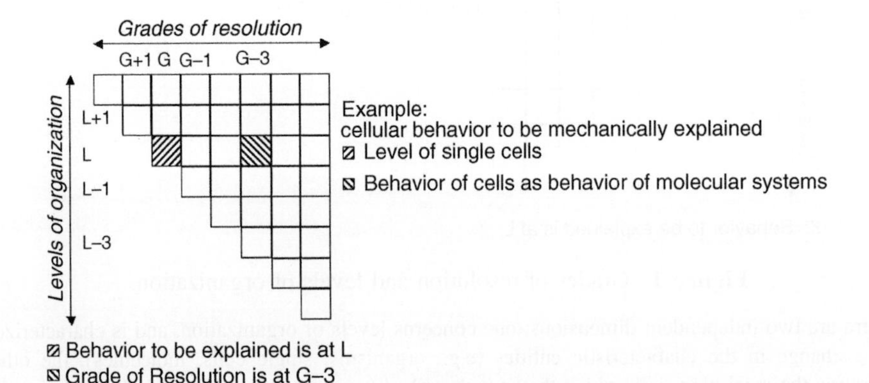

Figure 2 Changing grades of resolution, within a level of organization.

We may focus on some characteristic behavior of cells, describing cells as functional units; or we may think of cells as molecular systems. Other grades of resolution are possible.

with the sugar within the bacterium. We look at the enzymes which break down the sugars or at the metabolic pathways, but we look at those in toto; they are still pathways for the consumption of sugar of a certain sort within a cell. Now, to gain enhanced insight, we may look toward more detailed descriptions of the behavior; e.g., we might be interested in the rates at which various sugars are consumed. As we specify the behavior more exactly, we need an explanation of how the components function within the cell. If we think, e.g., of *E. coli* as a set of macromolecules, we need to know what the capacities of the enzymes are, and also how these enzymes behave *in vivo*. This is accomplished when we refer to lower level theories, which constrain the mechanisms that belong to higher levels of organization.

To know how the parts interact within the given system, we can compare a variety of different systems with a similar organization and constitution. We might, for example, compare protein synthesis in *E. coli* with protein synthesis in a pint of beer. This would be an inappropriate comparison, given the differences in the systems. We might also compare protein synthesis on different media or in differing temperature regimes. An enzyme has a specific affinity for its substrates, which can be assessed *in vitro*, quite apart from living organisms, which sometimes serves as a measure of the enzyme's affinities in the living organism (cf. Boogerd et al., 2005). This last means of analysis, if successful, would suggest a reduction to constituent behavior in simpler systems. In Fig. 3, we see that background theories refer to the behavior of non-interacting molecules, which are located at a lower level of organization L−3.

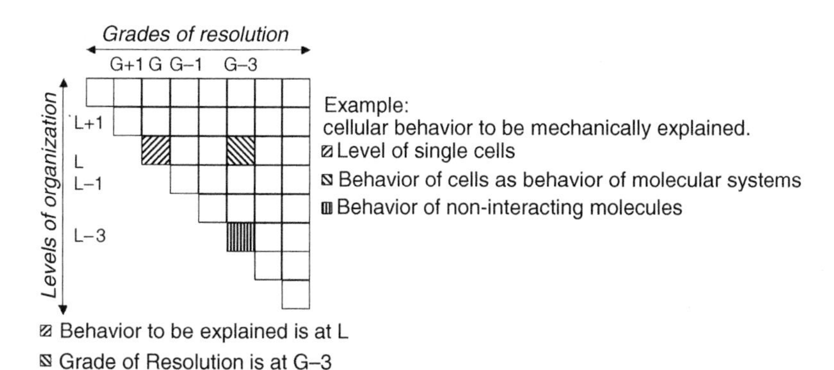

Figure 3 Changing levels of organization within a given grade.

We may focus, on, say, molecular details, with the aim of explaining the capacities of molecules, or the behaviors of cells. This is clearly a change in the characteristic entities, and so a change in the level of organization.

To see whether a mechanistic explanation is successful, two additional steps have to be completed. The first is a question of sufficiency. Quantitative models will yield definite predictions – if they are within the specified range of tolerance, they will count as sufficient:

(5) We have to know whether the constructed behavior is within the specified degree of tolerance (this is the sufficiency test).

The second question is one of adequacy for the explanatory problem at hand. A model will incorporate a variety of variables and parameters (e.g., the affinities of enzymes for its substrates, the amount of enzymes). We need to know, first of all, that the model postulates only parts and processes present in the complex system whose behavior is to be explained (e.g., a cell). It is possible to have a sufficient model, which predicts cell behavior, with fictitious parts; for example, a model that assumed the presence of an enzyme that *E. coli* cannot synthesize would be inadequate even if the model was empirically sufficient. An adequate model also requires that the parameter values incorporated in the model are realistic, that is, that they at least approximate the values in the cell. For example, if a model incorporates enzyme concentrations greater than those present in the cell, then that too would be an inadequate model.

(6) We have to check whether the mechanism is the actual one (this is the realism test).

3. THE DEVELOPMENT OF THE *LAC* OPERON AS A MECHANISTIC MODEL

In the early 1960s, Jacob and Monod published a series of ground-breaking papers focused on genetic regulation.[4] Though there were several alternatives to the Jacob and Monod model offered in the years which followed, their account of the *lac* operon is still widely accepted, at least in outline. The revolutionary aspect of the *lac* operon, within genetics, was that it introduced a distinctive class of genetic regulatory elements, governing the synthesis of enzymes and other products relevant to cell regulation. By roughly 1959, Jacob and Monod had established negative control as an important biological mechanism. In reflecting on the comparison between the bacterial model and phage, both of which exhibited the crucial regulatory phenomena, Jacob says this in his Nobel lecture:

[4] For a straightforward overview, see Beckwith (1967); and for a more comprehensive treatment, Beckwith & Zipser (1970).

> The most striking observation that emerged from the study of phage production by lysogenic bacteria and of induction of β-galactosidase synthesis was the extraordinary degree of analogy between the two systems. Despite the obvious differences between the production of a virus and that of an enzyme, the evidence showed that in both cases protein synthesis is subject to a *double genetic determinism*: on the one hand, by *structural genes*, which specify the configuration of the peptide chains; on the other hand, by *regulatory genes*, which control the expression of these structural genes. In both cases, the properties of mutants showed that the effect of a regulatory gene consists in inhibiting the expression of the structural genes, by forming a cytoplasmic product which was called the *repressor*. In both cases, the induction of synthesis (whether of phage or of enzyme) seemed to result from a similar process: *an inhibition of the inhibitor*.
>
> (Jacob, 1966, p. 1472)

The model explains genetic regulation through feedback, which is doubtless the reason Monod was inclined at one point to think of it as a cybernetic system. The molecularly characterized feedback relations supposedly brought cybernetic feedback within the scope of 'strong reduction'. We will describe the development and motivation for the model through the mid-1960s. In this context, that will be sufficient to see how it functions as a mechanistic model. In another paper, we intend to pursue the later development of the model.

In rough outline, in 1961, Monod and Jacob claimed that the *lac* operon consists of a regulatory region (the operator) and a set of structural genes (see Fig. 4). The operator, as they described it, consists of a promoter region, and a repressor. There are three structural genes regulated by the operator.[5] One of these structural genes codes for a permease that facilitates the uptake of lactose from the environment. One codes for β-galactosidase, an enzyme that cuts lactose (a dissacharide) into two simpler sugars, which can then be utilized in metabolism. In the presence of nutrient sources other than lactose, both the permease and the β-galactosidase are present but in extremely low concentrations. When lactose is present and other sugars are not, the concentrations of these products can increase by as much as a 1000-fold.

The crucial phenomenon, leading to the discovery of the *lac* operon, was the discovery of biphasic growth of *E. coli* bacteria in carbohydrate mixtures in the early 1940s by Monod (see Fig. 5). These results were associated by Andre Lwoff with what he called 'enzymatic adaptation'. In the late 1940s, Monod saw that even in the presence of lactose in the medium, there is no β-galactosidase present if there are other sugars, which we now know means there is no expression of the *lac* operon; that is, with glucose present, lactose fails to induce the *lac* operon, and the organism does not metabolize lactose.

[5] In their original work, Monod and Jacob recognized only two structural genes. That does not change the substance of our discussion, but there are clearly three.

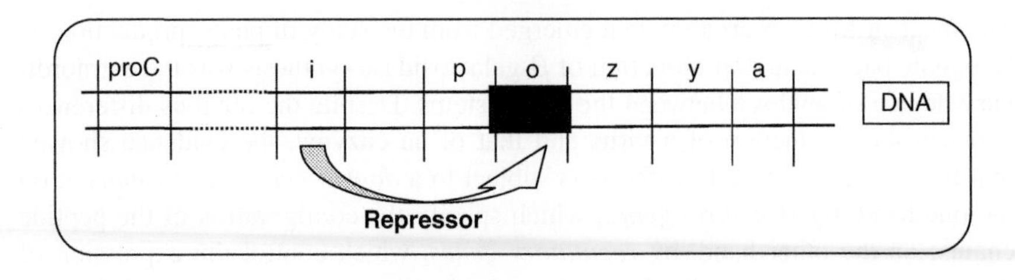

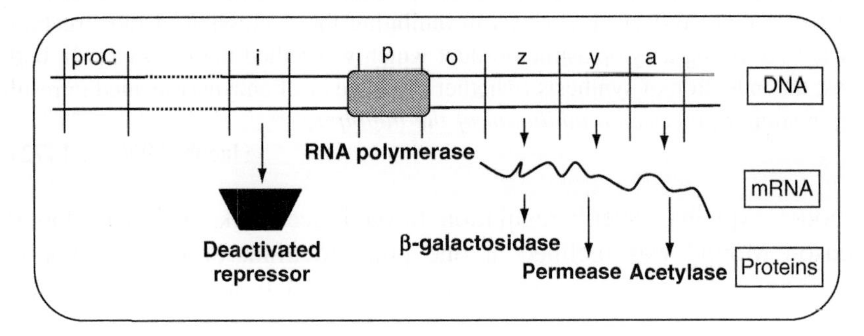

Figure 4 Control of expression in the lac operon.

We depict the three structural genes (z,y,a) and the control regions (i,o,p) on strands of DNA. The i region is the structural gene encoding the repressor, p is the promotor region, and o is the operator region bound by the repressor. In the top figure, the active repressor is present, and synthesis is inhibited. In the bottom figure, an inducer alters the shape of the repressor so that it no longer binds the operator, the RNA polymerase binds to the promotor region, and synthesis of the three structural genes proceeds. The z gene synthesizes mRNA for β-galactosidase, the y gene synthesizes mRNA for permease, and the a gene synthesizes mRNA for acetylase. The mRNA's are translated into proteins.

This switching behavior of *E. coli*, responding to the variation in sugar sources available in the environment is the fundamental behavior Jacob and Monod sought to explain. Strictly, of course, the *lac* operon model is a model of one component in the metabolic machinery of *E. coli*, though that component is crucially important in the behavior. In his Nobel lecture, Monod says that in 1947,

> . . . it became clear to me that this remarkable phenomenon was almost entirely shrouded in mystery. On the other hand, by its regularity, its specificity, and by the molecular-level interaction it exhibited between a genetic determinant and a chemical determinant, it seemed of such interest and of a significance so profound that there was no longer any question as to whether I should pursue its study.
>
> (1966, p. 476)

The problem Monod identifies is the 'respective roles of the inducing substrate and of the specific gene (or genes) in the formation and the structure of the enzyme' (ibid.).

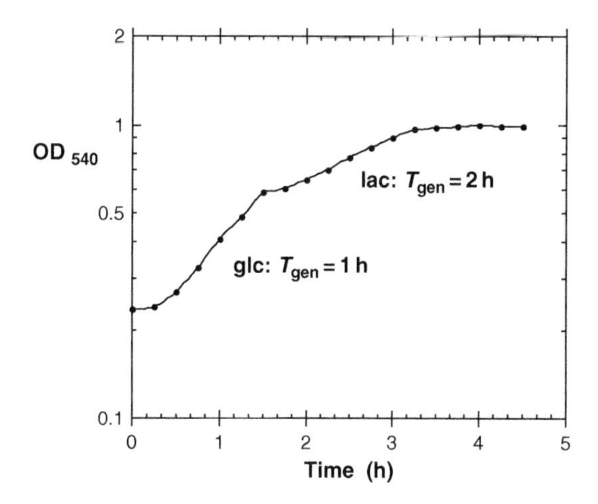

Figure 5 Biphasic growth of *E. coli* on a medium of glucose plus lactose.

The optical density (OD_{540}) (on a log scale) of a batch culture is depicted as a function of time. Difference in slopes reflects the difference in growth rate. In the initial phases of growth, *E. coli* grows more rapidly, consuming glucose. In later phases of growth, when glucose has been consumed, *E. coli* grows more slowly, relying on lactose; the generation time (T_{gen}) on lactose is two times longer that that on glucose.

By 1953, Jacob and Monod had two interesting mutants of *E. coli*. The 'wild type' (lac^+) is one in which we see the characteristic biphasic growth described above. It turns out that the two mutants involved genes with very different functions. It was a remarkable achievement to recognize the duality in the mechanism. The first mutant is a constitutive one (lac^-), in which lactose is always consumed by the organism. The second mutant is uninducible; that is to say, the uninducible mutant is one in which β-galactosidase is never synthesized, and so even if lactose is consumed it cannot be used in metabolism. This could be seen experimentally by indicators which could distinguish the presence of the *lac* mutants by different colors in populations. The lac^+ was either pink or white, depending on the specific medium on which the strains were grown. The lac^- was red. The presence of the lac^+ and lac^- mutant forms required two different mechanisms within the operator: one was involved with stimulation and the other with repression. The constitutive lac^- mutant is one in which a repressor is lost or dysfunctional and so one in which lactose is always consumed by the organism. The uninducible mutant is one in which activation is impossible, and so lactose is never consumed. As there are two genetic variants which are different from the 'wild type', we know that the operator in 'wild type' organisms has at least two components; otherwise, there would not be two different ways in which the mechanism could be 'broken'. This still left it unclear what these two components controlled. (We now know one controls the synthesis of β-galactosidase and the other regulates the importation of lactose into the cell. For Jacob and Monod this was not yet clear.)

Jacob and Monod found that some lac$^-$ organisms could be persuaded to produce β-galactosidase by inducers other than lactose, though this did not mean they used lactose in metabolism. This entailed that these lac$^-$ mutants lacked a factor the wild types had – one which would allow lactose to be brought into the cell. We now see this as the *lacY* gene. As a result, we know the structural genes regulated within the operon included at least two with very different functions. One was involved in bringing lactose into the cell, and one was responsible for producing β-galactosidase. To recapitulate, one mutant consumed lactose, taking it up from the environment, but could not metabolize it; the other could not take up lactose from the environment, and could have metabolized it if it had taken it in, as it produced β-galactosidase. The actual mechanism of control was still uncertain. More specifically, these results left open the question whether lactose consumption in the 'wild type' forms is triggered by stimulation or by inhibition of repression. Monod later described this as a 'double bluff', asking 'Why not suppose, then, . . . that induction could be effected by an anti-repressor rather than by repression by an anti-inducer' (1966, p. 479). This gave two different pictures of possible mechanisms behind biphasic growth. On the one hand, genetic control could be mediated by a switch that facilitates transcription of β-galactosidase. On the other hand, genetic control could be mediated by a switch that inhibits the production of β-galactosidase and can be turned off by a repressor.

The experimental solution was in bacterial sex (conjugation). Crossing lac$^+$ and lac$^-$ in some cases changed the acceptor lac$^-$ into a lac$^+$ organism, which meant that there had to be some factor available which repressed, or inhibited, expression of the gene. In other words, a bacterium which was incapable of limiting lactose consumption became capable of inhibiting lactose consumption, having received some factor from the lac$^+$ 'wild type' form. Shortly after this discovery, in 1960, Monod first introduced the term 'operon' to describe the coordinate expression of the *lacY* and *lacZ* genes. At this point, he knew only that the Y and Z genes were regulated by a promotor and an operator (POZY). Jacob and Monod, in the following years, promoted this as a general model of gene regulation (see Jacob 1966, Monod 1966, Jacob & Monod 1961, and Beckwith 1967).

What we have seen is the idea that there had to be a control mechanism to explain the ability of *E. coli* to switch food sources, which could (theoretically) have been subject to either stimulation or repression, and they settled on repression. Moreover, we have seen how they revealed the structure and organization of both regulatory elements and the structural genes they control, based on experimental evidence. What was most notably lacking was experimental evidence concerning the nature of the repressor.

Finally, we turn briefly to the issue of what we call 'realism'. Sufficiency of a proposed mechanism does not insure that the proposed mechanism is the one responsible for a given behavior. Even at the time that Jacob and Monod received the Nobel prize for their discovery of the *lac* operon, the evidence for it was indirect. There was good evidence that it was involved in the key phenomena (biphasic growth and lactose consumption), but the actual mechanism for repression had not been identified. Gilbert and Müller-Hill actually isolated the repressor in 1966, subsequently showing that the repressor binds directly to the DNA. Experimentally, it is interesting and difficult because the repressors only are present in very low concentrations. This is one of many avenues we will not follow out here.

4. MECHANISM AND EMERGENCE

The picture we have offered of mechanistic explanation is not inherently a narrowly reductionist picture, at least if we construe reduction in terms of theory reduction. It is, though, mechanistic in something like the sense advanced by Machamer, Darden, & Craver (2000); in Stuart Kaufmann's (1970) apt terms, it is an 'articulation of parts explanation'. It is an explanation of systemic behavior in terms of the behaviors of constituent parts within the systemic context.

This last qualification is not idle. A fully reductive explanation would explain systemic behavior in terms of the behaviors of constituent parts, where the range of behaviors is motivated in terms appropriate to the lower level. The constraints on a reductive explanation are substantial, provided it is intended to demonstrate the sufficiency of lower level models for explanatory purposes. It is not enough for a fully reductive explanation that we be able to redescribe some behavior, or some mechanism, in terms of constituents, or even that we can describe some state of the constituents sufficient for the behavior. This is something we can, in principle, do in any case. To lay claim to a reduction of one theory to another, it is necessary as well that we be able, at least, to show that a sufficient mechanism can be constructed relying only on the explanatory tools available within the lower level theory. This will sometimes require that we include information concerning the relevance of organization and concerning the behavior of constituents, neither of which are necessarily a function of constitution alone. These can be understood as twin constraints on the adequacy of reductionist models. Let us consider each briefly.

A description of a sufficient mechanism will often include information concerning organization and not simply constitution. In rare cases, simple constitution is sufficient, and organization does not matter or is not crucial. Organization will not matter when systemic behavior is an aggregative effect of constituent behaviors and will matter less insofar as systemic behavior is an aggregative effect of constituent behaviors (cf. Wimsatt, 1986, 1997; Bechtel & Richardson, 1993). The behavior of ideal gases satisfies this condition insofar as we are interested in explaining such things as the Boyle–Charles law. So, for example, in ideal gases the pressure on a container is simply the sum of the molecular momenta conveyed to the container, and the temperature becomes the mean molecular kinetic energy. These are simple aggregative conditions.[6] Genes are likewise sometimes treated as if their effects are largely or wholly additive, though this is certainly a misrepresentation of the usual case (cf. Wimsatt, 1980). Organization and interaction are the rules rather than the exception among genes. To take the opposite extreme, Stuart Kauffman (1993) has focused on the role that organization plays in understanding what he calls the 'origins of order'. In Kauffman's models, the constituents are treated as simple Boolean switches; he manages to extract a variety of surprising systemic effects from simple models whose only constituents are these Boolean switches and with randomly assigned connections. This is, as he says, 'order for free'. The result is a continuum of cases, depending on the extent to which interactions determine systemic behavior. At the extreme, represented by Kauffman, there is a sense in which the systemic behavior can be thought of as emergent, and contrary to reductionism, though still mechanistic (cf. Bechtel & Richardson, 1993).

The second constraint on reductionism is that there be a principled characterization of the behavior of constituents based on the lower level theory alone. There are many cases illustrating this constraint. One of the paradigms for interlevel reduction, the case of thermodynamics and statistical mechanics, illustrates the point precisely. In the 1860s, there was a movement toward understanding thermodynamics in terms of the kinetics of molecules. Maxwell was critical in this period, for the first time suggesting that the kinetic theory should be understood in a statistical or probabilistic form. Boltzmann took up the problems surrounding equilibrium distributions in the 1860s and nonequilibrium cases in the early 1870s. The mechanistic account they offered was often not well-received. One problem they confronted concerned what is called irreversibility, from J. Loschmidt in the middle of the decade. Newtonian mechanics (and for that matter, Quantum Mechanics) shows no temporal bias. With reversed motion, any process can be reversed without violating any dynamical principles. The second law of Thermodynamics requires a temporal direction, as Boltzmann acknowledged. The solution was probabilistic: the flow of heat from one source to a sink is the transfer of kinetic energy, and the most likely transition is from lower to higher entropy. Even if every microstate has the same probability, macrostates are associated with varying numbers of microstates; the most likely macrostates turn out to correspond to more frequent microstates. So the transition toward greater entropy is simultaneously a transition toward more frequent microstates.

---

[6] Historically, it was precisely because gases are *dis*organized that they were the focus for statistical mechanics.

One prominent objection to the idea, which Boltzmann clearly acknowledged, was the assignment of probabilities to microstates. This requires interpreting the distributions of probabilities among microstates and justifying those distributions. In equilibrium theories, the state of a system gets represented within a phase space defined in terms of the position and momenta of constituents. So the most probable macrostates are those that are associated more likely distributions. They become regions within the phase space. It is certainly possible to describe this transition from lower to greater entropy in kinetic terms. It is another task altogether to justify the description in terms of the Newtonian dynamics. That was, of course, the great accomplishment of the late nineteenth century. The point we want to draw from the example is that reduction is not simply a matter of redescription but of a principled redescription in terms motivated at the more basic lower level.

Reductive explanations are certainly mechanistic explanations, but not all mechanistic explanations are reductive explanations. In the mechanistic case we have highlighted in Section 3, there is certainly an explanation of systemic behavior (in this case, the switch to lactose consumption and biphasic growth), and this is certainly accomplished in terms of the organization and articulation of parts; but, crucially, the fundamental explanatory unit is the entire operon, described in a more detailed way (that is, the *lac* operon as a molecularly characterized device). That is, the system is the explanatory unit, and though it can be described with different degrees of detail, it is one and the same mechanism that is described. So the *lac* operon is a cybernetic device, a switching device, a regulatory gene, and a molecular mechanism. Describing the *lac* operon in molecular terms is not, as such, to descend to a lower level of organization, or a lower level of explanation. It is to describe the *lac* operon in a certain way, at a certain level of detail. The ability to do that does not insure that we are able to constructively represent, or predict, the behavior of the mechanism from molecular principles alone; it is a redescription of the mechanism in terms of 'parts and processes'. If we can not only describe the behavior in molecular terms but also construct its behavior on molecular principles alone, then we have a reductive explanation. It is this latter task, and not the simple redescription, that is the methodological heart and soul of reductionism.

Such a bottom up, constructive, process is not always something we can do, even when we have a molecular redescription available. We have argued elsewhere (Boogerd et al., 2005) that that is precisely what is missing in the case of cell biology, and that this constitutes a crucial case of emergence (see also Stephan, 1999, § 3.5). To carry through the constructive process, the behavior of constituents must suffice to explain the behavior of cells, where the behavior of the constituents is understood in terms of how those constituents behave outside the cellular context. We have shown that it is possible to give a fully mechanistic explanation of cellular behavior in terms of the behavior of molecular constituents in context. The key problem concerns understanding the behavior of the constituents, which are not context independent. It is only in context that we can specify the constituent behavior, and so only in context that we can derive the system behavior. So we are left with emergent behavior, indeterminate on the basis of constituent behavior in simpler systems or, we suspect, on the basis of constituent behavior in different systems (cf. Boogerd et al., 2005).

5. CONCLUSION: MECHANISTIC EXPLANATION AND SYSTEMS BIOLOGY

Mechanistic Explanations as proposed here make a strong case for systems biology. We have depicted mechanistic explanations as, essentially, more detailed redescriptions of system behavior. Not all explanations, even within functional biology, follow this form. Perhaps some mechanistic explanations do not follow this form. Still, many do, and within systems biology, they are more often a more appropriate type of explanation than any strictly reductive strategy that tries to pull down some single causal factor such as a genetic factor, and identify it as 'responsible for' some systemic behavior. Whether or not this is the only route to explain complex systems, genuine understanding is reached when we are able to redescribe a process, or a complex system, with a grade of resolution that allows us to see the relevant components 'at work'. We can then see how the system functions, and how components contribute to systemic behaviors. As Kitano says, this is a 'system level understanding'. It is also within this framework that we can reveal the general 'working capacities' of components; both experimental studies that investigate their capacities in context and behavior under systematically varied conditions can contribute to these models. That is what gives us the necessary background knowledge to understand the components' behavior within complex systems.

REFERENCES

Alexander S. *Space, Time, and Deity. The Gifford Lectures at Glasgow 1916–1918.* 2 Vol. Macmillan & Co, London, 1920.

Bechtel W & Richardson RC. *Discovering Complexity: Decomposition and Localization as Strategies in Scientific Research.* Princeton University Press, Princeton, 1993.

Beckner M. *The Biological Way of Thought.* Columbia University Press, New York, 1959.

Beckwith J. *Regulation of the* lac *Operon.* Science: 156, 597–604, 1967.

Bickle J. *Psychoneural Reduction: The New Wave.* MIT Press, Cambridge, 1998.

Boogerd FC, Bruggeman FJ, Richardson RC, Stephan A & Westerhoff HV. *Emergence and its Place in Nature.* Synthese: 145, 131–164, 2005.

Cartwright N. Ceteris Paribus *Laws and Socio-Economic Machines.* The Monist: 78, 276-295, 1995.

Craver CF. *Role Functions, Mechanisms, and Hierarchy*. Philosophy of Science: 68, 53–74, 2001.
Craver CF. *Structures of Scientific Theories*. In: Blackwell Guide to the Philosophy of Science. (Eds.: Machamer PK & Silberstein M), Blackwell, Oxford, 55–79, 2002.
Craver CF. *Explaining the Brain*. Oxford University Press, Oxford, in press.
Darden L & Maull N. *Interfield Theories*. Philosophy of Science: 44, 43–64, 1977.
Gilbert W & Muller-Hill B. *Isolation of the* Lac *Repressor*. Proceedings of the National Academy of Sciences USA: 58, 2415–2421, 1966.
Glennan S. *Mechanisms and the Nature of Causation*. Erkenntnis: 44, 49–71, 1996.
Glennan S. *Rethinking Mechanistic Explanation*. Philosophy of Science: 69, S342–S353, 2002.
Hooker C. *Towards a General Theory of Reduction*. Part I: *Historical and Scientific Setting*. Part II: *Identity in Reduction*. Part III: *Cross-Categorial Reduction*. Dialogue: 20, 38–59, 201–36, 496–529, 1981.
Jacob F. *Genetics of the Bacterial Cell*. Science: 152, 1470–1478, 1966.
Jacob F & Monod J. *Genetic Regulatory Mechanisms in the Synthesis of Proteins*. Journal of Molecular Biology: 3, 318–356, 1961.
Kauffman SA. *Articulation of Parts Explanation in Biology and the Rational Search for Them*. Boston Studies in Philosophy of Science: 8, 257–272, 1970.
Kauffman S. *The Origins of Order*. Oxford University Press, Oxford, 1993.
Kitano H. *Systems Biology: A brief overview*. Science: 295, 1662–1664, 2002
Kemeny JG & Oppenheim P. *On Reduction*. Philosophical Studies: 7, 6–19, 1956.
Machamer P, Darden L & Craver CF. *Thinking about Mechanisms*. Philosophy of Science: 67, 1–25, 2000.
McCauley R. *Explanatory pluralism and the coevolution of theories in science*. In: The Churchlands and Their Critics. (Ed.: McCauley R), Blackwell, Oxford, 1996.
Monod J. *From Enzymatic Adaptation to Allosteric Transitions*. Science: 154, 475–483, 1966.
Nagel E. *The Structure of Science*. Harcourt, Brace, & Co, New York, 1961.
Oppenheim P & Putnam H. *Unity of Science as a Working Hypothesis*. In: Minnesota Studies in Philosophy of Science. Vol. II. (Eds.: Feigl H, Scriven M & Maxwell G), University of Minnesota Press, Minneapolis, 3–36, 1958.
Richardson RC. *Reduction*. In: Encyclopedia of Cognitive Science (Ed.: Nadel L), Macmillan, New York, 897–904, 2002.
Richardson RC. *Reduction without the Structures*. In: The Matter of the Mind. Philosophical Essays on Psychology, Neuroscience and Reduction. (Eds.: Schouten M & Looren de Jong H), Basil Blackwell, Oxford. 2006.
Richardson RC & Stephan A. *Reductionism, anti-reductionism, reductive explanation*. In: Encyclopedia of Neuroscience (Eds.: Binder M, Hirokawa N, Windhorst U & Hirsch MC), Springer, Heidelberg, 2007.
Sarkar S. *Genetics and Reductionism*. Cambridge University Press, Cambridge, 1998.
Schaffner K. *Discovery and Explanation in Biology and Medicine*. University of Chicago Press, Chicago, 1993.
Simon H. *The Organization of Complex Systems*. In: Hierarchy Theory: The Challenge of Complex Systems (Ed.: Pattee HH), Braziller, New York, 1–27, 1973.
Skipper R & Millstein R. *Thinking about Evolutionary Mechanisms. Natural Selection*. Studies in History and Philosophy of Biological and Biomedical Sciences: 36, 327–347, 2005.
Stephan A. *Emergenz. Von der Unvorhersagbarket zur Selbstorganisation*. Dresden University Press, Dresden, 1999.
Wimsatt WC. *Complexity and Organization*. In: Boston Studies in the Philosophy of Science. Vol. 20. (Eds.: Schaffner KF & Cohen RS), D. Reidel, Dordrecht, 67–86, 1974.

Wimsatt WC. *Reductionism, Levels of Organization and the Mind-Body Problem*. In: Consciousness and the Brain (Eds.: Globus G, Maxwell G & Savodnik I), Plenum Press, New York, 205–267, 1976.

Wimsatt WC. *Reductionistic Research Strategies and their Biases in the Units of Selection Controversy*. In: Scientific Discovery, Volume II: Case Studies (Ed.: Nickles T), D. Reidel, Dordrecht, 213–259, 1980.

Wimsatt WC. *Forms of aggregativity*. In: Human Nature and Natural Knowledge. (Eds.: Donagan A, Perovich AN & Wedin MA), Dordrecht: Reidel, p. 259–291, 1986.

Wimsatt WC. *Aggregativity. Reductive heuristics for finding emergence*. Philosophy of Science: 64, S372–S384, 1997.

7

Theories, models, and equations in systems biology

Kenneth F. Schaffner

SUMMARY

This paper begins with a review of some claims made by biologists such as Waddington and von Bertalanffy, and others, that biology should seek general theories similar to those found in physics, for example in Newton's theory of gravitation and its elaboration in the Principia, some treatments of Maxwell's electromagnetic theory, or thermodynamics, quantum mechanics, and relativity theories. In these domains, broadly applicable differential equations system states and potential trajectories are found. I disagree with that view, and describe an alternative framework for biological theories as collections of prototypical interlevel largely qualitative causal models that can be extrapolated by analogy to different organisms. However, in the rare area of intersection between prototypical models which are largely qualitative and the equation-based models of physics, there are some significant accomplishments that may point the way toward important systems approaches in biology.

I look at two cases in particular in this intersection area: the development of the Hodgkin–Huxley giant squid model for action potentials, and at a more recent model of Ferrée and Lockery for worm (*Caenorhabditis elegans*) chemotaxis. The Hodgkin–Huxley strategy uses equations, but in specialized ways involving empirical curve-fitting and heuristic approximations, to build their model. In the worm example, model building proceeds from the organismal level down. It starts from a model of the nematode body, which captures the head and neck turning movements (head-sweep), then seeks a neural implementation of the head-sweep mechanism using tools from compartment theory.

Systems Biology
F.C. Boogerd, F.J. Bruggeman, J.-H.S. Hofmeyr and H.V. Westerhoff (Editors)
Isbn: 978-0-444-52085-2

The proponents of this model argue that their neural model is well based on the worm's neurophysiology but only weakly, at this point, on the organism's neuroanatomy. Though both approaches use some of the tools of biophysics and other mathematically sophisticated theories of physics, the manner of their implementation is quite different from physics, but may be generalizable as an approach for systems biology.

1. INTRODUCTION: THE STRUCTURE OF BIOLOGICAL THEORIES

Until quite recently, much of the analysis of theories in the biological and biomedical sciences had subscribed to what I term the 'Euclidean Ideal'. This notion assumes that the ideal structure of a scientific theory resembles Euclid's approach to geometry: a small number of fundamental definitions and axioms constitute the essence of a theory. The axioms are mathematically precise, and are then elaborated deductively in the form of theorems and applications that cover a broad (scientific) domain. This view of theory structure obtains fairly strong support in the physical sciences, and is exemplified by Newton's theory of gravitation and its elaboration in the *Principia* (1972 [1726]), by some treatments of Maxwell's electromagnetic theory (see Stratton, 1941), thermodynamics, and by quantum mechanics (see von Neumann, 1955). A similar orientation toward general theory in biology also can be found in the work of von Bertalanffy on 'general systems theory' and his sets of multiple partial differential equations.

Biologists – especially those biologists seeking a methodological unity with the physical sciences such as Waddington in his (1968) – and philosophers of biology, such as the early Michael Ruse (1973), have maintained that the laws and theories of biology have the exact same logical structure as do those of the physical sciences (though recently there have been some changes – see Kitcher (1984), Rosenberg (1985), Culp & Kitcher (1989), and van der Steen & Kamminga (1991)). This simple unity view is only supportable if one restricts one's attention to those few – but very important – theories in biology which in point of fact have a very broad scope and are characterizable in their more simplified forms as a set of 'laws' which admit of mathematically precise axiomatization and deductive elaboration. Examples are certain formulations of Mendelian genetics and of population genetics. I maintain that a deeper analysis of even these theories, however, will disclose difficulties with a strong methodological parallelism with the physical sciences (see Schaffner, 1980, 1986; Kitcher, 1984). I believe that a close examination of a wide variety of other biological theories in immunology, physiology, embryology, and the neurosciences will suggest that the typical theory in the biomedical sciences is a structure of overlapping interlevel causal temporal prototypical models.

The models of such a structure usually constitute a series of idealized prototypical mechanisms and variations (some of which may be mutants) that bear family or similarity resemblances to each other, and characteristically each has a (relatively) narrow scope of straightforward application to (few) pure types. The models are typically interlevel in the sense of levels of aggregation, containing component parts which are often specified in intermingled body part (e.g., head or tail), cellular (e.g., neuron or axon), and biochemical (e.g., receptor or ions) terms. Stages of temporal development in the models may represent either deterministic, causally probabilistic, random (Markovian), or even mixed connections. This probabilistic character of some causal connections (a failure of strict determinism) should be distinguished from the conceptually distinct failure of the exact match of a model to a nonpure type to which, nonetheless, it is closest given available knowledge. Such a match can be close, however, exhibiting a strong analogy between a model and an organism (or population of organisms). I argued at length (in my 1980) that this new type of theory, which I termed a 'theory of the middle range' (with apologies to R.K. Merton (1968) who first used that term in a somewhat different context), both is found and should be expected to be found in the biomedical sciences. The term 'middle range' seemed appropriate for two reasons: first, the theories were not broad sweeping general theories but they were not summaries of data either; they were midway between these extremes. Second, in terms of levels of aggregation of the entities in such theories, the theories were not about high-level populations evolving in evolutionary time and not about specific DNA sequences or specific enzymes functioning in well-defined biochemical pathways, but were at the level of the organelle, the gene as characterized by functional products, the cell, and the organ. Thus though interlevel, their levels of aggregation tended to concentrate in the 'middle range.'

Though the Waddington and von Bertalanffy programs have not been confirmed in the typical accomplishments and representations in molecular biology in general, and molecular genetics in particular, there are interesting advances that fall between those searches for broad theories couched in mathematically precise differential equation form, and the narrow classes of mechanisms, usually described in qualitative multilevel causal language, that constitute the vast majority of current biomedical explainers. In traditional population genetics, one important exception is the ability to develop a powerful axiomatization of the subject that does bear strong analogies to equation based theories of physics. (For a detailed example see the Jacquard axiomatization of population genetics summarized in Schaffner (1993), Chapter 8.)

There are several other theories that are equation-based which can be identified in contemporary biomedicine; and in the remainder of this paper I discuss two of these in detail. My view is that these can disclose some important ways that very general and quantitative principles can be applied fruitfully in biology and medicine. They also disclose the limitations of this kind of physics-oriented approach to biology, and a comparison of those areas where mathematical modeling works and at what points it begins to fail may indicate ways that systems biology can approach the issues of theories, models, and equations in this nascent area.

I will begin my discussion with a brief account of the development of the Hodgkin–Huxley Giant Squid Model for Action Potentials, a stunning accomplishment for which Hodgkin and Huxley shared the Nobel Prize in physiology or medicine in 1963. One of the current standard textbooks of neuroscience (Kandel et al. 2000) states that 50 years after its publication, 'the Hodgkin–Huxley model stands as the most successful quantitative computational model in neural sciences if not all of biology' (p. 156).[1]

2. THE DEVELOPMENT OF THE HODGKIN–HUXLEY GIANT SQUID MODEL FOR ACTION POTENTIALS AS A CLASSICAL EXAMPLE OF SYSTEMS BIOLOGY

Action potentials are waves of potential difference (or voltage) that move down nerve axons, communicating the effect of a stimulus from the receptors located near the beginning of the neuron to the termination of the nerve cell. To a first approximation, action potentials are the result of rapid (millisecond) changes in the membrane's permeabilities to sodium and potassium ions, changes that underlie the wave of potential difference. Hodgkin and Huxley's work on the action potential in nerve cells began from Hodgkin's earlier work on electric currents on the shore crab in the late 1930s (Hodgkin, 1964). He teamed up with Huxley, who was his student at Cambridge University, and they jointly turned their attention to the giant squid axon, which was a much more tractable experimental system to investigate the movement of specific ions, including sodium and potassium. Though their work was interrupted by World War II, they resumed their project in 1946, and in the late 1940s through to the early 1950s they conducted their classical experimental and theoretical investigations (Huxley, 1964). A series of papers culminated in their extraordinary 1952 article in the *Journal of Physiology* in which they systematically lay out the steps and their reasoning that culminates in the classical action potential model of nerve transmission (Hodgkin & Huxley, 1952).

The 1952 paper closely parallels their more historical account of their steps toward their quantitative model that appears in the two Nobel Prize lectures

[1] The philosophy of science literature has just recently begun to address the Hodgkin–Huxley action potential model as an important exemplar. Weber (2004) discusses it at some length in Chapter 2, and Bogen (2005) and Craver (2006) analyze it as well.

(Hodgkin, 1964; Huxley, 1964). They begin by first discussing their careful experimental results which had employed the voltage clamp apparatus, developed in 1949 by Kenneth Cole. This experimental device permits the establishment of a set of different potential differences across the squid nerve cell membrane, and recording of the effects that the different membrane potential have on the state of the cell. (A detailed description of the apparatus and technique can be found in the textbox on p. 152 of (Kandel et al. 2000).) Their earlier papers had indicated that the movement of currents based on ions across nerve cell membrane could be well represented by an 'equivalent circuit' involving a capacitor and three resistors, all in parallel, and with each resistor in series with a source of an electrical potential difference. This circuit captures the sodium (Na) and potassium (K) currents, as well as a small leakage current (*l*). This equivalent circuit adapted from their 1952 paper is shown in Fig. 1 (compare with Huxley, 1964).

The 'laws of working' (a term originally used by John Mackie, but see my discussion of the phrase in schaffner 1993, pp. 287, 306–307) that govern this circuit are the standard physical laws including Ohm's law as noted in the legend to Fig. 1. Additionally, the potential difference across the membrane established by differences in the Na and K ions is as required by the Nernst equation:

$$V_{\mathrm{ion}} - \frac{RT}{zF} \ln \left[\frac{X_{\mathrm{o}}}{X_{\mathrm{i}}} \right]$$

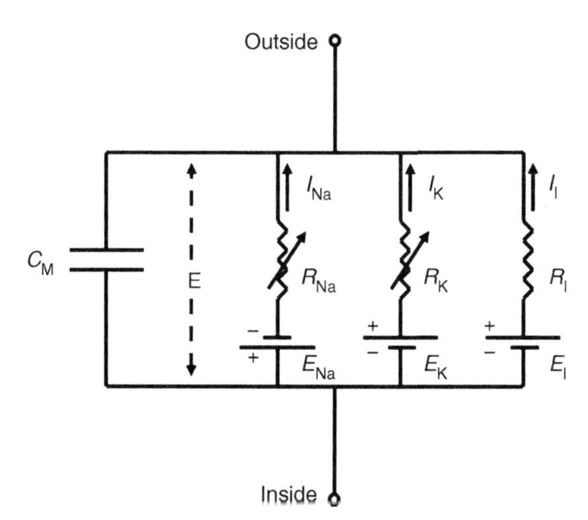

Figure 1 Equivalent circuit of a small area of membrane of the giant axon.

R_K and R_{Na} obey Ohm's law for rapid changes in the potential difference across the membrane, but change their values in times of the order of a millisecond if the membrane potential is held at a new value. *R* is constant.

where V is the potential difference (voltage), R and F the universal Boltzmann and Faraday constants , T the temperature, z the valence of the ion, and X_o and X_i are the concentrations of the ion outside and inside the cell. (Such laws are constraints and foundations, but are not the complete derivational source, for the later H and H equations I introduce further below (also see Bogen (2005) and Craver (in press) on this point.))

Part II of the Hodgkin–Huxley paper is a 'mathematical description of membrane current during a voltage clamp'. Equations for the sodium and potassium currents as conductances are developed. The equations do not come from 'first principles' but rather are empirical equations fitted from the voltage clamp data. They are typically chosen based on simplicity, with a first order equation being preferred over a second order, etc. A first order equation is satisfactory to represent a portion of the time course of nerve depolarization (a rapid change of voltage across the membrane), but a fourth order equation is needed to represent the beginning of the potassium depolarization process. The equation for potassium conductance, in the form that it could be compared with the empirical results, was chosen as

$$g_K = \left\{ (g_{K\infty})^{1/4} - \left[(g_{K\infty})^{1/4} - (g_{K0})^{1/4} \right] \exp\left(-\frac{t}{\tau_n} \right) \right\}^4$$

It is a theoretical equation, to use H and H's language, based on the equivalent circuit and the general empirically found form of the rise and fall of ion conduction during depolarization and repolarization. H and H doubt gives a 'correct picture' of the membrane, though they do provide a possible physical basis for the equation (see pp. 506–507 of the 1952 article). The equation contains a constant τ_n that can then be specified to be the best fit to experimentally determined depolarizations of different potential membrane differences. Hodgkin and Huxley found that there was reasonable agreement between theoretical and experimental curves. H and H then go on to develop the somewhat more complex reasoning leading to the equation for sodium conductance, which I shall not discuss, but which can be found on pp. 512–515 of their 1952 paper. They also develop equations for rate constants α and β, and the dimensionless proportions n, m, and h, of ions inside and outside the membrane, in Part II as well.

At the beginning of Part III of their 1952 paper, titled 'Reconstruction of Nerve Behavior,' H and H summarize the equations they have developed in Part II of that paper. The summary is from the H and H (1952) article and the numbering of the equations in brackets comes from their original equation numbers. The summary looks like this:

$$I = C_M \frac{\mathrm{d}V}{\mathrm{d}t} + \bar{g}_K n^4 (V - V_K) + \bar{g}_{Na} m^3 h (V - V_{Na}) + \bar{g}_l (V - V_l) \qquad (26)$$

$$\frac{\mathrm{d}n}{\mathrm{d}t} = \alpha_n(1-n) - \beta_n n \tag{7}$$

$$\frac{\mathrm{d}m}{\mathrm{d}t} = \alpha_m(1-m) - \beta_m m \tag{15}$$

$$\frac{\mathrm{d}h}{\mathrm{d}t} = \alpha_h(1-h) - \beta_h h \tag{16}$$

and

$$\alpha_n = 0.01 \frac{(V+10)}{\exp\left(\frac{V+10}{10}\right) - 1} \tag{12}$$

$$\beta_n = 0.125 \ \exp\left(\frac{V}{80}\right) \tag{13}$$

$$\alpha_m = 0.1 \frac{(V+25)}{\exp\left(\frac{V+25}{10}\right) - 1} \tag{20}$$

$$\beta_m = 4 \ \exp\left(\frac{V}{18}\right) \tag{21}$$

$$\alpha_h = 0.07 \ \exp\left(\frac{V}{20}\right) \tag{23}$$

$$\beta_h = \frac{1}{\exp\left(\frac{V+30}{10}\right) + 1} \tag{24}$$

The first four of these equations are the differential equations which govern the system's behavior. The many computer simulations of the H and H model involve programs that repeatedly step through those first four equations (see Fodor, 2005, for one example).

Equation [26] is then applied to the action potential. We are most interested in the 'propagated action potential,' as distinguished from a uniform membrane action potential. In the propagated action potential, the local circuit currents have to be provided by the net membrane current. At this point in their 1952 paper, H and H appeal to a well-known partial differential equation from cable theory (which is a variant of Laplace's well-known heat diffusion partial differential equation) relating the current to the second partial derivative of the potential difference (V) with respect to distance (x). This equation is given by the expression

$$i = \left(\frac{1}{r_1 + r_2}\right) \frac{\partial^2 V}{\partial x^2} \tag{27}$$

There are some simplifications then invoked, e.g., since $r_1 \ll r_2$, r_1 can be dropped. The expression for the current density for the fiber with a radius of a then allows the equation to be rewritten as

$$I = \left(\frac{a}{2R_2}\right) \frac{\partial^2 V}{\partial x^2} \tag{28}$$

This relation is then substituted into Eqn [26], which yields a partial differential equation that is 'not practical to solve as it stands' (p. 522). But a similarity is noted for the condition of steady propagation, one which permits the equation to be converted into an ordinary differential equation that can be solved numerically, if laboriously given the computational tools available in 1952. This is the propagated action potential equation and was written as

$$\frac{a}{2R_2\theta^2}\frac{\mathrm{d}^2V}{dt^2} = C_M\frac{\mathrm{d}V}{\mathrm{d}t} + \bar{g}_\mathrm{K}n^4(V - V_\mathrm{K}) + \bar{g}_\mathrm{Na}m^3h(V - V_\mathrm{Na}) + \bar{g}_l(V - V_l) \tag{30}$$

where θ is a parameter that has to be estimated numerically, based on the behavior of the equation at extreme boundary conditions (see p. 522 of H and H for details). A section on numerical methods of solution of such equations is interpolated in the 1952 article, and after a minor (abbreviational) substitution, Eqn [30] is rewritten as Eqn [31] (not shown here, but see p. 524 of the original article). This equation (either Eqn [30], or the equivalent Eqn [31], is solved numerically, and graphs of the membrane conductances during a propagated action potential are depicted. H and H's graphical results were shown in their Fig. 17 that were based on numerical solutions of Eqn [31] showing components of membrane conductances (g) during propagated action potential ($-V$). This figure is widely reproduced in standard neuroscience texts (also see their original article, p. 530). Readers will also recognize such graphs of the conductances as THE classical action potential result, which is represented, based on largely qualitative considerations, in typical neuroscience textbooks.

3. IMPLICATIONS OF THE HODGKIN–HUXLEY MODEL AND THEIR METHODOLOGY

3.1. One basic mechanism with many types of molecular realizations?

An examination of the form of the key equations, especially the batch summarized beginning with [26] above and then numbers [30–31] might suggest that H and H's accomplishment is not that different than, say, James Clerk Maxwell's articulation of the electromagnetic theory of light and Maxwell's derivation of the wave equation for an electromagnetic disturbance. (That disturbance importantly had the transverse wave features and the same velocity as light, which led Maxwell to postulate that light was an electromagnetic wave.) But the H and H equations are not universal equations as were Maxwell's – the H and H equations were empirically generated from curve fittings to the squid action potential changes read using the voltage clamp technique. Hodgkin and Huxley remarked on the limitations of their model a number of times during the course of their 1952 article, limitations that are well summarized by Bogen (2005) and Craver (2006).

Toward the very end of the 1952 paper, H and H wrote

> Applicability to other tissues. The similarity of the effects of changing the concentrations of sodium and potassium on the resting and action potentials of many excitable tissues (Hodgkin, 1951) suggest that the *basic mechanism* of conduction may be the *same* as implied by our equations, but the *great differences* in the shape of action potentials show that even if equations of the same form as ours are applicable in other cases, some at least of the parameters must have *very different values*.
>
> (p. 542) (my emphases)

In addition, toward the end of his Nobel lecture, Hodgkin returned to this issue and the related theme of a specific or 'definite' model of the membrane when he wrote

> To begin with we hoped that the analysis might lead to a definite molecular model of the membrane. However, it gradually became clear that different mechanisms could lead to similar equations and that no real progress at the molecular level could be made until much more was known about the chemistry and fine structure of the membrane. On the other hand, the equations that we developed proved surprisingly powerful and it was possible to predict much of the electrical behaviour of the giant axon with fair accuracy. Examples of some of the properties of the axon which are fitted by the equations are: the form, duration and amplitude of the action potential; the conduction velocity; impedance changes; ionic movements; and subthreshold phenomena including oscillatory behaviour.
>
> (1962, p. 42)

A review of contemporary molecular models of various ion channels capable of supporting action potentials suggests that H and H happened on a most remarkable level of abstraction/aggregation that would support very broad generalization in terms of the specificity of membrane currents, though not any specific molecular mechanisms. For example, the chapter by Koester and Siegelbaum on 'Propagated Signaling: The Action Potential' in Kandel et al. (2000) states somewhat 'teleologically' that

> The squid axon can generate an action potential with just two types of voltage-gated channels. Why then are there so many *different types* of voltage-gated channels found in the nervous system? The answer is that neurons with the expanded set of voltage-gated channels have much more complex information-processing abilities than those with only two types of channels.
>
> (p. 159) (my emphasis)

The number and types of ion channels are explained, and to an extent unified, by the underlying genetics (and epigenetics) of ion channel diversity, a topic to which I turn next.

3.2. Genetic and epigenetic diversity accounts for ion channel diversity

Hille recounts the history of ion channel research over the course of the last half-century following H and H's classic paper. The progress he writes has been 'phenomenal', and 'the field has become highly interdisciplinary, combining approaches of biophysics, pharmacology, protein chemistry, molecular and medical genetics, and cell biology' (Hille, 2001, p. 61). Several recent Nobel prizes have, in point of fact, been awarded for ion channel research, including to Neher and Sakmann in 1991, who developed the 'patch clamp method' that provided direct evidence of ion channels, and to MacKinnon in 2003, for structural and mechanistic studies of ion channels, including his pore model.

Genetic studies that began in the 1980s have indicated that there are three general genetic 'superfamilies' of ion channels, comprising ligand-gated, gap-junction, and the H and H type of action potential generating voltage gated channels. This last class, which is activated by depolarization, also contains three subclasses of channels selective for Na^+, and K^+, Ca^{2+} (Siegelbaum & Koester, 2000). Siegelbaum & Koester describe the similar architecture of this class of channels writing:

> They contain four repeats of a basic motif composed of six transmembrane segments [known as] (S1–S6). The S5 and S6 segments are connected by a loop, through the extracellular face of the membrane, the P-region, that forms the selectivity filter of the channel. A single subunit of voltage gated Na^+ and

Ca^{++} channels contains four of these repeats. Potassium channels are composed of four subunits, each containing one repeat.

(p. 119)

Additional ion family channels are in the process of being discovered and characterized, including a class of Cl^{-} channels. But already the number of different channel types is according to Siegelbaum & Koester 'enormous'. The diversity is accounted for in part because 'most channels are made up of multiple subunits that can be combined in different permutations to produce channels with different functional properties' (p. 119). Additionally, the variability is 'produced by differential expression of two or more closely related genes, by alternative splicing of mRNA transcribed from the same gene, or by editing of mRNA' (p. 120).

Some simplification of this extensive diversity occurs in the axonal region of the neuron where just the two major channel types, Na^{+} and K^{+}, are involved. However, even here, Hille also describes an extensive 'diversity of K channels' in different tissues and even within single cells. He sums up this 'microheterogeneity of K channels', noting that 'such results are typical of experimental discoveries today. The finer the method of analysis, the more apparent subtypes of channels are discovered' (Hille, 2001, p. 74). In spite of this extensive diversity and variation, genetics can provide a rationale for generalization, at least involving similarity modeling. On this point Hille writes: 'The Na, Ca, and K families of voltage-gated channels form a homologous gene superfamily, as may be expected from their broad apparent functional similarity. This means that many findings for one type of channel can be generalized to the others' (Hille, 2001; p. 85).

3.3. The H and H 'basic mechanism' as an emergent simplification

The account of the extensive diversity of specific mechanisms of ion channel types just summarized raises the question of how unity can be effectively achieved amid such natural variation. In a significant sense, H and H achieved that unification and simplification in advance of the more recent molecular knowledge by working at a higher level of abstraction. Their accomplishment suggests that in certain areas of biology, investigators can capture what might be termed 'emergent simplifications' that transcend the specific workings of the molecular details. In a way, a more abstract mechanism can be a 'basic mechanism', even if it is clearly realized that there are as yet unknown molecular details of the mechanism. Possibly such a basic mechanism is more like a 'prototypical' mechanism, which identifies and characterizes salient core features of a biological entity and its actions.

In some circumstances, the core features of those simplifications can be generated by quantitative investigations and represented by mathematical equations that are formally analogous to what we find in the Euclidean types of theories discussed in Section 1. But they lack that very broad universality, and instead serve their functions by being prototypes for analogical modeling to similar prototypes, albeit in this case, analogical modeling to other quantitative prototypes. In addition, they are not usually uni-level, but instead mix levels of aggregation. In the H and H work, the discussion is focused on current flows and potential difference changes because of ions and inferred ion channels, but as situated in an axon of a particular species. Further reflection of the H and H systems-level methodology may provide important generalizable heuristics that can inform biology pursued at the level of general systems.

4. A NEUROSCIENTIFIC ACCOUNT OF BEHAVIOR IN *C. ELEGANS*

An interesting comparison with the above H and H account can be found in a recent essay by Ferrée and Lockery. Whereas the typical study of the behavior of the model organism, *C. elegans*, tries to identify genes and molecular sequences that are characterized as 'causes' of behaviors, the example to be discussed in this section is more akin to the H and H inquiry and their mode of modeling. For an example of the more typical approach to worm behavior modeling, see Mario de Bono and Cori Bargmann's (1998) *Cell* paper with their focus on a DNA nucleotide change as the 'cause' of a behavioral phenotype involving social versus solitary feeding. Ferrée and Lockery, in contrast, provide an analysis that attempts to model the factors and interactions that govern the neurons not the genes. Ferrée and Lockery's general task was to determine 'the behavioral strategy for chemotaxis in *C. elegans*', and their specific approach was to 'derive a linear neural network model of the chemotaxis control circuit' in *C. elegans*, and then to 'demonstrate that this model is capable of producing nematode-like chemotaxis' (Ferrée & Lockery, 1999, p. 2). This then is a simulation study, but one based on a considerable amount of empirical work. The following account is adapted from Schaffner (2000), but here is updated and placed in a different context – that of exploring systems theory in biology and the extent and nature of mathematical equations in that area.

Ferrée and Lockery utilized a 'candidate neural network' based on Bargmann's earlier work on the worm (see Fig. 2). Lockery's own investigations (Goodman et al., 1998) have shown that the neural signals in *C. elegans* are encoded by graded electrical potentials (not by classic sodium action potentials). The individual neurons display nonlinear transfer functions, but Ferrée and Lockery

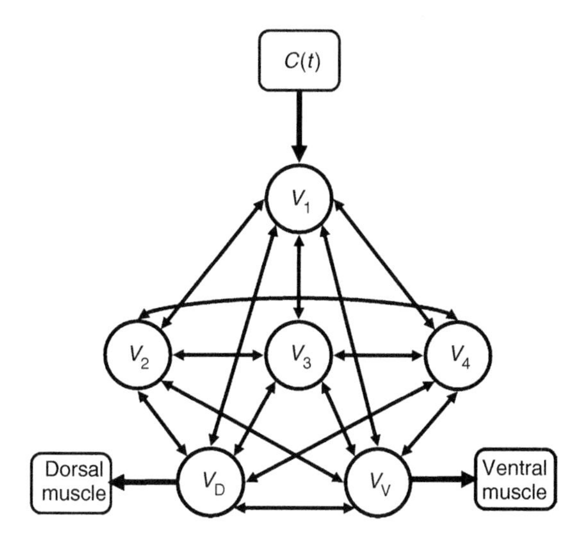

Figure 2 Neural network model for the chemotaxis control circuit of C. elegans.

The state variable of each neuron (circle) is voltage(V_i). The model contains one chemosensory neuron (V_1), three interneurons (V_2 – V_4) and two motor neurons (V_D, V_V). The chemosensory neuron receives input equal to the chemical concentration $C(t)$ at the tip of the nose, and the motor neurons innervate dorsal (D) and ventral (V) neck muscles. Based on a similar diagram from Feree and Lockery.

propose that one can look at a simplified linearization of the chemotaxis system that can give some insights about this behavior, albeit as a first approximation.

The model building proceeds from the organismal level down. It starts from a model of the nematode body, which captures the head and neck turning movements (head-sweep), then seeks a neural implementation of the head-sweep mechanism. Ferrée and Lockery argue that their neural model is based on the worm's neurophysiology, but only – at this point – weakly on the neuroanatomy. Citing Goodman et al., 1998, they suggest the neurons can be represented as single electrical compartments. (Compartment models, like the simpler cable theory models that backgrounded some of H and H's investigation, are one of the traditional strategies used in neuroscience; see Bower & Beeman, 1995.) An equation for the voltage V_i of the ith neuron can be written using standard compartment modeling as

$$C_i^{\text{cell}} \frac{\mathrm{d}V_i}{\mathrm{d}t} = -G_i^{\text{cell}}(V_i - E_i^{\text{cell}}) - I_i^{\text{elec}}(V) - I_i^{\text{chem}}(V) - I_i^{\text{sens}}(t) \tag{1}$$

where C_i^{cell} is the whole-cell capacitance, G_i^{cell} is the effective ohmic conductance associated with the linear region of the $I - V$ curve, and E_i^{cell} is the resting potential of an isolated neuron. Here $I_i^{\text{elec}}(V)$ and $I_i^{\text{chem}}(V)$ represent electrical and chemical synaptic currents, $V = (V_1, \ldots, V_N)$is an N-dimensional vector comprised of the voltages of all N neurons in the network, and $I_i^{\text{sens}}(t)$ represents chemosensory input (from Ferrée & Lockery, p. 14).

They then borrow from data on *Ascaris*, since as frequently noted in the worm literature, synaptic neurophysiological data are not yet available for *C. elegans*, so a closely related worm, *Ascaris*, is used. This data allows them to assert that the chemical synapses between cells i and j can be modeled by the a sigmoidal functional equation

$$I_i^{\text{chem}}(V) = \sum_{j=1}^{N} G_{ij}^{\text{chem}} \cdot \sigma(\beta_{ij}(V_j - \bar{V}_j)) \cdot (V_i - E_{ij}) \tag{2}$$

where G_{ij}^{chem} is the maximum conductance in the cell i because of synaptic connections from cell j and E_{ij} is the reversal potential for the corresponding postsynaptic current. Electrical synapses are similarly modeled by another slightly simpler third equation. Further, chemical inputs to the system are captured by

$$I_i^{\text{sens}}(t) = -\delta_{il}\kappa_{\text{sens}}C(t) \tag{3}$$

where $C(t)$ is the chemical concentration at the tip of the worm's nose, δ_{i1} is the standard Kronecker delta and κ_{sens} is a constant parameter.

The total synaptic model can be further simplified by representing only the chemical synapses. Equation (2), which is then governing, is nonlinear, but it can be linearized by using a Taylor series expansion (familiar to elementary calculus students) and retaining only the linear terms. This process yields the following set of equations:

$$\frac{\mathrm{d}V_i}{\mathrm{d}t} = \sum_{j=1}^{N} = A_{ij}V_j + b_i + c_i(t) \tag{4}$$

(The matrix A_{ij} and b_j are complicated functions of the G's, V's, and E's introduced in Eqns (1) and (2), and are not reproduced here; see Ferrée & Lockery, 1999, pp. 16–17.) This linearized equation and two quite simple body model equations are then combined with an equation representing the chemical environment, C, and the equations solved to yield a state trajectory $S(t)$ that begins from some specified initial state S_0. The simulation solutions were obtained by numerical integration, akin to H and H's work, though now using powerful computer tools, and some other tricks employed to eliminate transients. In their paper, Ferrée and Lockery show a comparison between real and simulated worms which provides a qualitative comparison of their model and actual worm pathways (see Ferrée & Lockery, 1999, p. 270).

Next, Ferrée and Lockery explored the linearized equation solution to develop a more intuitive result, since they note that 'distributed representations' often lack this property of intuitability. This part of their paper provides a 'simple rule for chemotaxis control which relates the body rate of turning ... to time derivatives of the chemosensory input . . . ' (p. 23). On the basis of the analysis, Ferrée and Lockery argue that their network uses strategies both of klinotaxis (alignment with a vector component of the stimulus field) and klinokinesis (change in turning rate in response to the scalar value of a stimulus field) to produce the behavior represented in their Fig. 3b. (Here the definitions of klinotaxis and klinokinesis follow Dunn, 1990.) These strategies also suggest seeking additional experimental worm stimulus and movement data to confirm or disprove the models.

Ferrée and Lockery's approach does not use genes, and it does not employ structural data from molecular biology. It does utilize physiology and neuroscientific compartment analysis to formulate a mathematical model of a neural network that qualitatively agrees with the worm's observed behavior. It is perhaps more similar to a biophysics approach such as H and H's action potential model than a biomolecular approach.

5. IMPLICATIONS OF THE FERRÉE AND LOCKERY MODEL FOR *C. ELEGANS* CHEMOTAXIS

The F and L model is a mathematical simulation of *C. elegans* chemotaxis relying on a simplified neural circuit, and on generally accepted model-building strategies found in the neurosciences. Like H and H, it seeks to identify an appropriate level of abstraction from the much more complicated details that might constitute the specific mechanisms of neural interplay. But thus far it has not enjoyed the broad acceptance and heuristic fertility of the H and H model. Why that is the case needs further thought, since the general strategies appear to be similar. Possibly, what seems to be the idiosyncratic aspects of the model arise because F and L are not dealing with a constituent low-level mechanism that could be found in multiple instances and more easily generalized to similar ion channel activities in related type of cells. Rather F and L deal with an intentionally particular wiring diagram that may restrict the generalizability of the results of their model. However, the more methodological modeling and equation building and solving strategies may have broader applicability, such as the reliance on an empirically confirmed circuit, and the application of compartment modeling to such interactive networks.

6. EIGHT IMPLICATIONS OF THE TWO EXEMPLARS FOR SYSTEMS BIOLOGY

I can think of eight implications of the above discussion for philosophical issues in systems biology; other may see additional implications.

(1) In biology, the roles that general theories have in physics (as explainers, organizers of domains of inquiry, and experimental fertility) is carried out by a series of prototypes (think of these as models or as mechanisms) which are causal–temporal multilevel systems and are analogically related to other prototypes.
(2) Those prototypes may be formulated in quantitative terms, though typically they are not, and in their quantitative variants these may even appear in a mathematical form that is very much like that found in general physical theories, such as in Maxwell's equations or the axioms of quantum mechanics. But the equations describing the prototypes are not universal ones, rather they are tied closely to the specific organisms on which they are based, though they can be extended. Such equations are also typically approximations rather than exact equations.
(3) Biological prototypes are applied, in the sense of being used as explainers or as extending the application to another biological organism, more by analogical reasoning than by determining the mathematically expressed initial conditions (or boundary conditions) and proceeding deductively, by inserting those conditions into equations to particularize and thus apply the general system.
(4) Biological prototypes need not be only gene based. The H and H and the F and L exemplars are not gene based, but do their work well, though the H and H model is much more generalizable, for reasons speculated on in the text above. Genetic information may assist in specifying highly particular variants of mechanisms, such as receptors, and even in identifying classes of control mechanisms, but a genetic dimension is not always needed.
(5) Biological prototypes incorporate critical structural information. This structural information is 'biological' in nature, as opposed to simple physical and chemical structural information. In an important sense, the explanation for biological structure requires an implicit appeal to billions of years of evolution, but working biologists need to assume that structure is an 'emergent' given in investigating any biological system, and need to characterize the prototype being studied at the appropriate levels with such structure assumed. We saw this in the H and H example in terms of accepting a cable model for the giant squid axon, constructed in part out of a structured membrane permitting sodium and potassium ion passage. In the F and L *C. elegans*

example, we encountered a pre-existing wiring diagram of connected neurons, in addition to pre-existing compartments characterizing the neurons. See Kitano (2001) for the importance of taking structural information into account in systems biology.

(6) Biological prototypes may be most useful when they display dynamical or behavioral features, in addition to structural organization, whether these dynamical/behavioral aspects are presented as temporal–causal sequences or as a dynamics that can be captured mathematically and in simulations, as in the H and H and F and L exemplars discussed in this paper. The importance behavior and dynamics for systems biology is also stressed by Kitano (2001).

(7) In the case where the systems are characterized quantitatively, as in the H and H and F and L exemplars, simulations can be constructed fairly easily, and these can be valuable both in testing a prototype and in possibly extending it, by allowing for variation of parameters in a precise manner and their application to experimental systems. These simulations, however, need to be controlled both by specific data and by general biological principles, and not be purely speculative exercises in mathematical model building, as in much of von Bertalanffy's (1968) writings.

(8) The two exemplars discussed above go some way toward identifying some potentially useful philosophical issues in systems biology, but they need to be supplemented with other exemplars, to provide a more comprehensive picture. One such additional area might involve gene-based systems together with high-throughput data. A valuable proof of a principal paper along these lines is the Ideker et al.'s (2001) galactose metabolism model for yeast, one that points toward additional features, such as gene–protein and protein–protein interactions, that are likely to be important in a philosophy of systems biology.

REFERENCES

Bogen J. *Regularities and causality; generalizations and causal explanations.* Studies in the History and Philosophy of Science C: 36, 397–420, 2005.

Craver CF. [forthcoming]. *Explaining the Brain.* Oxford: Oxford University Press, 2006.

Culp S & Kitcher P. *Theory Structure and Theory Change in Contemporary Molecular Biology.* British Journal for the Philosophy of Science: 40, 459–483, 1989.

de Bono M &Bargmann C. *Natural Variation in a Neuropeptide Y Receptor Homolog Modifies Social Behavior and Food Response in C. elegans.* Cell: 94, 677–689, 1998.

Dunn GA. *Conceptual problems with kinesis and taxis.* In: Armitage JP & Lackie JM (Eds.) Biology of the Chemotactic Response, Cambridge University Press, Cambridge, pp. 1–14, 1990.

Ferrée TC & Lockery SR. *Computational Rules for Chemotaxis in the Nematode C. elegans.* Journal of computational neuroscience: 6(3), 263–77, 1999.

Fodor, Anthony A. (2005). http://www.afodor.net/HHModel.htm (Accessed October 29, 2006).

Goodman MB, Hall DH, Avery L & Lockery SR. *Active Currents Regulate Sensitivity and Dynamic Range in C. elegans Neurons*. Neuron: 20, 763–772, 1998.

Hille B. *Ion channels of excitable membranes*. Sunderland, Mass., Sinauer, 2001.

Hodgkin AL & Huxley AF. *A Quantitative Description of Membrane Current and its Application to Conduction and Excitation in Nerve*. Journal of Physiology: 117, 500–544, 1952.

Hodgkin AL. *The Ionic Basis Of Nervous Conduction*. Science: 145, 1148–54,1964.

Huxley AF. *Excitation and Conduction in Nerve: Quantitative Analysis*. Science: 145, 1154–9, 1964.

Kitcher P. *1953 and all that: A tale of two sciences*. Philosophical Review: 18, 335–73, 1984.

Ideker T, Thorsson V, Ranish JA, Christmas R, Buhler J, Eng JK, Bumgarner R, Goodlett R, Aebersold R & Hood L. *Integrated genomic and proteomic analyses of a systematically perturbed metabolic network*. Science: 292(5518), 929–34, 2001.

Kandel ER, Schwartz JH & Jessell TM. *Principles of neural science*. New York, McGraw-Hill, Health Professions Division, 2000.

Kitano H (editor). *Foundations of Systems Biology*, Cambridge: MIT Press, 2001.

Lockery S. Lockery's home page: http://chinook.uoregon.edu/index.html, accessed October 29, 2006.

Merton RK. *On sociological theories of the middle range*. In: Social Theory and Social Structure (ed. Merton RK). New York: Free Press, 1968.

Newton I. *Principia*. 3rd ed. 2 vols. Cambridge, MA: Harvard University Press, 1972 [1726].

Rosenberg A. *The structure of biological science*. Cambridge: Cambridge University Press, 1985.

Ruse M. *Philosophy of biology*. London: Hutchinson, 1973.

Schaffner KF. *Theory structure in the biomedical sciences*.The Journal of Medicine and Philosophy: 5, 57–97, 1980.

Schaffner KF. *Exemplar reasoning about biological models and diseases: A relation between philosophy of medicine and philosophy of science*. The Journal of Medicine and Philosophy: 11, 63–80, 1986.

Schaffner KF. *Discovery and explanation in the biomedical sciences*. Chicago: University of Chicago Press, 1993.

Schaffner KF. *Behavior at the Organismal and Molecular Levels: The case of* C. elegans. Philosophy of Science: 67, PSA-1998, Vol. 2 Proceedings, S273–S288, 2000.

Siegelbaum SA & Koester J. *Ion channels*. In: Kandel ER, Schwartz JH & Jessell TM (eds.), pp. 105–123, 2000.

Stratton JA. *Electromagnetic Theory*. 1st ed. New York: McGraw-Hill, 1941.

Van der Steen WJ & Kamminga H. *Laws and Natural History in Biology*. British Journal for the Philosophy of Science: Vol. 42, No. 4, 445–467, 1991.

von Bertalanffy L. *General System Theory*, Braziller, New York, 1968.

von Neumann J. *Mathematical foundations of quantum mechanics*. Tr. Beyer RT, Princeton: Princeton University Press, 1955.

Waddington CH. *Introduction*. In: Towards a theoretical biology, edited by Waddington CH, Vol. 1. Chicago: Aldine Press, 1968.

Weber M. *Philosophy of Experimental Biology*. New York: Cambridge University Press, 2004.

8

All models are wrong

. . . some more than others

Olaf Wolkenhauer and Mukhtar Ullah

SUMMARY

Can systems biology be a 'holistic' approach as suggested by some authors? The fact that we are dealing with complex systems means that we must 'reduce' biological reality. The whole purpose of (mathematical) modelling is abstraction, i.e, a reduction of complexity. The success of differential equation modelling in the physical and engineering sciences might give a misleading impression that models in systems biology can provide accurate or replica representations of the physicochemical reality in question. Although most models of cellular processes cannot be accurate, they are useful because understanding arises from reducing one type of reality into another. It is therefore the complexity of intra- and intercellular processes that motivates modelling as a means to reduce complexity. The present text is to discuss Robert Rosen's critique of mathematical modelling of complex (biological) systems. The aim is to provide a concise summary of his distinction between analytical and synthetic models and to renew the interest in Rosen's work. He taught us to carefully (re)consider the modelling process. With an appreciation of the uncertainty involved in modelling complex natural systems, he demonstrated that not all models are equally good – for our understanding of living systems.

> *"As the complexity of a system increases, our ability to make precise and yet significant statements about its behavior diminishes until a threshold is reached beyond which precision and significance (or relevance) become almost exclusive characteristics."*
>
> (Lotfi Zadeh)

Systems Biology
F.C. Boogerd, F.J. Bruggeman, J.-H.S. Hofmeyr and H.V. Westerhoff (Editors)
Isbn: 978-0-444-52085-2

1. INTRODUCTION

We model to explain observations, aspects of complex systems. Modelling is necessarily an abstraction, that is, a reduction of complex interrelations to essential features, without loosing the ability to predict. Figure 1 illustrates the predictive power of abstract modelling using the drawing of a bird. Despite its simplicity, this drawing allows us to identify/predict a real male lapwing if we see one. There are various possibilities to improve this model: using a finer pen, showing feathers, depicting the bird in flight etc. The message is that there is no correct or wrong model but models with different purposes, models that address different aspects of a complex system and that have different accuracies. Mathematical modelling in systems biology is the art of making appropriate assumptions, thereby choosing an appropriate level of abstraction. Mathematics is the art that makes us realise reality, and as Picasso commented, art is a lie that makes us realise truth.

The original conception of systems biology as a merger of control theory and molecular and cell biology has as its central dogma the observation that it is system dynamics and organising principles that give rise to the functioning and function of cells (Wolkenhauer & Mesarovic, 2005). Cell function, including growth, differentiation, division and apoptosis, are temporal processes and we will only be able to understand them if we model them as dynamic systems. While the areas of genomics and bioinformatics are identifying, cataloguing and characterising the components that make up a cell, systems biology focuses on an understanding of functional activity from a systems perspective. The biological agenda of systems biology can subsequently be defined by the following two

Figure 1 This is not a bird – it is a model of a bird, an abstraction. The drawing illustrates the power of abstraction in modelling and the importance of choosing an appropriate level and resolution for a model. Despite its simplicity, this picture allows an unambiguous identification of the species and sex of the bird within a context (e.g. some location and time).

questions related to intra- and intercellular processes within a cell and in cell populations:

How do the components within a cell interact, so as to bring about its structure and realise its functioning?

How do cells interact, so as to develop and maintain higher levels of structural and functional organisation?

A defining feature of systems biology is the role that mathematical modelling plays. As indicated above, for mathematical modelling to be employed we require a certain level of complexity that is necessary to convince the nonmathematician of its usefulness. In systems biology, complexity emerges as

- A property of an encoding, e.g. a larger number of variables that determine the behaviour of a system.
- An attribute of the natural system under consideration, e.g. the connectivity, nonlinearity of relationships.
- As related to the technologies by which we take measurements and generate experimental data, e.g. limited precision and accuracy.
- Associated with uncertainty arising from the methodologies employed to investigate and model the system. For instance, the choice of a conceptual framework, i.e. whether differential equations, automata, etc. are employed.

It is clear that a model of a cell, cell function or pathway, is conceptual – not reality. All theoretical results are derived within a chosen conceptual framework, thus formulated to correspond in some useful way to the real world, as it presents itself to us in experiments. A pathway is an interpretation of observable facts, in the light of biological and mathematical concepts that we ourselves construct using natural language, drawings, images and formal, mathematical methods.

The definition of a pathway usually starts with a selection of proteins and their modified forms. This choice should be determined by the function the pathway realises in the cell but is often also determined by the costs and logistics of time-consuming experiments that can be conducted to investigate the pathway. In any case, the selected components of a pathway should display some autonomy or robustness in the sense that their interactions and the resulting behaviour is to a large degree independent of other components that are not included in its definition. By defining a pathway, we therefore isolate it from other components of the cell. This suggests that the concept of 'cross-talk' among pathways is contradicting the common definition of a pathway, because if two pathways interact to an extent that they cannot be looked at in isolation, all components should be considered one pathway. With the complexity of cellular processes, the uncertainty that arises from our limited ability to observe these processes and take measurements, it becomes clear that the modelling process itself is of interest.

In his book *Life Itself*, Robert Rosen (1991) argues that mathematical modelling in the Newtonian realm of physics – the world of mechanisms – is inadequate to describe biological systems (organisms). The basis for his argument is a distinction between two approaches for mathematical modelling of cellular systems: synthetic vs. analytic modelling. We are going to present the mathematical foundation for this argument and show formally that there are analytical models which are not synthetic models. In other words, the conventional way in which we model cells, cell function or pathways has a principal limit in what we can know about the cell as a living organism. Rosen's discussion takes place at a high level of abstraction, occasionally leaving the reader to fill in details. We provide a concise summary of a formal framework in which one can formalise the process of mathematical modelling of complex natural systems. We maintain that such a framework is relevant for systems biology and hope the present paper makes the work of Rosen more accessible.

2. MODELLING THE MODELLING PROCESS

Robert Rosen (1934–1998) was a theoretical biologist who focussed his career on the question what it is that makes an organism alive. Rosen came to realise that mathematical modelling in the Newtonian realm of physics – the world of mechanisms – was inadequate to describe biological systems. On the basis of the argument presented below, he introduced a class of metaphorical, relational paradigms for cellular activity, called (M,R)-systems (Rosen, 1971, 1991; Wolkenhauer, 2001). Using concepts from category theory, he demonstrated the minimal requirements that would have to be in place for a cell to be 'alive'. Linear systems interpretation of (M,R)-systems was assumed by Casti in the context of engineering systems (Casti, 1988a,b). Further generalisation were considered in the context of *Artifical Life* (Nomura, 2004) and in systems biology (Cho et al., 2005). Letelier and coworkers related (M,R)-systems to autopoietic systems, as well as metabolic networks (Letelier et al., 2003). Considering the fundamental conclusions that can be drawn from Rosen's work, it is not surprising that the concept has been challenged (e.g. Landauer & Bellman, 2002).

The discussion of how we try to understand intra- and intercellular dynamics, using mathematical modelling and computer simulations, remains an important question in the context of systems biology (Wolkenhauer & Mesarovic, 2005). The purpose of the present text is, however, not a discussion of the validity of Rosen's arguments but to provide a concise summary of the formal distinction between 'analytical' and 'synthetic' modelling, which is at the root of Rosen's distinction between organisms and mechanisms and his critique of the 'physical approach' to model biological systems. Rosen's discussion takes place at a high level of abstraction, occasionally leaving the reader to fill in details. In this situation, a clear mathematical notation is important and typographical errors can easily mislead the reader.

The outline for the remaining text is as follows: We introduce analytical models and synthetic modelling in separate sections, before establishing a relationship between the two. Finally, we point out some of the conclusions Rosen himself has drawn from the presented material and we give links to the literature that take up some of the issues relevant to the present text.

3. ANALYTICAL MODELLING

We investigate a natural system by considering it an abstract set M of abstract states $p \in M$. For the analysis of complex natural systems, and in particular biological systems, Rosen argued that it is most important that we should not make any assumptions about the structure of M. Observations or measurements we make on M are encoded by observables

$$\begin{aligned} \xi : M &\to \xi[M] \\ p &\mapsto \xi(p), \end{aligned}$$

where we use square brackets to denote the range or set image values of a map. In a complex system and with limited means for observation or measurement of the system, it may happen that more than one $p \in M$ maps into the same value. In other words, every map ξ on M induces an equivalence relation

$$E_\xi(p, p') \quad \text{if and only if} \quad \xi(p) = \xi(p')$$

and hence equivalence classes for which elements in M are indistinguishable w.r.t. ξ

$$[p]_\xi = \{p' : \xi(p') = \xi(p)\}.$$

The set of equivalence classes on M forms a quotient set or partition, denoted M/E_ξ (Fig. 2). For complex systems, what we observe is not M directly but a reduced state space M/E_ξ. The image set $\xi[M]$ and the partition M/E_ξ are isomorphic, that is, there is a one-to-one mapping, which is denoted by

$$\xi[M] \cong M/E_\xi. \tag{1}$$

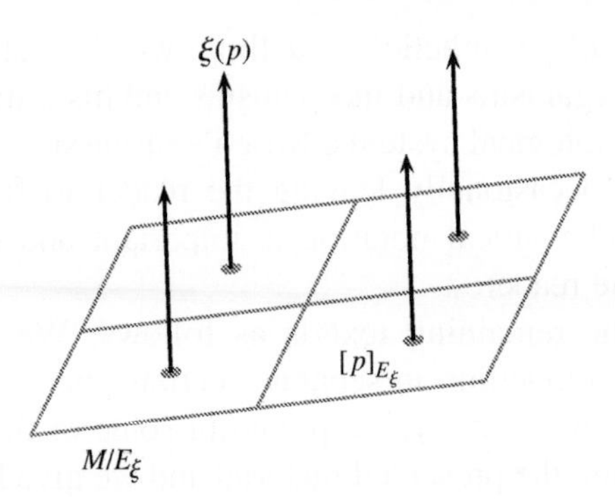

Figure 2 Equivalence classes, $[p]_{E_\xi}$, induced by an observable ξ, form a partition M/E_ξ. The values $\xi(p)$ in the range $\xi[M]$ act as *labels* to the equivalence classes.

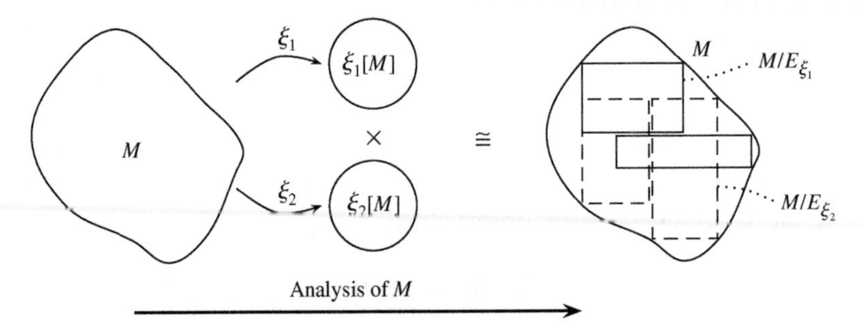

Figure 3 The analysis of a natural system, represented by M, leads to an analytical model $\mathcal{S}_a(M)$ which describes M in terms of the direct product of image sets and partitions induced by observables ξ_i.

As shown in Fig. 3, two observables can provide more information than one,

$$M \to \xi_1[M] \times \xi_2[M]$$
$$p \mapsto (\xi_1(p), \xi_2(p))$$

such that

$$\xi_1[M] \times \xi_2[M] \cong M/E_{\xi_1\xi_2}$$

where

$$E_{\xi_1\xi_2} = E_{\xi_1} \cap E_{\xi_2}$$

means that for any two values p and p', $\xi_1(p) = \xi_2(p')$ and $\xi_1(p) = \xi_2(p')$. That is, the equivalence classes of $E_{\xi_1\xi_2}$ are the intersections of all the equivalence classes of E_{ξ_1} with all equivalence classes of E_{ξ_2}. $E_{\xi_1\xi_2}$ has more equivalence classes than E_{ξ_1} or E_{ξ_2}, and may thus be considered a refinement of E_{ξ_1} and E_{ξ_2}. With only two unlinked observables we have three analyses of M, namely M/E_{ξ_1}, M/E_{ξ_2} and $M/E_{\xi_1\xi_2}$, where the latter is a refinement of the former. The more refined a partition, the more equivalence classes define it. Each class of $M/E_{\xi_1\xi_2}$ is a subset of M/E_{ξ_1}, respectively M/E_{ξ_2}, which means we can relate the partitions by some mapping that is defined in terms of a subset relation. In other words, the set of models $\{M/E_{\xi_1}, M/E_{\xi_2}, M/E_{\xi_1\xi_2}\}$ is a partially ordered set (poset).

Note that for Eqn 1 to hold, every E_ξ-equivalence class intersects every other $E_{\xi'}$-equivalence class, and vice versa. In this case, the two observables are said to be totally unlinked. The partition M/E can be considered a base space and an equivalence class $[p]_E$ of this partition as a fibre, consisting of all $p' \in M$ that project onto p. For any number of observables on M we define an analysis of M in terms of a direct product:

$$\mathcal{S}_a(M) = \prod_i \xi_i[M], \tag{2}$$

where

$$\prod_i \xi_i[M] \cong M/\cap E_{\xi_i}.$$

For the discussion that follows, we shall denote the set of all observables as

$$H(M, \mathbb{R}) = \{\xi \quad \text{for which } \xi : M \to \mathbb{R}\}.$$

Note that we can only unambiguously decode back from the product of the partitions if the observables are unlinked, i.e. if the value of one observable is not in some way determined by the other. Fig. 4 illustrates this point by considering four observables and their partitions. We find in case of ξ_1 and ξ_2 whatever pair $(r, r'), r \in \xi_1[M], r' \in \xi_2[M]$, is chosen, always every E_{ξ_1}-class intersects every E_{ξ_2}-class; the pair is thus unlinked. The same is true for (ξ_1, ξ_4), (ξ_2, ξ_4), (ξ_1, ξ_3) and (ξ_2, ξ_3) but not for (ξ_3, ξ_4), if the circle of M/E_{ξ_4} is a subset of the inner rectangle of M/E_{ξ_3}.

Considering the partition M/E_{ξ_i}, an analytical model of M, as a single mathematical object, we can form a category of models. In this setting, the product of factors

$$P = \prod_i M/E_{\xi_i}$$

is an object for which there exists a family of morphisms that are surjective projections onto the factors

$$\{\pi_i : P \to M/E_{\xi_i}\}$$

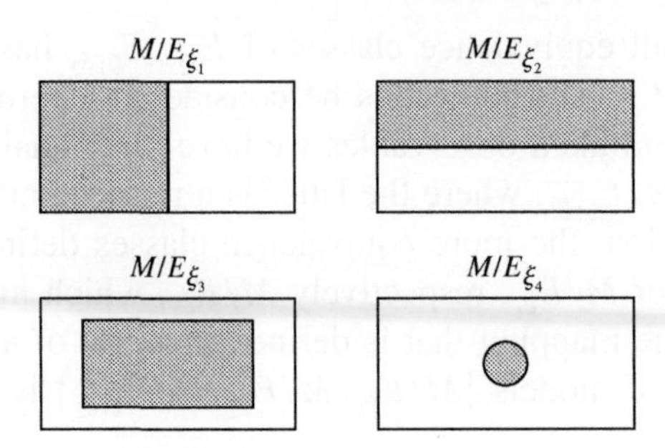

Figure 4 Four partitions induced by observables. All observables are mutually unlinked with the exception of ξ_3 and ξ_4.

such that for every object O and every family of morphisms

$$\{f_i : O \to M/E_{\xi_i}\}$$

there is a unique morphism, that is a map, $f : O \to P$ such that for the composition of the maps π_i and f

$$\pi_i \circ f = f_i$$

for all i, and

$$f(q) = (f_1(q), f_2(q), \ldots, f_i(q), \ldots),$$

for every q in O. We can summarise the analysis $\mathcal{S}_a(M)$ of M as a direct product in form of a commutative diagram

$$\begin{array}{ccc} O & & \\ {\scriptstyle f}\,\vdots & \searrow^{f_i} & \\ P = \prod_i M/E_{\xi_i} & \xrightarrow[\pi_{\xi_i}]{} & M/E_{\xi_i} \end{array}$$

In the category of abstract sets and morphisms, named **Set**, the direct product is the cartesian product; in the category of groups, **Grp**, it is the group direct product.

Relating the direct product in a general category to the category of analytical models, we find that for totally unlinked observables there exists an isomorphism between M and $\prod_i M/E_{\xi_i}$. Considering only two observables, ξ_1 and ξ_2, the commutative diagram takes the form

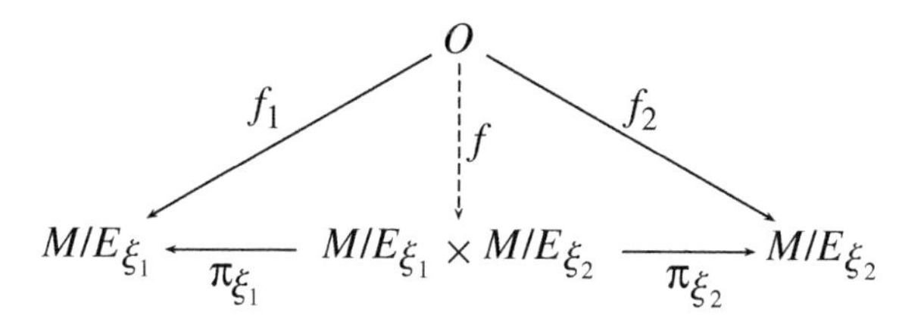

In the above diagram, replacing O by the partition $M/(E_{\xi_1} \cap E_{\xi_2})$, we can always define a map

$$f: M/(E_{\xi_1} \cap E_{\xi_2}) \to M/E_{\xi_1} \times M/E_{\xi_2}.$$

This map associates with each equivalence class in partition $M/(E_{\xi_1} \cap E_{\xi_2})$ the pair of classes of which it is the intersection. The map f is onto $M/E_{\xi_1} \times M/E_{\xi_2}$ if the two observables are totally unlinked, i.e. if each E_{ξ} equivalence class intersects with every $E_{\xi'}$-class, and conversely. If in addition the intersection of an E_{ξ}-class with an $E_{\xi'}$-class contains exactly one element, then f is one-to-one and M is said to be isomorphic to $M/(E_{\xi_1} \cap E_{\xi_2})$. If f is an isomorphism, we obtain via individual observations of ξ_1 and ξ_2 information about M itself. If ξ_1 and ξ_2 are not related in that way, f is only onto (surjective) and what we obtain is a representation of the quotient set of M; an embedding of M into a product, and not a representation of M such a product.

An important observation is that we are now in a position to consider more than one analysis of M, by the mutual refinement of equivalence relations, leading to a partial order of analyses $\mathcal{S}_a(M)$. The category of analytical models, denoted $\mathcal{A}(M)$, that is the set of all analytical models, is shown in Fig. 5. As can be seen from the figure, the objects of the category are analytical models, whereas the morphisms in that category are defined in terms of an inclusion relation. One should not fail to notice that with the partial

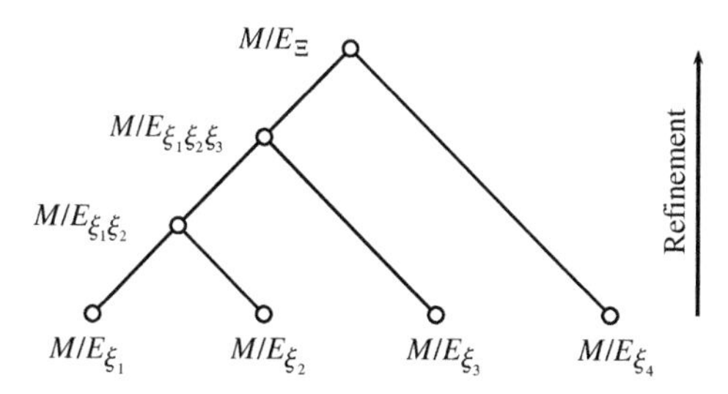

Figure 5 Extract from the category of analytic models, $\mathcal{A}(M)$, generated by the product of partitions that are induced from three totally unlinked observables, $\Xi = \{\xi_1, \xi_2, \xi_3\}$. Each node in which two nodes ξ, ξ' below merge is defined as a product $M/E_{\xi} \times M/E_{\xi'}$, realising the mapping $f: M/(E_{\xi} \cap E_{\xi'}) \to M/E_{\xi} \times M/E_{\xi'}$. A refinement model is said to be *'bigger'* than the one shown.

order of models, there are maps between models, and we are thus in a position to establish a modelling relation between models. In other words, there is a means to discuss the modelling process itself! The totality of analytical models can be identified with the totality of equivalence relations on M, as well as the totality of sets of observables, the set of all subsets of $H(M, \mathbb{R})$.

4. SYNTHETIC MODELLING

The key to analytical modelling was that we do not make any assumptions about the abstract set of pure states of the natural system, represented by M. This is clearly not what most projects in systems biology do. For most cases we assume, construct or synthesise M but subsequently forget that this is no more than a hypothesis. The traditional way of modelling is discussed in this section before we compare the two approaches.

In synthetic modelling, we synthesise M from subsets O_j in terms of the disjoint union

$$\biguplus_j O_j \doteq M, \tag{3}$$

where the equality is by construction. The process is shown in Fig. 6.

In the setting of category theory, we obtain a synthetic object as the coproduct of the family of objects $\{O_j\}$

$$C = \coprod_j O_j$$

for which there exists a family of injections

$$\{\iota_j : O_j \to C\}$$

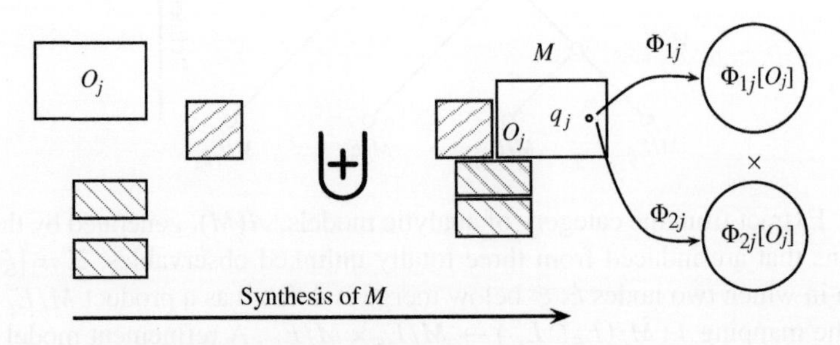

Figure 6 The synthesis of a natural system as a direct sum of subspaces.

such that for every object X and every family of morphisms

$$\{f_j : O_j \to X\}$$

there is a unique morphism $f : C \to X$ such that

$$f \circ \iota_j = f_j$$

for all j. In the form of a commutative diagram, for every object X in the category, for every j, there exists a unique morphism f that makes the following diagram commute for all j:

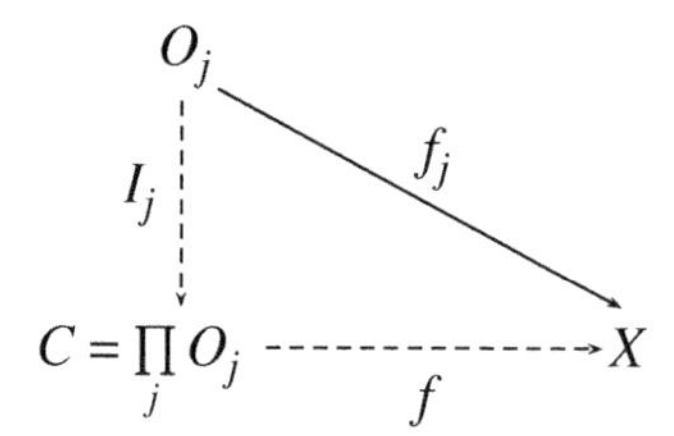

The direct sum is thus a dual concept to the direct product with the arrows being reversed in the commutative diagrams. In the category of sets, **Set**, the co-product is the disjoint union $C = \biguplus_j O_j$ and $\iota_j : O_j \to C$ is the inclusion. In the category of abelian groups, the co-product is the group direct sum, often denoted

$$C = \bigoplus_j O_j$$

and $\iota_j : O_j \to C$ are the injections of the jth summand. In the category of groups, **Grp**, the coproduct is the free product of groups. When the objects, from which the direct product/sum is formed, are abelian, e.g. vector spaces (modules over a field), or abelian groups (modules over integers) then the direct sum is the same as the direct product. We return to this question further below in the context of dynamic pathway modelling.

Next, we define for each O_j a fixed analytical model, using Eqn 2 as a template

$$\mathcal{S}_a(O_j) = \prod_i \Phi_{ij}[O_j] \tag{4}$$

where Φ_{ij} is an observable or state variable on O_j

$$\begin{aligned} \Phi_{ij} : O_j &\to \Phi_{ij}[O_j] \\ q_j &\mapsto \Phi_{ij}(q_j), \end{aligned}$$

and

$$\prod_i \Phi_{ij}[O_j] \cong O_j / \bigcap_i E_{\Phi_{ij}}.$$

Note that the O_j can be identical but in any case they are considered separate subspaces for which the observables are totally unlinked. We can now define the synthetic model as an encoding of M in terms of the direct sum of subspaces O_j

$$\mathcal{S}_s(M) = \biguplus_j \mathcal{S}_a(O_j). \tag{5}$$

We observe that in the synthetic model an abstract state $p \in M$ is encoded by an ordered tuple of single-valued elements $(q_1, q_2, \ldots, q_j, \ldots)$, with $q_j \in O_j$. Each q_j may, however, also be understood as an ordered set or vector

$$\bar{q}_j = (\ldots, 0, 0, q_j, 0, 0, \ldots),$$

such that we can write for the p of a synthesised M

$$p = \bar{q}_1 + \bar{q}_2 + \cdots.$$

This evidently describes an observable on M,

$$\xi(p) = \sum_j (\Phi_{1j}, \Phi_{2j}, \ldots, \Phi_{ij}, \ldots) \tag{6}$$

where $(\Phi_{1j}, \Phi_{2j}, \ldots \Phi_{ij}, \ldots)$ is an element of $\prod_i \Phi_{ij}[O_j]$. More specifically,

$$\xi : M \to \mathcal{S}_s(M) = \biguplus_j \mathcal{S}_a(O_j)$$

$$p \mapsto \xi(p) = \sum_j \Phi_j(q_j),$$

where we view Φ_j as an operator, that is a map, with

$$\Phi_j = (\Phi_{1j}, \ldots, \Phi_{ij}, \ldots).$$

The sum of observables on O_j induces therefore an observable on M.

5. SYNTHETIC VS. ANALYTIC MODELLING

With an observable Eqn 6, defined on M but arising from a synthetic model, we can now compare this to the analytic approach. Towards this end let us generalise

Eqn 6 to define an observable η as an operator that is a linear combination of maps Φ_j

$$\eta = \sum_j \theta_j \Phi_j \tag{7}$$

where θ_j is some real-valued parameter:

$$\begin{aligned} \eta : M &\to \eta[M] \\ p &\mapsto \sum_j \theta_j \Phi_j. \end{aligned}$$

To relate an observable η on M defined in an analysis of M with an observable η arising from a synthetic model, we can define an equivalence of the two,

$$\eta(p) = \left(\sum_j \theta_j \Phi_j \right)(p) \doteq \sum_j \theta_j \Phi_j(q_j), \tag{8}$$

only because of the involved linearity and the disjointness of summands O_j, which implied that the observables Φ_{ij} on O_j, by definition of the synthetic approach, are *unlinked*.

Note however that η is an element of the set of all maps from M to $\mathbb{R}$, denoted $H(M, \mathbb{R})$. There are, thus, many more observables than those that are linear combinations of the kind Eqn 7. It would, therefore, be possible to find an observable $\xi \in H(M, \mathbb{R})$ in an analysis of M, where for any two q and q' in the same equivalence class of O_j/E_{Φ_j},

$$\Phi_j(q) = \Phi_j(q')$$

holds but where ξ splits this equivalence class such that

$$\xi(q) \neq \xi(q').$$

In other words, although every synthetic model is an analytical model, there are analytical encodings of M that are not synthetic models; that do not possess a synthetic refinement.

6. DYNAMIC PATHWAY MODELLING

The vast majority of mathematical models and computational simulations in systems biology are developed within the framework of time-invariant differential equations (rate-equations). Regardless of whether one derives the equations from the law of mass action, using Michaelis–Menten kinetics or justifies these equations using power-law representations, such as S-systems, these models take the form

$$\text{state-transition:} \qquad \frac{\mathrm{d}}{\mathrm{d}t}x = V(x(t), u(t))$$

$$\text{response map:} \qquad y(t) = h(x(t)),$$

where $x \in X$ denotes the state, $x = (x_1, \ldots, x_n)$, of the system and X is called the state-space. One might consider V a mapping $V: X \times U \rightarrow \mathbb{R}^n$, which in the geometric setting of nonlinear systems theory becomes a vector field. The mapping V defines for a particular state x and stimulus $u = (u_1, \ldots, u_m)$ at time t how the system changes, given initial condition $x(t_0) = x_0$. Mathematically, these models are usually described in terms of finite-dimensional vector spaces.

The discussion above centred around the difference between analytical and synthetic modelling. Formally, the main difference between these two approaches is linked to the difference between the direct product and direct sum. Direct products and sums differ for infinite indices. An element of the direct sum is zero for all but a finite number of entries, while an element of the direct product can have all nonzero entries. In other words, the direct sum consists of the elements of the direct product which have only finitely many factors. For finite-dimensional vector spaces, the direct sum is the same as the direct product. Finite dimensional vector spaces are the basis for virtually all dynamic pathway modelling projects in systems biology. In dynamic pathway modelling, we are using (nonlinear) ordinary differential for the following reason:

> In modelling cells, cell function or pathways, causation is the principal of explanation of change; a relation not between things but changes of states. For anything to be different (changed) to anything else, either space or time, or both have to be supposed.

We noted above the central dogma of systems biology that systems dynamics are at the root of cell function. Although the areas of genomics and bioinformatics are identifying, cataloguing and characterising the components that make up a cell, this static information will not be sufficient to understand cellular processes such as growth, differentiation or apoptosis. The processes are not arbitrary but coordinated, controlled and regulated. To control, regulate or coordinate whatever, means to adapt, maintain or optimize; but this does mean that there must exist a goal, objective or function. To induce a change towards this goal, information must be fed back. Feedback implies a before and after, which is why we have to understand cell function as a dynamic system (Wolkenhauer et al., 2005b). Feedback loops are the basis for control and regulation in dynamic systems. In regulation, the system tries to maintain a variable or its level;

regulation is thus understood as a form of disturbance rejection, leading to robustness of the system against external perturbations. Control on the other hand, is related to sensitivity, the systems response to stimuli in an efficient way (Wolkenhauer et al., 2005a).

Differential equations are *the* natural language of change but with ordinary differential equations we 'only' capture time but not space. Apart from interactions through dynamics, spatial organisation is the second key aspect to biological systems. Partial differential equations are an option if spatial processes, like diffusion, cannot be ignored. The theory of partial differential equations is, however, rather complex and in many situations it will not be easy to identify parameter values from experimental data. Similarly, the translocation of proteins, for example nucleo-cytoplasmic shuttling in cell signalling, leads to transport delays or dead-times that have a significant influence on the dynamics of a system. This effect is widely ignored in systems biology, because the theory of delay differential equations is nontrivial. Processes would therefore be more accurately described by partial differential equations or delay difference equations. Infinite-dimensional systems theory is concerned with the formulation of such systems in state-space form, analogous to those for those described by ordinary differential equations. In this setting, one introduces a suitable infinite-dimensional state space and suitable operators instead of the usual matrices.

The differences of different approaches, the assumptions implied and consequences to prediction is therefore particularly important in systems biology.

7. ALL MODELS ARE WRONG, SOME ARE USEFUL

We presented a concise summary of Rosen's conceptual framework on the basis of which he discusses the difference between organisms and mechanisms, and which is also at the root of his critique of conventional mathematical modelling of cellular systems. In this setting, the synthetic approach of mathematical modelling as tightly associated with the direct sum, while the notion of an analytical model is tied to the notion of a direct product. In the context of category theory, these two concepts are dual and only in special circumstances (which Rosen argues involves linearity and finiteness) the two can be equivalent.

The most important aspect of the conceptual framework presented here is that it provides a means to discuss the modelling process itself, formally. Starting with basic considerations about the way we observe a natural system, we are able to define categories of models. It is then possible to relate models and investigate abstractions not just of natural systems but also of models.

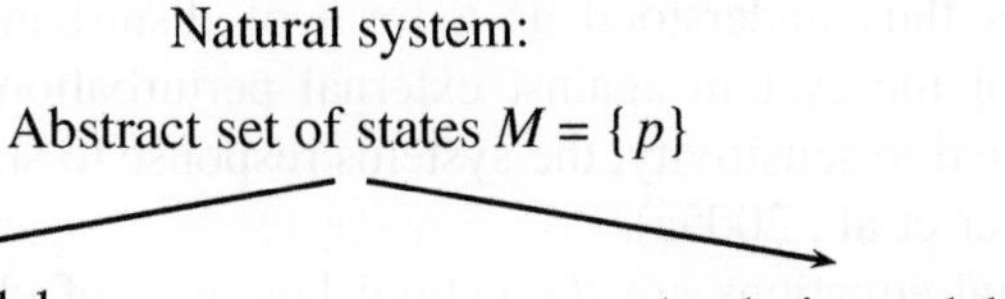

On the basis of the material presented here, Rosen defines a natural system as a 'simple system' or mechanism if all of its models are simulable. In Chapters 7 and 8 of his book *Life Itself*, he then argues that 'there exist modes of *organisation* whose material realizations cannot be simple'. (. . . organisms). While simulation is what machines do, more often than not a biological system is a 'complex system'. In Chapter 8, Rosen summarises his argument: If a natural system M is a mechanism, then the category of all analytical and synthetic models of M has a unique largest model $\mathcal{S}^{\max}$. This model captures all that is knowable about M. If a natural system is a mechanism, then there is a necessarily finite set of minimal models $\mathcal{S}^{\min}$ and the maximal model $\mathcal{S}^{\max}$ is equivalent to the direct sum of the minimal ones

$$\mathcal{S}^{\max} = \sum_i \mathcal{S}^{\min}$$

which means the maximal model is a synthetic model.

The uncertainty principle of systems biology states that as the complexity of a system increases, our ability to make precise and yet general statements diminishes. If we wish to understand the general principles of cell functions (say cell differentiation), independent of cell type or organism, then we need to generalise rate equations or automata and move a level up in abstraction. The relationship between state-variables and observable is also the subject of the recent book of the Netsruev group (Netsruev, 2003). The underlying idea of Rosen's discussion of observables (Rosen, 1991), as well as more recent texts on smooth manifolds (Netsruev, 2003), is to show whether it is possible to replace a set of states, M, for a system with a set of real-valued mappings from M to $\mathbb{R}$ (observables) on M. In (Netsruev, 2003) the Netsruev group discusses the possibility that every statement about M can be turned into a statement about the set of observables. Applied to modelling of dynamic systems, this work is closely related to recent developments in the setting of hybrid systems (Haghverdi et al., 2003; Tabuada et al., 2004; Tabuada & Pappas, 2005). Progress in this area is going to be important for dynamic pathway modelling in systems biology.

ACKNOWLEDGEMENTS

The authors acknowledge the support received as part of the EU FP6 STREP COSBICS and the German Federal Ministry for Education and Research (BMBF) as part of the SMP Protein within NGFN II. Mukhtar Ullah received for his contributions support from the UK Department for the Environment, Food and Rural Affairs (DEFRA) as part of a collaborative project with the Veterinary Laboratory Agency (VLA), Weybridge.

REFERENCES

Casti J. *Linear Metabolism-Repair Systems.* Int J General Systems: 14, 143–167, 1988a.

Casti J. *The Theory of Metabolism-Repair Systems.* Applied Mathematics and Computation: 28, 113–154, 1988b.

Cho KH, Johansson KH & Wolkenhauer O. *A hybrid systems framework for cellular processes.* Biosystems: 80, 273–287, 2005.

Haghverdi E, Tabuada P & Pappas G. *Bisimulation Relations for Dynamical and Control Systems.* In: Electronic Notes in Theoretical Computer Science. (Eds.: Blute R & Selinger P), Elsevier Science B.V., volume 69, 2003.

Landauer C & Bellman K. *Theoretical Biology: Organisms and Mechanisms.* In: AIP Conference Proceedings, American Institute of Physics, volume 627, 59–70, 2002.

Letelier J, Marin G & Mpodozis J. *Autopoietic and (M,R) systems.* Journal of Theoretical Biology: 222, 261–272, 2003.

Netstruev J. *Smooth Manifolds and Observables.* Springer Verlag, 2003.

Nomura T. *Formal Description of Autopoiesis for Analytic Models of Life and Social Systems.* In: Proceedings of the 8th International Conference on Arti cial Life (ALIFE VIII). (Eds.: Standish RK, Bedau MA & Abbass HA), MIT Press, 15–18, 2004.

Rosen R, *Some Realizations of (M,R)-Systems and their Interpretation.* Bulletin of Mathematical Biophysics: 33, 303–319, 1971.

Rosen R. *Life Itself.* Columbia University Press, 1991.

Tabuada P & Pappas G. *Quotients of Fully Nonlinear Control Systems.* SIAM Journal on Control and Optimization: 43, 1844–1866, 2005.

Tabuada P, Pappas GJ & Lima P. *Compositional Abstractions of Hybrid Control Systems.* Journal of Discrete Event Dynamical Systems: 14, 203–238, 2004.

Wolkenhauer O. *Systems Biology: The reincarnation of systems theory applied in biology?* Briefings in Bioinformatics: 2, 258–270, 2001.

Wolkenhauer O & Mesarovic M. *Feedback Dynamics and Cell Function: Why Systems Biology is called Systems Biology.* Molecular BioSystems: 1, 14–16, 2005.

Wolkenhauer O, Ullah M, Wellstead P & Cho KH. *The Dynamic Systems Approach to Control and Regulation of IntraCellular Networks.* FEBS Letters: 579, 1846–1853, 2005a.

Wolkenhauer O, Sreenath SN, Wellstead P, Ullah M & Cho KH. *A Systems- and Signal Oriented Approach to IntraCellular Dynamics.* Biochemical Society Transactions: 33, 507–515, 2005b.

9

Data without models merging with models without data

Ulrich Krohs and Werner Callebaut

SUMMARY

Systems biology is largely tributary to genomics and other 'omic' disciplines that generate vast amounts of structural data. 'Omics', however, lack a theoretical framework that would allow using these data sets as such (rather than just tiny bits that are extracted by advanced data-mining techniques) to build explanatory models that help understand physiological processes. Systems biology provides such a framework by adding a dynamic dimension to the merely structural 'omics'. It makes use of bottom-up and top-down models. The former are based on data about systems components, the latter on systems-level data. We trace back both modeling strategies (which are often used to delineate two branches of the field) to the modeling of metabolic and signaling pathways in the bottom-up case and to biological cybernetics and systems theory in the top-down case. We then argue that three roots of systems biology must be discerned to account adequately for the structure of the field: pathway modeling, biological cybernetics, and 'omics'. We regard systems biology as merging modeling strategies (supplemented by new mathematical procedures) from data-poor fields with data supply from a field that is quite deficient in explanatory modeling. After characterizing the structure of the field, we address some epistemological and ontological issues regarding concepts on which the top-down approach relies and that seem to us to require clarification. This includes the consequences of identifying modules in

Systems Biology
F.C. Boogerd, F.J. Bruggeman, J.-H.S. Hofmeyr and H.V. Westerhoff (Editors)
Isbn: 978-0-444-52085-2

large networks without relying on functional considerations, the question of the 'holism' of systems biology, and the epistemic value of the 'systeome' project that aspires to become the cutting edge of the field.

1. INTRODUCTION

New scientific theories are not made out of nothing. Although departing and emancipating themselves from their predecessors more or less radically, they typically incorporate elements from predecessor theories such as definitions, interpretations of phenomena, and models. Similarly, the wider scientific fields in which theories are embedded remain to some extent tributary to older fields, e.g., by retaining problems, explanatory strategies, experimental techniques, or data sets. A scientific field often (but not always) also comprises laws and theories that aim to realize its explanatory goals. Although we will refer to many of these elements throughout this chapter, our emphasis is more narrowly on theory dynamics, leaving the treatment of other aspects of the emerging field of systems biology for another occasion.[1] Systems biology is a particularly interesting candidate for investigating the dynamics that underlie the formation of new scientific fields because it assimilates models from various older theories as well as special kinds of data sets from data-driven research programs ('omics') that grossly lack a theoretical perspective. Systems biology promises to merge all these elements into a new framework that continuously extends and modifies the models that originally inspired it.

As a science in the making, systems biology stirs debates in which conceptual and empirical issues are so entangled that progress would seem to require the combined efforts of scientists and philosophers acting as conceptual analysts (see Callebaut, 2005 and Krohs, 2006a, for reflections on the nature of philosophy of biology). The 'new' philosophy of biology that originated some three decades ago has long focused almost exclusively on issues in evolution and systematics (see, e.g., Sober, 2006). As a positive result of this intellectual investment, the current interaction between evolutionary biologists or systematicists and 'their' philosophers can be described without much exaggeration as symbiotic: both parties truly benefit from their collaboration. One major drawback has been the neglect, with few exceptions, of far 'hotter' areas of biological research and

[1] See Darden & Maull (1977) and Darden (2006) for a specification of the notion of field and the roles theories play in them, and, e.g., Bechtel & Richardson (1993, Chapter 10) and Dougherty & Brago-Neto (2006) for a discussion of (and more references to the by now huge literature on) the interrelations between models and theories of complex biological systems.

development,[2] in particular molecular biology, (gen)omics and, most recently, systems biology. This volume is a most welcome first attempt to correct this bias. However, we should be aware from the outset that articulating a full-fledged philosophy of systems biology will be a formidable task indeed. The challenge is for a community of philosophers with the necessary critical mass to learn to collaborate with systems biologists (who will have to accept them as 'partners' of some kind!) and master the relevant technical literature, which is quickly growing, in sufficient detail. This cannot be expected to happen overnight. But the challenge is certainly a seducing one. A cursory look at the literature in this burgeoning field[3] suffices to conclude that with a few notable exceptions, systems biology remains largely uninformed of the results of contemporary philosophy of science, although many of the issues – e.g., data- vs. hypothesis- or theory-driven research, reduction or emergence, realism vs. simplification or idealization regarding models and theories, mere aggregates vs. systems – cry out for philosophical analysis. The good news for philosophers is that they will not have to start from scratch: many if not most of the relevant issues will not take them by surprise as they will already be more or less familiar from other contexts (cf. Boogerd et al., 2005; O'Malley & Dupré, 2005).

Obvious first questions for philosophers to ask about systems biology concern its 'topography.' It is generally accepted that systems biology not only has several roots, but also consists of two branches: 'bottom-up' and 'top-down' systems biology. After a preliminary sketch of the layout of the field (Section 2) and a survey of its main roots – models of metabolic and signaling pathways (Section 3), biological cybernetics and mathematical systems analysis (Section 4), and 'omics' (Section 5) – our main concern will be to reconstruct the theoretical links between the different branches of systems biology in order to present the current structure of the field (Sections 6 and 7). In addition, we

---

[2] Development and application of scientific results in general are 'messy' areas that have traditionally been disregarded by philosophers (and the scientists echoing them) operating under the Platonist/Cartesian illusion that 'pure' – i.e., physical (see, e.g., Kitano, 2002b) and ultimately mathematical – foundations are to be privileged; see Wimsatt (in press) for a forceful rebuttal. More generally, it would seem that the philosophy of a field like systems biology, which relies heavily on massive data acquisition and has 'emergence' written on its banner, could benefit from adopting an account of explanation and unification that centers on "cooperation and communication among theoretical and phenomenological equals, rather than on imperialism and competition for primacy and fundamentality, which reduces or replaces one theory by another or trivializes one explanandum as epiphenomenal to another" (Griesemer, 2006, p. 5). In this vein, we interpret the search for 'foundations' suggested in the title of this volume as a search for conceptual filiations (resulting, hopefully, in conceptual clarification) rather than for epistemic priority.

[3] Including the rhetoric of promise that accompanies its incipient stage, which is often much more daring than one would expect after the disillusionment that followed the announcement of the results of the Human Genome Project (see, e.g., Lewontin, 1997; Newman, 2003). To guarantee their empirical adequacy, philosophical reflections on systems biology will soon have to be complemented by both professional historical research into its antecedents and investigations of the social, institutional, and economic framing of the field by researchers in science, technology, and society studies.

will address some ontological and epistemological issues regarding functionality, holism, and model realism that arise in the top-down branch of systems biology, which breaks most radically with tradition (Section 8).

2. PRELIMINARY TOPOGRAPHY OF THE FIELD

Systems biological models aim at a detailed account of the dynamics of complex biological systems, where 'detailed' means that not only basic qualitative characteristics of the system but also more specific properties of a dynamical network are reproduced by the model; in particular, the ontology of the model should refer to the systems' physical components. It is well established that such models can be separated into two classes with different historical roots (Westerhoff & Palsson, 2004; O'Malley & Dupré, 2005). The two classes of models are often characterized as bottom-up and top-down (Kitano, 2002b; Bray, 2003; O'Malley & Dupré, 2005; Palsson, 2006). Although we view the roots of systems biology in a somewhat more differentiated way, we will adopt these labels as denoting different ways of modeling. The idea behind this nomenclature is that some models are built from data concerning well-characterized components of the system – the 'bottom-level' of the biological system – so that the system- or top-level is reassembled mathematically from the components' contributions, whereas other models are built from data concerning the system as a whole and try to break down the system mathematically into modules and components. Bottom-up models typically consider data on a few (less than ten) components only, while top-down models may be based on data about as many as hundreds or thousands of components. Therefore, bottom-up models are classified as data-poor models, while top-down models are considered as data-rich (Westerhoff & Palsson, 2004). A similar difference in data richness is characteristic of the forerunner approaches on which systems biology is based, although in this case the borderline is not between bottom-up and top-town but between modeling and the acquisition and analysis of complete data sets with respect to a particular domain, such as in genomics. The huge amounts of data produced by the genome projects were in fact collected almost free of any theoretical burden; as could have been expected, they turned out to explain next to nothing. For Baconian induction from theoretically unorganized observations is bankrupt, as population geneticist Lewontin (1997, p. 30) reminds us: "Before sense experiences become 'observations' we need a theoretical question, and what counts as a relevant observation depends upon a theoretical frame into which it is to be placed" (see also Snoddy et al. (2000) on the challenge of developing computerized information systems that 'can help understand' complex biological systems). The first part of the title we chose for our chapter refers to this circumstance. The second part of the title, in contrast, is a gross exaggeration, but it casts some light on the situation. The models incorporated into the endeavor of systems biology are certainly based on sound empirical data. However, model building in this tradition has, for most of the time, either suffered from a relative paucity of data or was not able to integrate even available data for technical reasons.

Although there is less consensus regarding the roots of systems biology than with respect to the classification of systems biological models, the table which O'Malley and Dupré (2005) have compiled from several sources nevertheless offers a clear picture. Different theories of systems dynamics (among them cybernetics and systems theory) are mentioned as belonging to the root of top-down systems biology, whereas classical molecular biology, being a basis of modeling of metabolic pathways, and genomics are univocally regarded as forming the root of bottom-up systems biology. Thus, authors agree that two roots gave rise to the two respective classes of systems biological models and disagree only on details concerning which forerunner approaches exactly constitute these roots. We challenge the 'two-roots' view and want to propose three roots instead because we believe that only a delineation that is fine-grained enough allows us to adequately reconstruct the structure of the field. Consequently, our main focus will be on those aspects of different forerunner models that are retained in systems biology and on the interrelationship between the two classes of models.

Each of the three roots we discern contributes to both classes of systems biological models, and we intend to show that this is a more adequate reconstruction of the field than an account in terms of two roots giving rise to two classes of models in a one-to-one manner. These three roots are (i) modeling of metabolic and signaling pathways, (ii) biological cybernetics and systems analysis, and (iii) genomics as well as other 'omics' projects such as proteomics (the analysis of the proteins expressed in an organism or in a cell during its life cycle), transcriptomics (the cell-scale study of the genes transcribed and of the RNA transcripts), and metabolomics (the study of small-molecule metabolite profiles on the cellular level). A first reason to view 'omics' as a third root rather than classifying it together with pathway modeling relates to the richness or poverty in structural and other data of the different roots and branches of systems biology. Data on the structure of a biological system, mainly on its different components, is what is meant when models are classified as data-rich or data-poor. Richness in structural data is found neither in pathway modeling nor in biological cybernetics, which both focus on a few parameters per system only; it stems exclusively from 'omics', which thus contrasts with both of the other roots (Table 1). Note that the latter, however, are rich in kinetic data (though not in the number of independent kinetic parameters). In pathway modeling, these are data on the dynamics of the components of the system, whereas in biological cybernetics mainly data on system-level dynamics are used. Similarly, rich kinetic data can be found in the models of the two branches of systems biology, but here the divide is less strict with respect to all three kinds of data.

Table 1 Data richness of roots and branches of systems biology.

Kind of data	Pathway modeling	Biological cybernetics	Omics	Bottom-up systems biology	Top-down systems biology
Structural	−	−	+	+/−	+
Systems dynamics	−/+	+	−	+/−	+
Components dynamics	+	−	−	+	+/−

Note: For the five different roots and branches of systems biology we discern the typical richness (allowing for exceptions) in data of a certain kind, '+' indicating high richness, '−' poverty of data.

3. THE FIRST ROOT OF SYSTEMS BIOLOGY: MODELS OF METABOLIC AND SIGNALING PATHWAYS

Enzymes are the biocatalysts that enable the metabolic transformation of organic molecules under thermodynamical conditions relevant for life on earth. Enzyme kinetics, including its regulation by all kinds of effectors, is therefore one of the main fields that help explain metabolism and its regulation. Most anabolic and catabolic biochemical processes involve several reaction steps that lead, say, from a molecule that is taken up (or stored) either to one that is incorporated into a functional structure; or to one with less free energy, allowing part of the released reaction energy to be used to drive other, energy-consuming reactions; or from a degraded structure to an excretable molecule. Such multireaction processes are thus traditionally regarded as metabolic pathways, leading from one molecule species to another, each enzyme having its specific role within the pathway.

3.1. Regulatory metabolic and signaling systems

All known metabolic pathways are regulated. Those that are simple enough and sufficiently separable from the embedding system were often chosen for analysis through mathematical modeling and simulation. The theoretical background of such models is to be found in enzyme kinetics, the theory of dynamical systems, and nonlinear thermodynamics (Westerhoff & Palsson, 2004). Regulation may involve both negative and positive feedback and feed-forward control within the pathway, and also regulation by external effectors. As pathways are usually not completely separated from each other, regulation may also occur by direct 'crosstalk' with other pathways through shared metabolites (see, e.g., Michal,

1982). Also, the presence of the enzymes that are involved in a certain pathway may be controlled by regulation of the expression of the related genes by a transcriptional regulatory network. Most prominent among genetic network models is the *lac* operon of *Escherichia coli*. Here, the transcription of genes for enzymes that are part of a catabolic pathway is regulated by the carbon and free-energy sources lactose and glucose; in short, lactose upregulates and glucose suppresses the expression of the *lac* operon (Pardee et al., 1959; Jacob & Monod, 1961; see Weber, 2005 and Richardson & Stephan, this volume). Signaling systems are the other domain of pathway modeling.[4] Among these are the visual cascade (Chabre & Deterre, 1989) and other instances of G-protein signaling (Dohlman & Thorner, 2001). Often rhythms on a short time scale were under investigation, e.g., in the regulation of the swimming behavior and chemotactic response of *E. coli* (Spudich & Koshland 1975; Bourret & Stock, 2002). All these systems are conceived as networks that process some kind of signal or stimulus that elicits a specific response. In chemotaxis of *E. coli*, the stimulus is a change in the concentration of certain chemicals, mostly nutrients like glucose and amino acids; the response is a modification of the swimming behavior, enabling the bacterium to swim, by means of a modulated random walk, along a chemical gradient. In the G-protein cascades, a photon or a transmitter molecule may act as stimulus or as signal, respectively. The response is a change in the conductivity of certain ion channels leading to a transient depolarization of the cell membrane or activation of other enzymes. These cascades are self-regulated. In many cases, the response decays even if the changed conditions pertain, allowing so-called perfect adaptation to the background stimulus level.

3.2. Modeling regulatory networks and gathering data

Mathematical models of metabolic pathways usually rely on the kinetic parameters of the enzymes, determined with isolated active enzyme, and on estimates of enzyme and metabolite concentrations. Other regulatory networks include parameters also for, e.g., protein–protein interaction. Most models are based on basic chemical reaction kinetics and enzyme kinetics of the Michaelis–Menten kind. They may be given in terms of a set of ordinary differential equations that take into account the main in- and outflows of metabolites, including intermediate products of the pathway. Models of this kind often describe the smallest system that qualitatively shows the same dynamics as the biological pathway. As such models proceed from data about components to a description of a complex system, pathway modeling can be characterized as bottom-up. This makes it a candidate for an important

[4] Both domains are not clearly separated, since any regulation of a metabolic pathway may be regarded as signal processing as well.

root of the bottom-up branch of systems biology. We will discuss in Sections 6.1 and 7 the differences between pathway models and the related kind of systems biological models.

This sketch regards model building as a process starting from given kinetic data. However, the data must be collected first. Biologists often start by singling out a metabolic or sensory capacity that is to be analyzed and eventually modeled. The next step, after the characterization of the capacity, is the biochemical or physiological analysis of the pathway, yielding the components of the pathway. Third, these components have to be isolated, characterized, and their kinetic parameters must be determined. This yields the data set from which mathematical modeling can start. While modeling itself is bottom-up, the whole process of gathering the data is a top-down process, starting from a higher-level capacity and proceeding to the contributions of the lower-level components of the system. (For an analysis of a historical case see Darden & Craver, 2002.)

To summarize, the process of model building in the tradition of metabolic and signaling pathway research typically consists of the following steps: (i) singling out a capacity of the organism or cell; (ii) experimental breakdown of the underlying system into its components, providing the data for the modeling; (iii) synthesizing a model from subsets of the data, which may involve several steps of simplification and refinement; (iv) analyzing the model analytically or numerically, i.e., by running simulations; and (v) assessing the reliability of the results that the model generates.

4. THE SECOND ROOT OF SYSTEMS BIOLOGY: BIOLOGICAL CYBERNETICS AND MATHEMATICAL SYSTEMS ANALYSIS

Biological cybernetics concentrates on the modeling of regulatory processes that occur in complex systems. Like pathway modeling, it starts from the description of systemic capacities, most often ones that are related to signal and stimulus processing. But in contrast to pathway modeling, it does not aim for an empirical analysis of the investigated capacity at the molecular level before modeling the system. Molecular and cellular processes are black-boxed and empirical data are collected at the systems level by input–output analysis. Biological cybernetics thus tries to identify the regulatory processes that are required to produce the observed systemic response at an abstract level. Relating the systems output to controlled stimulus input allows inferences with respect to frequency filtering, signal delay, and the amplification characteristics of the system. The regulatory instances found by such an analysis are often depicted as equivalent electric circuits as they are also known from electrophysiology (where they are also often based on data at the molecular level; cf. Schaffner, this volume).[5]

4.1. Cybernetic models in biology

Though cybernetic approaches have been applied at the molecular level in systems biology, many of the prominent models from biological cybernetics deal not with molecular but with nervous control systems, often related to the visual system. Among such nervous control systems is the regulation of the diameter of the human pupil in response to light conditions. The iris sphincter muscle contracts in response to stimulation of the retina by light, thus decreasing the diameter of the pupil and hence the intensity of illumination of the retina. Both pupils are regulated separately, but not completely independent from each other. This allows for a cybernetic analysis of the network that brings the regulation about. As a result, the order of nonlinear transformation of the stimulus and integration of stimuli from both retinas could be determined and an equivalent circuit of the system assigned to anatomical structures of the nerve pathways of the visual system (Stark, 1959; Varjú, 1967, 1969).

That modeling in these cases does not refer to the molecular level of the system could explain why several authors who have proposed reconstructions of the history of systems biology do not refer to cybernetics at all. However, biological cybernetics has investigated other than nervous systems as well, and well-known cases combine analyses of molecular processes and neural signaling. This is the case, say, for Hodgkin and Huxley's model of nervous signal propagation in the giant axon of squid (Hodgkin et al., 1952; Hodgkin & Huxley, 1952; see Krohs, 2004 and Schaffner, this volume), which today still provides the basis for electrophysiological modeling. It also holds for the analysis of the lateral inhibition of visual neurons in a field of receptor neurons, which has resulted in a model of synaptic processes involving the molecular level (Furman, 1965).

An important instance of an application of systems analysis that refers directly to the (biochemically unidentified) molecular level is the early model of the circadian clock by Arthur Winfree (1967, 1977). The circadian clock regulates sleep–wake rhythm even in the absence of any *Zeitgeber*. The phase of the rhythm can be shifted by light stimuli and by certain chemical stimuli. This led Winfree to postulate an endogenous biochemical oscillator of unknown

[5] Norbert Wiener was one of the inventors of cybernetics as an approach applicable to the description of regulatory processes in biology as well as in engineering (Rosenblueth et al., 1943; Wiener, 1948) and is regarded as one of the forerunners of systems biology (e.g., Kitano, 2002a). Ludwig von Bertalanffy's General System Theory (Bertalanffy, 1932–1942; 1968) was to some extent incorporated in later cybernetic and systems-theoretical accounts (see, e.g., Mesarovic, 1968), which today inspire a hierarchical approach to systems biology (Mesarovic et al., 2004). Other systems theoreticians that became important for biological modeling are discussed in O'Malley and Dupré (2005).

molecular constitution, which he described by a mathematical model accounting for the data obtained by input–output analysis.

4.2. Contents of the black boxes and the 'direction' of cybernetic modeling

Biological cybernetics uses data from input–output analysis, i.e., from systems dynamics under controlled conditions, to infer to regulatory instances within the black-boxed system. These instances are specified primarily on an abstract level, without reference to their physical realization. On this level, then, the content of the black box is analyzed as the functional structure of the system that the model describes – more precisely, a possible functional structure. What is searched for is a model of the lowest possible mathematical complexity that can explain the behavior of the system. This criterion is ambiguous; it neither needs to single out exactly one possible model, nor can one decide in all cases which one of two models is the less complex (Varjú, 1967). The biological realization might be more complex than the minimal model anyway. By beginning the analysis of the systems dynamics from data on systems-level behavior and inferring the functional structure on a level below, biological cybernetics follows a top-down modeling strategy.

Although cybernetic models only describe the functional structure of a system, cybernetics does not ignore the components of a system; it aims to relate the low-level regulatory units to cellular and molecular structures. This is done in a second step that supplements the regulatory analysis of a system. Only anatomical, physiological, and biochemical analyses can reveal the actual physical structure that realizes the regulatory functions. As mentioned, cybernetic and molecular analysis went hand in hand in the case of the squid axon model. Another case in which such a mixed strategy was applied is the biological clock. Initially, the molecular basis of the oscillator was completely unknown; Winfree's model stated minimal requirements for the regulatory mechanism. With increasing biochemical knowledge, molecular data were obtained, allowing the model to be refined. Modeling is continued within a systems-biological framework, the latest version by Leloup and Goldbeter (2003) being a prime example of the success of this research strategy. Similarly, an early cybernetic model of *E. coli* chemotaxis (Spudich & Koshland, 1975) was supplemented later on by the mechanistic models cited in Section 3.1. Having in mind such combinations of cybernetic modeling and molecular analysis, it is worth considering the direction of modeling again. Initially it was top-down, from the system-level response to the functional structure of the system. At this point, molecular analysis came in; its results could now be related to the model, the molecules being regarded as instantiations of the functional components of the system.

We thus have bottom-up modeling as well, since the model can be built from data about its material components. Pathway models of processes that are going on in the system may supplement the top-down cybernetic model. The top-down systems model and the bottom-up pathway models may be refined and adapted to each other, usually by iterated steps, so that in the end the model is built by a blending of both basic strategies. This shows that biological cybernetics and pathway modeling are not completely isolated (which was to be expected, as pathway modeling makes use of cybernetic control concepts when referring to, e.g., feedback regulation). Nevertheless, both strategies are clearly discernible by the reliance of paradigmatic cases on bottom-up and on top-down strategies, respectively. These different strategies being the criterion for discerning two branches of systems biology, they should also suffice to discern different roots of the field.

Model building in the tradition of biological cybernetics and systems analysis embraces the following steps: (i) singling out a capacity of the organism or cell; (ii) experimental input–output analysis of the underlying system, providing the data that go into modeling; (iii) analysis of the data for filtering, delay, amplification, etc. that is going on in the black box between input and output; (iv) analyzing the model analytically or numerically; and (v) assessing the reliability of the results that the model generates. Steps (ii) and (iii), hence the models themselves, differ from the case of pathway modeling.

5. THE THIRD ROOT OF SYSTEMS BIOLOGY: 'OMICS'

In the 1990s, large-scale sequencing projects were set up as a new field of biology, which received funding on a large scale. Meanwhile, the application of new high-throughput technology yielded the complete sequence of the human genome. Similar projects to compile comprehensive catalogues of cell contents are run with proteins, metabolic reactions, etc. In addition to genomic projects, there are projects in proteomics, metabolomics . . . – in short, a new biological discipline called 'omics.' All 'omic' projects have their specific methods of data analysis and presentation, but they also all lack modeling strategies. There does not seem to be much to explain with respect to large collections of structural data of the 'omic' kind. Only in the systems biological perspective do 'omic' data sets become interesting for modeling purposes. But these projects were not created out of nothing either. Though they rely crucially on newly developed high-throughput methods for data acquisition, they have forerunners in other projects that catalogued cellular components. We review some of these early projects in addition to genomics proper as a basis for our discussion of the transformation of such projects into systems biology.

5.1. Early genome projects: Chromosome maps

Chromosomes, when carefully stained, show a banding pattern. This pattern is usually fairly coarse, yet extremely rich in the giant chromosomes that are found in the salivary glands of the larvae of diptera (e.g., flies). Their banding pattern provided a coordinate system to localize genes on a chromosome long before it was found that a chromosome consists of a long strain of DNA (coiled on histones) or, in the case of the giant chromosomes, of a bundle of parallel strains.[6] Morgan and his coworkers, mainly Sturtevant and C. B. Bridges, managed to assign genes to certain *loci* on the chromosomes by analyzing partial linkage between different Mendelian genes. Linkage of an eye color mutation of *Drosophila melanogaster* to sex gave the starting point to relate genes to chromosomes, since the sex chromosome could be easily identified under the microscope. A large amount of data was collected during more than 20 years. Eventually, fairly detailed chromosome maps were obtained (C. B. Bridges, 1935, 1938; P. N. Bridges, 1942). The compilation of the chromosome maps of *D. melanogaster* can be regarded as an early genome project to the extent that it was an attempt to catalogue the complete inventory of structural features of the chromosome and relate all known genes to these structures. Similar maps were compiled later on for species lacking giant chromosomes, including *H. sapiens* (McKusick, 1998).

After the discovery of the DNA structure, gene mapping could move on to the molecular level. The circular chromosome molecule of *E. coli* was the first genome that was tackled. Again a coordinate system was needed. A measure was found in the duration of bacterial conjugation required for the transfer of a particular gene. A 100-min scale allows addressing any place on the chromosome. Thus, the genes – as identified by phenotypic effects – can be mapped on the chromosome (Jacob & Wollman, 1958). Again, the aim of the project was to "relate information on [*E. coli*] K-12 genetic material gained by genetic and by physical methods" (p. 92).

5.2. Molecular genome projects

The refinement of the gene map of *E. coli* by analyzing the sequence of many of its genes may count as the first genome project with molecular resolution. Later on, with the development of quicker and partially automated DNA sequencing techniques, scientists began to sequence not only single genes but also the whole genome.[7] This was a serious shift in perspective. The older genome projects

---

[6] In addition to the bands, there are larger morphological landmarks, dubbed, e.g., dog-collar, onion, and ballet skirt, that help identify individual (parts of) chromosomes even when these are strongly altered (Bridges, 1935).

[7] As the perspective had shifted to gathering complete sequence information, *Haemophilus influenzae*, a bacterium with a much smaller genome than *E. coli*, was a more promising object to finish such a task. Its

mapped known genes on the physical structure of a chromosome, i.e., they related physiological roles to genomic structures. Sequencing those parts of the DNA that were identified as gene loci and consequently equaled with genes mainly refines knowledge of the structure to which the physiological role is mapped. Whole-genome sequencing, in contrast, focuses exclusively on the structure of the genome. The assignment of physiological roles to substructures is now an additional task. Some of the assignments can be read from gene maps. Others, so the promise, can be inferred from the sequence (we will address the limitations of this in the next paragraph).

With eukaryote genome projects, researchers did not concentrate on the easy task, as initially with the prokaryotes, but shifted quickly to one for which large-scale funding could be raised: the sequencing of the genome of *Homo sapiens*. The human genome project promised to turn medicine into a straightforward technological application of biology within a short time (Drell & Adamson, 2000). Considerable parts of the sequence were published simultaneously by the International Human Genome Sequencing Consortium (2001) and Venter et al. (2001). Three years later, the Consortium presented 99% of the human euchromatic sequence (International Human Genome Sequencing Consortium, 2004). The sequence has by now become an indispensable tool in different fields of biology. Comparison with the genomic sequence of chimpanzee and gorilla genomes as well as with the *D. melanogaster* and *Caenorhabditis elegans* genomes provides important insights into evolution, though not into the evolution of, e.g., cognitive capacities.[8] However, initial expectations that this project would yield medical applications ('genes to drugs') within short time were soon frustrated, not really to the surprise of biologists who were aware that disease biology is complex, and that drug development must be driven by insights into biological responses (Butcher et al., 2004). At present, the complete knowledge of genomic data provides only very limited insights into the physiology of an organism. Hence, it is of limited relevance for explaining, say, diseases or the ontogeny of organisms. One reason is of course that the physiological roles of DNA segments cannot be assumed to be easily readable from the sequence, especially in eukaryotes, though the toolbox of bioinformatics provides some help.

5.3. Early proteomic projects

Proteins, in contrast to genes, are individual macromolecules; hence structural data need not be supplemented with knowledge about functions to compile

complete genomic sequence was published in 1995 (Fleischmann et al., 1995). The sequence of *E. coli* followed two years later (Blattner et al., 1997).

[8] For a philosophical analysis of the way in which results from the projects concerning non-human organisms are used to interpret human genome data see Ankeny (2001).

a catalogue. The protein species contained in a cell under certain conditions could thus be made visible as soon as protein analytical methods allowed for high enough resolution. In a sense, then, proteomics began with the first series of experiments that could resolve most of the proteins of a cell, presented by Patrick O'Farrell in a seminal methodological paper on two-dimensional gel electrophoresis of proteins. He resolved 1100 proteins from *E. coli* and estimated a maximal resolution of the method of 5000 proteins (O'Farrell, 1975). His method allows to detect many mutations and, of course, changes in the protein expression pattern. Less resolution was needed initially in the analysis of the proteomic subset of heat shock proteins in *Drosophila*, in which early genomic and proteomic data were already combined (Tissières et al., 1974; Ashburner & Bonner, 1979). However, this 'early proteomic project' was not primarily interested in the dynamics of a proteomic network as is systems biology. The structural data that could be obtained were mainly used within the framework of analysis of already known functions of individual proteins.

6. THE BRANCHES OF SYSTEMS BIOLOGY: MERGERS OF THE DIFFERENT ROOTS

Systems biology aims to model biological systems in ways that combine aspects that we have addressed in the discussion of its three roots: (i) detailed models of genetic and metabolic networks that are built from molecular components in the tradition of bottom-up pathway modeling; (ii) models that describe the overall dynamics of large networks and divide large (sometimes cell-encompassing) networks into modules (nearly separable regulatory subunits), a modeling strategy related to cybernetic top-down modeling; and (iii) grounding of many of its models in 'omic' data sets, which is clearly the case with the top-down approach, but to some extent also with cases from the bottom-up tradition. The availability of high-throughput methods and the 'omic' data sets gathered by using them can be seen as crucial for the transformation of pathway modeling and biological cybernetics into systems biology. The new richness of structural data allowed the refinement of models that were based on kinetic data of only a few components, leading to detailed models of the dynamics of complex networks.

6.1. The first branch of systems biology: Detailed bottom-up regulatory models

A fairly continuous development leads from the tradition of modeling regulatory networks to more recent models of similar networks that run under the unifying label of systems biology. Among these are the current models of motor regulation in *E. coli*, signaling pathways, and complex biosynthetic and other metabolic pathways (Kitano 2002c, Part IV; Kholodenko, 2006). Like the earlier models, they start from a physiological capacity that is analyzed by molecular–biological methods, the model being built bottom-up from these data. The main difference compared to the earlier version is that the models provide more and more details. In most cases, this does not mean that the models become more fine-grained with respect to components of pathways already considered in the older models. As discussed in Section 3, those models already referred to the single species of molecules involved in a reaction chain and accounted for their kinetic parameters. The main way in which more details are considered, then, is that additional components at the periphery of the metabolic pathways are integrated into the model so that the reaction networks grow. This is the case, for instance, with signaling networks, which were formerly restricted to the size of, say, a single G-protein signaling pathway, and perhaps its coupling to the activation of the Ca^{2+}-pathway. The new models of signaling pathways represent a network that incorporates about four times as many components (Kholodenko, 2006; Palsson, 2006, p. 85). In the case of chemotactic motor response of *E. coli*, models now consider the gene regulatory pathways in addition to the processes that account for the response in a certain physiological state (Bourret & Stock, 2002). With respect to metabolic pathways, models now combine within one network what were previously considered separate pathways among which at best some crosstalk might have existed (Thiele et al., 2005). The model of the mammalian circadian clock was refined and now represents a network of five genes and their gene products (Leloup & Goldbeter, 2003).

At first glimpse, and keeping in mind that the detailed *Biochemical pathways* map was already available in 1974, it is astonishing that more detailed models of the sort described have been developed only recently. In its first edition, the map depicted 760 enzymes and the metabolic pathways they are involved in. This was about one third of the enzymes known at the time and covered most of the core metabolism of the eukaryotic cell (Michal, 1982). However, there is a crucial difference. Although the old maps of metabolic pathways depicted all the known enzymes, reactions, and regulatory relationships, this was a static depiction of the metabolic network that did not allow to calculate its dynamics. In contrast, systems biological models of such networks aim for computability of the network dynamics. In this sense, they are the successors of dynamic models of restricted metabolic pathways. These models are still created bottom-up and are compiled from data about single enzymes with known kinetic constants. What is new is the large amount of structural data that are considered in the model. In this sense, these systems biological models merge the old rationale of modeling regulatory pathways that were comparatively data-poor with richer molecular data.

This branch of systems biology is regarded as being predominant in Europe. Nevertheless, the Japanese and US–American traditions of top-down modeling (see Westerhoff & Palsson (2004) and Cassman et al. (2005) for the geographic assignment) until recently also relied mostly on exemplars from the bottom-up branch when presenting convincing results of systems biology (see Kitano, 2002c).

6.2. The second branch of systems biology: Making sense of 'omic' data by top-down modeling

The second branch of systems biological modeling proceeds in top-down fashion. The change with respect to the tradition of top-down modeling of cybernetics and systems analysis seems to be more dramatic than the change within the bottom-up tradition. As argued above, modern 'omic' projects, in contradistinction to their predecessors from pre-genomic days, do not allow much biologically relevant interpretation and are hardly accessible to mathematical modeling of any physiological interest. This is overcome in top-down systems biology by a combination of two strategies. The first strategy is to use data that show not only one state of a network but also its *dynamics*. To do this, multiple 'omic' data sets of the same sort, but of different states of a cell, must be collected. This may be done, e.g., in successive physiological states, or at different times during the development of an organism, or by monitoring 'perturbed' systems (Ideker et al., 2001). Perturbation may be effected by altering the system's environment or by changing its components, which in the case of a genetic network means modifying or deleting genes. Given the availability of high-throughput methods and their further improvability, this is within the range of feasibility. These data provide a basis for modeling. Different ways were found to depict a network in terms of the interaction of its components, which are not further characterized in functional terms, and to account for its dynamics, e.g., Difference-Based Regulation Finding (Onami et al., 2002) and stoichiometric matrices (Palsson, 2006). However, data on the dynamics of a whole network have only limited explanatory power as long as the network cannot be structured in a way that makes its internal dynamics understandable. Therefore, the second strategy involved in the systems biological evaluation of 'omic' data sets is structuring the network into manageable subnetworks ('modules'). To accomplish this, the problem must be solved as to how subnetworks that are physiologically relevant may be singled out.

Systems biologists decompose a network into modules in two different ways. The first way to decompose a network of 'omic' scale isolates the subnetworks in terms of knowledge about network capacities and the contributions of components to these capacities. Such physiological modules are modeled separately, often in a bottom-up manner. Reassembling different modules is another bottom-up step, but may not result in a model of the complete network. The 'omic' scope still requires top-down modeling that embeds the modules into the cellular processes. Such mixed modeling was successful with different metabolic, genetic-regulatory, and signaling networks (see Kitano, 2002c; Palsson, 2006). Relying on known functional modules, however, may distort the model of the network to the extent that modules that are individuated functionally (often on the basis of intuitive reasoning) may not be as independent from the processes in which they are embedded as the separate bottom-up models may suggest. Functional modules need not coincide with dynamical modules of a network, hence an approach that first models functional modules separately and then assembles these models into a network might end up with a distorted network structure. To avoid such distortion, a second technique of network decomposition uses criteria gained from the structure and dynamics of the network itself. Different mathematical methods are used to individuate so-called unbiased modules, i.e., modules as delineated from the embedding network by criteria for strong internal and weak external interaction rather than by functional considerations (Rohwer et al., 1996; Koza et al., 2002; Friedman, 2004; Papin et al., 2004). We leave the analysis of the technical details and pitfalls of 'unbiased modularization' to the specialists, but will return to some of its consequences in Section 8.1.

This kind of top-down modeling shows certain similarities with biological cybernetics, although in contrast to the latter it does account for the molecular components of a network. The strategy of top-down systems biology is to identify modules within a network, each of which displays a characteristic dynamics. In biological cybernetics, regulatory instances are postulated within the system, which are also autonomous to a certain degree, yet interconnected. However, in systems biology the system is given in terms of data about its components, while biological cybernetics does not necessarily refer to any component within the black-boxed system.

7. THE STRUCTURE OF THE FIELD

Having described some of the characteristics of models that can be attributed to the different roots and branches of systems biology, we now want to use this material to elaborate the relationships that hold between the different fields. Although it should be possible in principle to reconstruct a theory net from the related models and to analyze all the links between them, this would require reconstructive efforts of almost 'omic' dimensions. We therefore draw a coarser-grained picture of the links that hold between the different fields, concentrating on questions about the principal relationships between the various approaches.

The field we are looking at may itself be conceived as a network of the five approaches that we label pathway modeling, biological cybernetics, 'omics,' bottom-up systems biology, and top-down systems biology, respectively. The first three of those we have described as the roots of systems biology, the latter two as its branches. So we have to look how each particular root feeds into the branches.

It seems obvious that pathway modeling hands down its bottom-up way of modeling and reliance on detailed kinetic data to the bottom-up branch of systems biology. Both approaches build models in a way that may be characterized as functional composition from detailed kinetic and some structural data. We see richer structural data, i.e., incorporation of more components into the model, on the side of systems biology. To the extent that it is achieved by the application of high-throughput methods, this gain in richness of structural data can be seen as an influence of 'omics'. We therefore regard 'omics' as another, though less important, root of bottom-up systems biology (see Fig. 1). The other main difference between both approaches is the way the functionality of the systems components is dealt with. Pathway modeling sometimes assigns regulatory functions to particular components of a pathway, referring to a rate-limiting step at which the flux through the pathway is controlled. This notion was already given up in metabolic control analysis (Kacser & Burns, 1973; Heinrich & Rapoport,

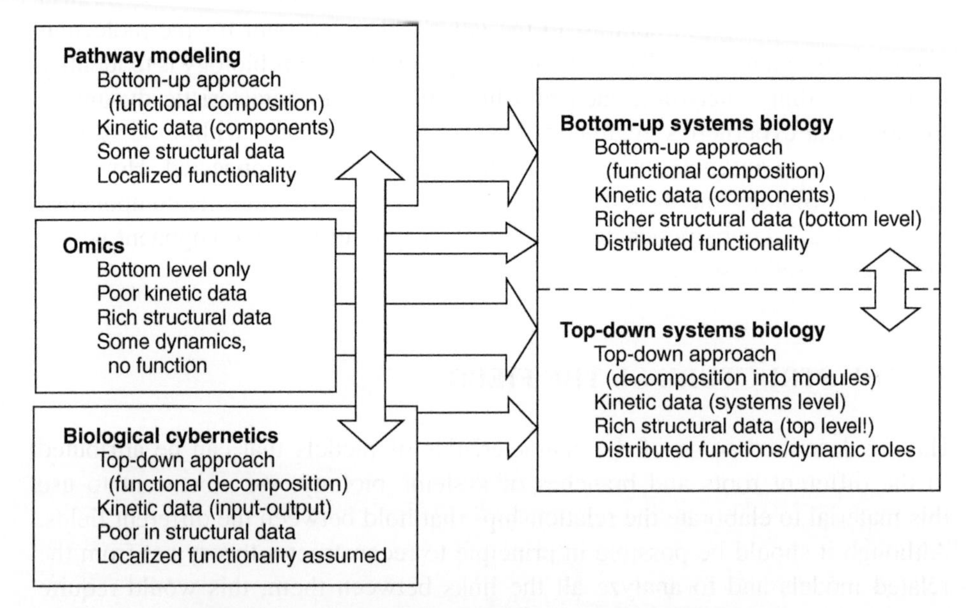

Figure 1 The structure of the field of systems biology.

Thickness of arrows indicates roughly the strength of the links between the root disciplines (left) and the branches of systems biology (right). The vertical arrows indicate interchange between pathway modeling and biological cybernetics, and between both branches of systems biology, respectively. The dotted demarcation line between the two branches of systems biology shall indicate that both kinds of modeling are often combined within one model.

1974) as control turned out not to depend on any single step in most metabolic pathways, but to be distributed. In the larger detailed models of systems biology, the notion of distributed functionality becomes indispensable.

Pathway modeling and the bottom-up branch of systems biology may be viewed as constituting a continuous tradition. The change from biological cybernetics and systems analysis to top-down systems biology is more dramatic. Cybernetic modeling proceeds by functional decomposition of a system, whereas in top-down systems biology this is only one, the less favored, strategy. Decomposition of a network into modules by dynamical criteria is favored. In addition, kinetic data of the input–output type, most important for cybernetics, are not so important for systems biology, where the dynamics of the system is studied by an analysis of 'omic' data sets. The latter also provide the rich structural data on which top-down modeling in systems biology relies. We therefore estimate the link between 'omics' and top-down systems biology, its second root, to be much stronger than the one between 'omics' and bottom-up systems biology (Fig. 1). This result is astonishing in a way, because 'omic' data seem to be data about components rather than systems. But this classification holds within the framework of 'omics' only. In top-down systems biology, structural 'omic' data as well as kinetic data obtained from them should be regarded as top-level data (Palsson, 2006, pp. 63–65).[9] These components are not considered with respect to their individual role, they merely represent nodes of the network. 'Omic' data, then, define primarily the state and dynamics of the network as a whole and are only secondarily analyzed with respect to lower-level modules. A move away from localized functionality, as was mentioned with respect to bottom-up systems biology, occurs in the top-down branch as well. Cybernetics assumes localized functionality, although as a result of the black-boxing of regulatory pathways the functions are not ascribable to any component if the approach is not combined with pathway analysis. Depending on the individuation of models by functional or dynamical criteria, localized functionality is replaced in top-down systems biology by either distributed functionality or assigning distributed dynamical roles within the network.

Like the cases from Section 4, and in particular that of the biological clock model shown, pathway modeling and biological cybernetics are not completely separable. Modeling here is top-down, but then plausible molecular regulatory mechanisms are substituting the control elements that are standardly used in cybernetics. Cybernetic models thus often refer to the molecular basis of regulatory circuits and hence incorporate bottom-up models. This means that in fact both fields are not completely separated but linked to each other, as indicated by the vertical arrow in the left part of Fig. 1. By this link, pathway modeling also

[9] Palsson speaks about top-down data types instead of top-level data, already referring to the direction of modeling from the systems level down to its (in this case functionally individuated) modules.

has an indirect influence on top-down systems biology, and biological cybernetics accordingly has an indirect influence on bottom-up systems biology. 'Omics', in contrast, is not strongly linked to the two other roots of systems biology.[10] It stands itself in the traditions of molecular biology and enzyme kinetics. However, since molecular biology is also an important source for pathway modeling (which is clear in particular from early models: Krebs, Warburg), we do not trace back the tradition of 'omics' to molecular biology as one discernable root of systems biology (as, e.g., Westerhoff and Palsson (2004) do), but refer to 'early omic projects' as the onset of this root of systems biology.

Up to this point, we followed the habit of separating systems biological models into two classes that are clearly discernable by the 'direction' of modeling. However, as we already noted, bottom-up and top-down approaches are often combined. Palsson, for one, points out the need to combine bottom-up and top-down strategies in the modeling of transcriptional regulatory networks (Palsson, 2006, p. 66) and some convergence is stated with respect to the two branches of the field (Westerhoff & Palsson, 2004). The top-down approach tries to decompose large networks into modules. These modules may be described separately by smaller top-down models, but also by bottom-up models. This overcomes the strict demarcation between both branches of the field, which is indicated by the dashed line in Fig. 1. Systems biology, although certainly not a homogeneous field, can be regarded as a unity if the aspect of the direction of modeling is given less weight. A shared perspective can be found in regarding functionality as distributed over the components of complex networks. Another commonality of both branches of systems biology is the search for detailed models, though the degree of detail aspired at by researchers is higher with respect to top-down than with respect to bottom-up models. Most important in this respect might be the attitude to model 'whole' systems in such a way as to describe networks embracing entities that are in some sense 'wholes': cells, organs, organisms, etc.

8. EPISTEMOLOGICAL AND ONTOLOGICAL ISSUES REGARDING TOP-DOWN SYSTEMS BIOLOGY

Up to now, we have dealt with systems biology from a reconstructive perspective to analyze the structure of the merger of different roots into systems biology and the changes in basic modeling strategies that took place. As we have seen, systems biology of the bottom-up branch is very much in line with its predecessor theories. The top-down branch, however, displays some drastic novelties in data

[10] There are, of course, other links that represent shared methods such as different kinds of radioactive marking and diverse analytic techniques. We regard these as too general to be relevant to the structure of the field.

handling and modeling. In the remainder, we will therefore concentrate on the top-down approach and discuss some questions pertaining to the epistemological, conceptual, and ontological foundations of this young biological discipline. We do not aspire to solve these questions here.

8.1. Decomposition of large networks and functionality

The dynamics of a large network is neither explainable from an account of its components alone, nor does an account of the overall dynamics of a network count as its own explanation. In contrast, an explanation may refer to the roles of subnetworks or modules in the dynamics of the whole network. An important step in systems biological top-down modeling is, therefore, the decomposition of a system into interacting subsystems. As already mentioned, there are two modularization strategies, one using criteria of functionality to identify modules and the other proceeding by a mathematical analysis of network structure (Section 6.2). The first strategy to decompose a network starts from a biological capacity that one can attribute to a network, e.g., regulation of a developmental pathway or processing of some signal. This capacity may be analyzed further by singling out the functional contributions of sub-networks to the capacity. A sub-network to which a function can be ascribed, then, is considered a functional module. (Within the module, functionality need not be, and usually is not, localizable at any particular component but can be distributed.) The concept of function, however, is controversial within philosophy of science, and different accounts compete with each other (for a review, see, e.g., Wouters, 2005). We therefore must specify which notion of function is applicable here. An approach that fits the strategy described well is Cummins' functional analysis (see Boogerd et al., 2005) according to which the function of a system's component is its systemic role in contributing to a capacity of that system (Cummins, 1975).[11] Since most functions are not performed permanently, 'role' is to be understood dispositionally here. Functional analysis may be based on the results of physiological experiments, as in the biochemical analysis of metabolic and signaling pathways (whose results may be further used for bottom-up pathway modeling), or, to some extent, be performed mathematically, like in the analysis of input-output data, as long as the results can in some way be related to structural data.[12] In

[11] Other accounts of function refer to the evolutionary history of a biological trait and single out the role a trait was adapted for as its proper function. These accounts are barely applicable to systems biology, which mostly lacks an evolutionary perspective and in most cases investigates a network as it is now without taking into account its history (cf. also note 16).

[12] One point should be added to our rendering of Cummins's account: Functional analysis not only refers to a system and its capacities, but also includes a description ('analytic account') of this capacity. Therefore, the capacity and functions by which the system's components contribute to it cannot be read from the system alone, but are always relative to the structuring of the capacity by the description given by scientists. Functions

more classical physiological disciplines, Cummins-style functional analysis has been performed successfully. Applied to large networks, however, the structuring becomes somewhat problematic because as a result of the distributed nature of the functions in these cases, the limits of a functional subunit are not necessarily sharp. Instead of following structural borders that may exist in the network, functional analysis cuts the network into modules according to the functions the descriptions (the analytical account) can address. The position of these cuts, however, depends on how functional contributions can be conceptualized. The cuts need not coincide with the actual structure of a network and in most cases the functional modules will not be as independent from each other as they appear under this decomposition. Functional decomposition may consequently distort the picture of the network.

The question is, then, whether functional decomposition is an adequate heuristic for the decomposition of large-scale networks in cases where the goal is a faithful representation of a whole network in a model. The problem of decomposing a network faithfully is claimed to have been solved by various formal methods for obtaining 'unbiased modules' (see references in Section 6.2). These methods do not rely on a functionality criterion but refer to structural and dynamical properties of networks instead. Here, networks are viewed as 'nearly decomposable' into structurally and dynamically largely autonomous components or modules sensu Herbert Simon (1969), providing 'natural joints' along which a network can be decomposed to yield modules. We note, however, and again in the vein of Simon's pragmatic 'bounded rationality' assumption ("any real-world system is too complex to study in all of its complexity, so we must make simplifications"), that there are no good reasons to expect these methods to yield a (unique) 'best' decomposition, whatever that should be; the 'best' one is likely to get its robust results based on triangulation, which will always reflect the researchers' particular aims with the modeling enterprise at hand (see, e.g., Wimsatt, 1981; Simon, 1982; cf. also Section 8.3).[13,14]

The structural/dynamical decomposition strategy is supposed to give a less distorted picture of a network than (often 'intuitive') functional decomposition (Bechtel and Richardson, 1993; Schaffner, 1998; Papin et al. 2004).[15] Rather than pursuing the question whether the various methods for 'unbiased modularization' that have been proposed may deliver the goods, we here want to point

therefore depend on the concepts by which the contributions to a capacity are classified. This relativization, which could in principle be almost arbitrary, may be overcome, however, by referring to the ontogenetically effective fixation of components of functionally organized systems (Krohs, 2004 and submitted).

[13] See Callebaut and Rasskin-Gutman (2005) for state-of-the-art applications of near-decomposability to complex biological (and other) systems and discussion of the epistemological issues they raise.

[14] Introducing an evolutionary perspective may also help structuring a network in cases where evolution results in recurring, dynamic organizational principles (de Atauri et al., 2004).

[15] Ideally for the modeler, of course, the two sets of modules arrived at on the different routes would coincide, but in general they will not.

to the change in perspective associated with this move. A network decomposed in one of these ways is considered not in a functional, but in a connectionist perspective, as known from cognitive science (neural networks composed of large numbers of units together with weights that measure the strength of connections between the units but without particular functions ascribed to the units; see, e.g., Rumelhart & McClelland, 1986; McClelland & Rumelhart, 1986). Functionality may be brought in again in 'unbiased' top-down systems biology if functions are ascribed to the modules in a later modeling step. This, however, is neither required nor does it change the 'unbiased' picture of a network once it is obtained.

Abstaining from function ascriptions might be regarded as unusual in biology (see, e.g., Rosenberg, 1985 on the ubiquity of functional explanation at the level of macromolecules). We want to point out an important consequence of this move: Abandoning the functional perspective on biological systems amounts to 'physicalizing' them, as reflected in the – welcome – improvement of the mathematical tractability of a system that is decomposed according to explicit formal criteria. On the other hand, such physicalization renders problematic the application of a whole range of concepts that are related to the concept of functionality and that are used, within the realm of natural sciences, exclusively in biology. In particular, this move eliminates the only reference point for what might be regarded as the biological 'meaning' of a structure within an organism, viz. its function within the system or disposition to fulfill a functional role. 'Meaning' has to be put in quotes here anyway, but we cannot think of a weaker adequate interpretation than that as a systemic function.[16] An approach that refrains from function ascriptions should not attempt, then, to find anything like an account of biological 'meaning' in its models (unless having explicated what 'meaning' could be instead). Yet finding biological 'meaning' in 'omic' data is regarded as one of the main challenges in systems biological modeling (Huang, 2004; Joyce & Palsson, 2006).

---

[16] An evolutionary perspective might allow a different interpretation of the concept of biological meaning, but none of the systems-biological models we have considered in this chapter make claims to having evolutionary import. Although at present still somewhat marginal within the field, work that profitably combines a systems-biological perspective with evolutionary considerations is actually undertaken. Thus, e.g., Huang (2004) suggests that as "structures midway between genome and phenome", molecular network topology opens a new window to study evolution in complex living systems because it sheds new light on the old debate on the relative contributions of natural selection and self-organization ('intrinsic constraints'), respectively. Another example is Eric Davidson and his co-workers' research at the interface of systems biology and evolutionary developmental biology ('EvoDevo'): their research on the control of the development of the animal body plan by large gene regulatory networks (GRNs) documents how the evolution of body plans must depend upon change in the hierarchical architecture of developmental GRNs and suggests that the conservation of phyletic body plans may have been due to the retention since pre-Cambrian time of 'GRN kernels' underlying development of major body parts (Davidson et al., 2002; Davidson & Erwin, 2006; cf. Callebaut et al., in press, for a critical assessment of the GRN approach in the larger framework of EvoDevo).

This tension between the justification of the 'unbiased' strategy and the status of the result of the procedure needs to be resolved. This could be done in different ways. First, there is the aforementioned secondary ascription of functions to the 'unbiased' modules. This is a reasonable strategy, which, from an epistemological point of view, does not differ much from any other ascription of functions to known sub-structures of an organism in other physiological disciplines, including molecular biology. On this account, the unbiased modularization helps to identify, by top-down analysis, the physiological subunits of a network, which cannot be identified bottom-up merely from knowledge of its components without bias. As soon as functions can be ascribed to these modules with respect to an organismic capacity, 'meaning' is bestowed on the modules with respect to the organism as a whole. A second way of reading 'meaning' from top-down models could refer to capacities of the network itself rather than of the organism. The network has, e.g., (self-)regulatory properties that may be regarded as its capacities. These may be analyzed functionally, without requiring reference to capacities of the organism. On this account, both the organism and the network itself must be considered as ontological wholes because functional decomposition cannot be applied satisfactorily to incomplete entities (otherwise one would end up with ascriptions like the function of the wound of a severely injured animal to be squirting blood and the like). The strategy of relating biological function and hence 'meaning' to 'omic' biological systems as wholes, however, required to explicate the concept of wholeness with respect to networks, a topic to which we turn next.

8.2. The 'holism' of systems biology

Systems biology shifts the focus of physiological inquiry from the analysis of (molecular) subsystems that contribute to certain capacities to 'whole system' accounts. What systems biology lacks, however, is a clear ontology of systems: It is neither clear how systems are to be individuated nor what it means for a whole system, in contrast to an incomplete one, to be under consideration (Mesarovic et al., 2004; O'Malley & Dupré, 2005; Krohs, 2006a). In systems biology, 'holism' is often used as a mere buzzword to oppose 'reductionism', which is often pejoratively connotated.[17] Thus, e.g., when Mendes (2002) diagnoses the "trend away from extreme reductionism to systems descriptions and analyses"

[17] To complicate matters, reduction(ism) is also often opposed to emergence (emergentism), which invites the further question how holism and emergentism are to be related (see, e.g., Williams, 2002 and Boogerd et al., 2005 on emergent properties at the molecular and cellular levels, and Thalos, 2006 on emergent physical processes). These issues cannot be pursued in this chapter. Let us only note that on at least some authoritative current accounts, (certain kinds of) emergence, rather than implying the failure of reduction, presupposes reducibility; see, e.g., Stephan, 2002; Boogerd et al., 2005: Wimsatt, in press.

as a positive outcome of the emerging 'omic' disciplines, 'reductionism' for him simply means a focus on specific organismal components, which he contrasts with 'whole-organism measurement'. Since it is blatantly unclear how one is to 'measure' a whole organism, we assume that the point is that parameters of a certain kind (related to, e.g., genes, proteins, or metabolic processes) are collected on an organismic or cellular scale in an attempt at completeness (cf. Butcher et al., 2004 on integrating 'reductionist data'). Wholeness in this sense can be regarded as completeness of data of a certain kind with respect to a given compartment, which may be an organelle, a cell, an organ, an organism, or even an ecosystem.[18]

Though it initially seemed plausible, using compartmentalization as an indicator for the delineation of a whole system needed justification for the different kinds of 'omic' systems that may be subjected to systems biological investigation. 'Whole' cannot mean 'closed', since any biological system that interacts with its environment by exchange of mass and energy is an open system. A reasonable criterion would therefore be the strength of internal and external interactions. Wholeness should not be ascribed to a system whose components interact with the environment as strongly as with each other. On this account, a whole system can be individuated in very much the same way as unbiased modules from a network by its partial autonomy with respect to its environment. This, of course, is not yet an applicable criterion but a meta-requirement that any criterion should meet. The compartmentalization criterion, on the other hand, could probably be used as a substitute in the case of cellular genomes, but even in this case one should discuss whether in eukaryotes the whole system includes only nuclear or also extranuclear DNA, whether the DNA of an organelle is really a whole system in itself, and whether it would be consistent to regard nuclear DNA as a systemic whole but not mitochondrial and plastidal DNA. It would seem to be even more questionable whether the borders of proteomes, metabolomes, and interactomes coincide with the borders of compartments. If not – and there is strong evidence that '-omes' should be conceived as spanning different levels and compartments (Finkelstein et al., 2004; Thiele et al., 2005) – a top-down systems biological model that relies on 'omic' data for a certain compartment represents an incomplete system. 'Omic' data on whole organisms, on the other hand, distort the picture of network dynamics as there are in fact different processes going on in different compartments. An ontology of systems and an epistemology of systems-as-wholes is required if top-down systems biology is to find criteria for structuring its models adequately.

[18] In systems biological parlance, a compartment is a container of finite size for substances. A substance requires a unique compartment to contain it. Compartments need not correspond to actual structures in- or outside of, say, a cell. For a discussion of the problems arising with species that can cross compartment boundaries see Hattne (2004).

8.3. 'Realistic' representation and systems biological 'models of everything'

The reliance of top-down systems biology on genomic data is often associated with the promise that this allows for a 'realistic' representation of biological processes in mathematical terms (e.g., Kitano, 2002b, 2006). The most realistic model, Kitano believes, is a model of all genomic and proteomic interactions and dynamics of an organism. The project to build such a model runs under the title of 'systeome', though what is envisaged is also a simulation model, not only an 'omic' data set. Kitano proposes to include research on the systeome from five different organisms as the grand challenge for systems biology. He regards the human systeome project (which, if funded, would certainly become the grand cornucopia of the field) as most important. The promise is to have realized the systeome of the human cell by the year 2020, with 20% error margin, and to have included all genetic variations, drug responses, and environmental stimuli ten years later (Kitano, 2002b, p. 25). We are not in a position to assess the feasibility of this project, nor can we judge its scientific and medical value in the long run; but we do want to ask what kind of a model such an envisaged 'realistic' systeomic model is, how it relates to the kinds of models we have discussed in this chapter, and, most importantly from our point of view, what explanatory power it might have.

Calling an envisaged systeomic model, 'realistic' suggests that it can give a reliable, undistorted picture of biological processes. This would seem to require that the picture ought to be as detailed as the processes going on in nature, and that the structure of the model, i.e., the components and their relations, is isomorphic to the structure found in nature. (Given that we have no direct access to nature, it is far from obvious how such a comparison between nature and our models of it could ever be made.) It does *not* mean that the model has to consist of material components like an organism, as it does not represent the material nature – the embodiment – of the system. The model is 'realistic' in the sense that it does not simplify by neglecting parameters. A realistic model, then, is maximally detailed and unbiased. It is not obvious that any model can be realistic in this sense, and certainly no model referred to in this chapter meets this requirement. Most, if not all, bottom-up models are not realistic in this sense. Even detailed models usually neglect, e.g., that organismic cells are spatially inhomogeneous, i.e., that chemical gradients are present and consequently transport phenomena occur. Ordinary differential equations that are used in most models describe homogeneous ('well-stirred') systems only. Using partial differential equations may lead to increased realism, but quickly results in intractable systems.[19] Top-down models are not realistic either. As we

[19] Prokaryotes, at least, may be described in a phenomenologically adequate way as homogeneous systems.

have seen, they decompose the system into modules. Structuring according to functional criteria distorts the picture anyway, and structuring into 'unbiased' modules still treats a nearly decomposable system as a completely decomposable one and therefore represents the interactions of the modules in an idealized way. We conclude that the systeomic models envisaged by Kitano are top-down models that distinguish themselves from the models of all classes described so far by the perspective of not simplifying the system.

It is noteworthy that simplification by data reduction, though it looks like a deficiency, is usually intended. One reason is that models that involve too many variables become easily intractable, not only analytically but also numerically, so that they cannot even be used for running simulations. There is another reason for simplification: We can learn more about the main processes within a system if we look at them without being influenced by too many minor sideways that only blur the picture of the core processes. Models have to be simplified or idealized to be informative (Wimsatt, 1987 and in press; Wolkenhauer & Ullah, this volume). A simplified model that still reproduces the relevant part of the dynamics of the modeled system can explain the principles of the dynamics, since one can read from the models how the dynamics of the biological systems is brought about. The (fictitious) 'realistic' model, in contrast, simply reproduces the dynamics. It allows the faithful simulation of all processes of the considered kinds that take place in an organism. It has predictive power, but little or no explanatory value (see, e.g., Cartwright, 1989; Lewontin, 1997). Its main value will be as an easily tractable analogue of organismic processes on which virtual experiments may be performed. The processes of such *in silico* experiments are measurable without technical restrictions that hold with respect to the natural systems, but the dynamics of the simulation demands itself for explanation by simplifying models (Krohs, 2006b). Anyhow, data about the biological system may be obtained vicariously from the simulation. This follows a long and interesting tradition that also has learned about limits of such a transfer (Morgan, 2003). As it is highly unlikely that a nonmaterial system could reproduce the dynamics of a material system exactly (Griesemer, 2005), and since 'omic' data sets provide dynamical data of poor resolution anyway, it seems reasonable that Kitano does not aim for better than 20% error.[20]

The systeome project aims to collect data without providing a strategy to arrive at explanatory models. Though coming under the label of systems biology, it turns out to be a purely 'omic' project, as is also made clear in its name. The only improvement with respect to other 'omic' projects is that it integrates a dynamic perspective, but instead of taking explanatory advantage from this perspective as

[20] Although this might not be relevant to the success of the systeome project, it would be interesting to learn how many of the results concerning human and mouse systeomes can be expected to be distinguishable within this error margin.

systems biology proper does, the systeomic project degrades network dynamics to another source of large data sets. This critique, however, applies only to attempts of 'realistic' modeling that aim at doing without any simplification. All extant systems-biological models do justice to the explanatory attitude of model building. They gain an explanatory status by idealization of the modeled system, which allows an exploration of which components and relations actually bring about the dynamics of the network under investigation.

9. CONCLUSION

We have reviewed the field of systems biology and its variegated roots to characterize its structure, and have addressed some epistemological and ontological issues regarding concepts on which the field relies and that seem to us to require clarification. Systems biology is largely tributary to genomics and other 'omic' disciplines that generate vast amounts of structural data. 'Omics', however, lack a theoretical framework that would allow to use these data sets as such (rather than just tiny bits that are extracted by advanced data-mining techniques) to build explanatory models that help understand physiological processes. Systems biology provides such a framework by adding a dynamic dimension to merely structural 'omics', making use of bottom-up and top-down models. The former are based on data about systems components and the latter on systems-level data. We traced back both modeling strategies (which are often used to delineate two branches of the field) to the modeling of metabolic and signaling pathways in the bottom-up case, and to biological cybernetics and systems theory in the top-down case. We identified three roots of systems biology: pathway modeling, biological cybernetics, and 'omics'. The data richness one encounters in both branches of systems biology stems from the 'omic' disciplines. Both other roots were found to be comparatively data-poor but strong in model building. Therefore, we regard systems biology as merging modeling strategies (supplemented by new mathematical procedures) from data-poor fields with the data supply from a field that is quite deficient in explanatory modeling.

An epistemologically important problem of the top-down approach to systems biology arises from the fact that the systems under investigation are large and must be decomposed into subsystems in order to model them in a way that allows to explain their dynamics. Decomposition into modules is performed according to either functional or structural criteria. While the latter way, usually called 'unbiased modularization', distorts the picture of the system to a lesser degree than functional decomposition, it gives up the established link between physiology and function. We have argued that functionality had to be reintroduced if one is interested in statements about the biological 'meaning' of the dynamics of a system. As we view scientific explanation not as mere depiction of some structure and dynamics but as making it understandable (which in the case of a large network boils down to giving a simplified picture of it), we criticized the project of a 'realistic' representation of all metabolic processes in a 1:1 manner as lacking explanatory power and, more generally, as being epistemologically misguided. It regresses to the 'omic' approach by once again offering 'complete' but physiologically uninterpreted data sets. In addition to epistemological issues, we addressed the ontological question of how to define the 'wholeness' of the systems investigated by top-down systems biology and confirmed the diagnosis that a clear concept of wholeness is still missing. We had to confine ourselves to discussing why it is insuffient to ground this concept in reference to (sub)cellular compartments. We regard it as a primary demand for systems biology to establish a sound concept of wholeness.

REFERENCES

Ankeny RA. *Organisms as models: understanding the 'lingua franca' of the Human Genome Project.* Philosophy of Science: 68 (Proceedings), S251–261, 2001.

Ashburner M, Bonner JJ. *The induction of gene activity in drosophila by heat shock.* Cell: 17, 241–254, 1979.

Bechtel W, Richardson RC. *Discovering complexity: decomposition and localization as strategies in scientific research.* Princeton University Press, Princeton, NJ, 1993.

Bertalanffy L von. *Theoretische Biologie*, 2 vols. Berlin, 1932; 1942.

Bertalanffy L. von. *General system theory: foundations, development, applications.* Braziller, New York, 1968.

Blattner FR, Plunkett G III, Bloch CA, Perna NT, Burland V, Riley M, Collado-Vides J, Glasner JD, Rode CK, Mayhew GF, Gregor J, Davis NW, Kirkpatrick HA, Goeden MA, Rose DJ, Mau B, Shao Y. *The complete genome sequence of* Escherichia coli *K*-12. Science: 227, 1453–1474, 1997.

Boogerd FC, Bruggeman FJ, Richardson RC, Stephan A, Westerhoff HV. *Emergence and its place in nature: a case study of biochemical networks.* Synthese: 145, 131–164, 2005.

Bourret RB & Stock AM. *Molecular information processing: lessons from bacterial chemotaxis.* The Journal of Biological Chemistry: 277, 9625–9628, 2002.

Bray D. *Molecular networks: the top-down view.* Science: 301, 1864–1865, 2003.

Bridges CB. *Salivary chromosome maps.* The Journal of Heredity: 26, 60–64, 1935.

Bridges CB. *A revised map of the salivary gland X-chromosome.* The Journal of Heredity: 29, 11–13, 1938.

Bridges PN. *A new map of the salivary gland 2L-chromosome.* The Journal of Heredity: 33, 403–407, 1942.

Butcher EC, Berg EL & Kunkel EJ. *Systems biology in drug discovery.* Nature Biotechnology: 22, 1253–1259, 2004.

Callebaut W. *Again, what the philosophy of biology is not.* Acta Biotheoretica: 53, 93–122, 2005.

Callebaut W, Müller GB & Newman SA. *The organismic systems approach: EvoDevo and the streamlining of the naturalistic agenda.* In: Integrating evolution and development: from theory to practice. (Eds.: Sansom R & Brandon RN), MIT Press, Cambridge, MA, in press.

Callebaut W & Rasskin-Gutman D, eds. *Modularity: understanding the development and evolution of natural complex systems.* MIT Press, Cambridge, MA, 2005.

Cartwright N. *How the laws of physics lie.* Oxford University Press, Oxford, 1989.

Cassman M, Arkin A, Doyle F, Katagiri F, Lauffenburger D & Stokes C. *WTEC Panel report on international research and development in systems biology.* World Technology Evaluation Center, Baltimore, 2005.

Chabre M & Deterre P. *Molecular mechanism of visual transduction.* European Journal of Biochemistry: 179, 255–266, 1989.

Cummins R. *Functional analysis.* The Journal of Philosophy: 72, 741–765, 1975.

Darden L. *Reasoning in biological discoveries: mechanisms, interfield relations, and anomaly resolution.* Cambridge University Press, New York, 2006.

Darden L & Craver CF. *Strategies in the interfield discovery of the mechanism of protein synthesis.* Studies in History and Philosophy of Biological and Biomedical Sciences: 33, 1–28, 2002.

Darden L & Maull N. *Interfield theories.* Philosophy of Science: 44, 43–64, 1977.

Davidson EH et al. *A genomic regulatory network for development.* Science: 295, 1669–1678, 2002.

Davidson EH & Erwin DH. *Gene regulatory networks and the evolution of animal body plans.* Science: 311, 796–800, 2006.

de Atauri P, Orrell D, Ramsey S & Bolouri H. *Evolution of 'design' principles in biochemical networks.* Systems Biology: 1, 28–40, 2004.

Dohlman HG & Thorner JW. *Regulation of G protein-initiated signal transduction in yeast: paradigms and principles.* Annual Review of Biochemistry: 70, 703–754, 2001.

Dougherty ER & Brago-Neto U. *Epistemology of computational biology: Mathematical models and experimental prediction as the basis of their validity.* Journal of Biological Systems: 14, 65 90, 2006.

Drell D & Adamson A. *Fast forward to 2020: what to expect in molecular medicine.* TNTY Futures: 1(1), s.d. (2000). http://www.tnty.com/newsletter/futures/index.html.

Finkelstein A, Hetherington J, Li L, Margoninski O, Saffrey P, Seymour R & Warner A. *Computational challenges of systems biology.* Computer: 37 (5, May), 26–33, 2004.

Fleischmann RD et al. *Whole-genome random sequencing and assembly of* Haemophilus influenzae *Rd.* Science: 269, 496–512, 1995.

Friedman N. *Inferring cellular networks using probabilistic graphical models.* Science: 303, 799–805, 2004.

Furman GG. *Comparison of models for subtractive and shunting lateral-inhibition in receptor-neuron fields.* Kybernetik: 2, 257–274, 1965.

Griesemer JR. *The informational gene and the substantive body: on the generalization of evolutionary theory by abstraction.* In: Idealization xii: Correcting the model. Idealization and abstraction in the sciences. (Eds.: Jones MR & Cartwright N), Rodopi, Amsterdam, 2005.

Griesemer JR. *Theoretical integration, cooperation, and theories as tracking devices.* Biological Theory: 1, 4–7, 2006.

Hattne J. *MesoRD user's guide.* http://mesord.sourceforge.net/manual/docbook/book1.html, 2004.

Heinrich R & Rapoport TA. *A linear steady-state treatment of enzymatic chains.* European Journal of Biochemistry: 42, 89–95, 1974.

Hodgkin AL & Huxley AF. *A quantitative description of membrane current and its application to conduction and excitation in nerve.* The Journal of Physiology: 117, 500–544, 1952.

Hodgkin AL, Huxley AF & Katz B. *Measurement of current-voltage relations in the membrane of the giant axon of Loligo.* The Journal of Physiology: 116, 424–448, 1952.

Huang S. *Back to biology in systems biology: what can we learn from biomolecular networks?* Briefings in Functional Genomics and Proteomics: 2, 279–297, 2004.

Ideker T, Galitski T & Hood L. *A new approach to decoding life: systems biology.* Annual Reviews of Genomics and Human Genetics: 2, 343 372, 2001.

International Human Genome Sequencing Consortium. *Initial sequencing and analysis of the human genome.* Nature 409, 860–921, 2001.

International Human Genome Sequencing Consortium. *Finishing the euchromatic sequence of the human genome.* Nature: 431, 931–945, 2004.

Jacob F & Monod J. *On the regulation of gene activity.* Cold Spring Harbor Symposia on Quantitative Biology: 26, 193–211, 1961.

Jacob F & Wollman EL. *Genetic and physical determinations of chromosomal segments in Escherichia coli.* Symposia of the Society for Experimental Biology: 12, 75–92, 1958.

Joyce AR & Palsson BØ. *The model organism as a system: integrating 'omics' data sets.* Nature Reviews Molecular Cell Biology: 7, 198–210, 2006.

Kacser H & Burns JA. *The control of flux.* Symposium of the Society for Experimental Biology: 27, 65–104, 1973.

Kholodenko BN. *Cell-signaling dynamics in time and space.* Nature Reviews Molecular Cell Biology: 7, 165–176, 2006.

Kitano H. *Systems biology: a brief overview.* Science: 295, 1662–1664, 2002a.

Kitano H. *Systems biology: toward system-level understanding of biological systems.* In: Foundations of systems biology. (Ed.: Kitano H), MIT Press, Cambridge, MA, 1–36, 2002 (2002b).

Kitano H, ed. *Foundations of systems biology.* MIT Press, Cambridge, MA, 2002 (2002c).

Kitano H. *Computational cellular dynamics: a network–physics integral.* Nature Reviews Molecular Cell Biology: 7, 163, 2006.

Koza JR, Mydlowec W, Lanza G, Yu J & Keane MA. *Automated reverse engineering of metabolic pathways from observed data by means of genetic programming.* In: Foundations of systems biology. (Ed.: Kitano H), MIT Press, Cambridge, MA, 95–121, 2002.

Krohs U. *Eine Theorie biologischer Theorien: Status und Gehalt von Funktionsaussagen und informationstheoretischen Modellen.* Springer, Berlin, 2004.

Krohs U. *Philosophies of particular biological research programs.* Biological Theory: 1, 182–187, 2006a.

Krohs U. *A priori measurable worlds.* Proceedings (Models and Simulations, London 2006), Philosophy of Science Archive, Pittsburgh, ID code 2787, 2006 (2006b). http://philsci-archive.pitt.edu/archive/00002787/.

Krohs U. *Functions as based on a concept of general design.* Submitted.

Leloup JC & Goldbeter A. *Toward a detailed computational model for the mammalian circadian clock.* Proceedings of the National Academy of Sciences of the USA: 100, 7051–7056, 2003.

Lewontin RC. *Billions and billions of demons.* The New York Review of Books: January 9, 28–32, 1997.

McClelland JL, Rumelhart DE & the PDP Research Group. *Parallel distributed processing: explorations in the microstructure of cognition* Volume 2: *Psychological and biological models.* MIT Press, Cambridge, MA, 1986.

McKusick VA. *Mendelian inheritance in man*, 3 vol. The Johns Hopkins University Press, Baltimore, 1998.

Mendes P. *Modeling large biological systems from functional genomics data: parameter estimation.* In: Foundations of systems biology. (Ed.: Kitano H), MIT Press, Cambridge, MA, 163–186, 2002.

Mesarovic MD, ed. *Systems theory and biology.* Springer, New York, 1968.

Mesarovic MD, Sreenath SN & Keene JD. *Search for organizing principles: understanding in systems biology.* Systems Biology: 1, 19–27, 2004.

Michal G. *Biochemical pathways.* Boehringer, Mannheim, 3rd ed. 1982.

Morgan MS. *Experiments without material intervention: model experiments, virtual experiments, and virtually experiments.* In: The philosophy of scientific experimentation. (Ed.: Radder H), University of Pittsburgh Press, Pittsburgh, 216–235, 2003.

Newman S. *The fall and rise of systems biology: recovering from a half-century gene binge.* GeneWatch: 8–12, July-August 2003.

O'Farrell PH. *High resolution two-dimensional electrophoresis of proteins.* The Journal of Biological Chemistry: 250, 4007–4021, 1975.

O'Malley MA & Dupré J. *Fundamental issues in systems biology.* BioEssays: 27, 1270–1276, 2005.

Onami S, Kyoda KM, Morohashi M & Kitano H. *The DBRF method for inferring a gene network from large-scale steady-state gene expression data.* In: Foundations of systems biology. (Ed.: Kitano H), MIT Press, Cambridge, MA, 59–75, 2002.

Pardee AB, Jacob F & Monod J. *The genetic control and cytoplasmic expression of 'inducibility' in the synthesis of β-galactosidase by* E. coli. Journal of Molecular Biology: 1, 165–178, 1959.

Palsson BØ. *Systems biology.* Cambridge University Press, Cambridge, MA, 2006.

Papin JA, Reed JL & Palsson BO. *Hierarchical thinking in network biology: the unbiased modularization of biochemical networks.* Trends in Biochemical Sciences: 29, 641–647, 2004.

Rohwer JM, Schuster S & Westerhoff HV. *How to recognize monofunctional units in a metabolic system.* Journal of Theoretical Biology: 179, 213–228, 1996.

Rosenberg A. *The structure of biological science.* Cambridge University Press, Cambridge, 1985.

Rosenblueth A, Wiener N & Bigelow J. *Behavior, purpose and teleology.* Philosophy of Science: 10, 18–24, 1943.

Rumelhart DE, McClelland JL & the PDP Research Group. *Parallel distributed processing: explorations in the microstructure of cognition* Volume 1*: Foundations.* MIT Press, Cambridge, MA, 1986.

Schaffner KF. *Genes, behavior, and developmental emergentism: one process, indivisible?* Philosophy of Science: 65, 209–252, 1998.

Simon HA. *The sciences of the artificial.* MIT Press, Cambridge, MA, 3rd edition 1996 (1st edition1969).

Simon HA. *Models of bounded rationality and other topics in economics: collected papers* (2 Vols.). MIT Press Cambridge, MA, 1982.

Snoddy J, Schmoyer D, Fischer K, Chen G-L, Land M, Petrov S, Martin S, Michaud E, Barry B, Rinchik G, Hoyt P, Doktycz M & Uberbacher E 2000. *Information systems to support experimental and computational research into complex biological systems and functional genomics: several pilot projects.* Research Abstract #60, DOE Human Genome Program Contractor-Grantee Workshop VIII, Santa Fe, NM, 2000. http://www.ornl.gov/sci/techresources/Human_Genome/publicat/00santa/60.html.

Sober E, ed. *Conceptual issues in evolutionary biology.* 3rd edition. The MIT Press, Cambridge, MA, 2006.

Spudich JL & Koshland DE Jr. *Quantitation of the sensory response in bacterial chemotaxis.* Proceedings of the National Academy of Sciences of the USA: 72, 710–713, 1975.

Stelling J, Gilles ED & Doyle FJ III. *Robustness properties of circadian clock architectures.* Proceedings of the National Academy of Sciences of the USA: 101, 13210–13215, 2004.

Stark L. *Stability, oscillations, and noise in the human pupil servomechanism.* Proceedings of the IRE: 47, 1925–1939, 1959.

Stephan A. *Emergentism, irreducibility, and downward causation*, Grazer Philosophische Studien: 65, 77–93, 2002.

Thalos M. *Non-reductive physics.* Synthese: 149, 133–178, 2006.

Thiele I, Price ND, Vo TD & Palsson PO. *Candidate metabolic network states in human mitochondria.* The Journal of Biological Chemistry: 280, 11683–11695, 2005.

Tissières A, Mitchell HK & Tracy UM. *Protein synthesis in salivary glands of* D. melanogaster. *Relation to chromosome puffs*. Journal of Molecular Biology: 84, 389–398, 1974.

Varjú D. *Nervöse Wechselwirkung in der pupillomotorischen Bahn des Menschen* I and II. Kybernetik: 3, 203–226, 1967.

Varjú D. *Human pupil dynamics*. In: Processing of optical data by organisms and machines. (Ed.: Reichardt W), Academic Press, New York, 1969.

Venter JC, et al. *The sequence of the human genome*. Science: 291, 1304–1350, 2001.

Weber M. *Philosophy of experimental biology*. Cambridge University Press, Cambridge, 2005.

Westerhoff HV & Palsson BO. *The evolution of molecular biology into systems biology*. Nature Biotechnology: 22, 1249–1252, 2004.

Wiener N. *Cybernetics: or control and communication in the animal and the machine*. Wiley, New York, 1948.

Williams RJP. *Emergent properties of biological molecules and cells*. In: Reductionism in the biomedial sciences (Eds.: Van Regenmortel MHV & Hull DL), Wiley, Chichester, 15–34, 2002.

Wimsatt WC. *Robustness, reliability, and overdetermination*. In: Scientific Inquiry and the social sciences. (Eds.: Brewer MB & Collins BE), Jossey-Bass, San Francisco, 124–163, 1981.

Wimsatt WC. *False models as means to truer theories*. In: Neutral models in biology. (Eds.: Nitecki MH & Hoffman A), Oxford University Press, Oxford, 23–55, 1987.

Wimsatt WC. *Re-Engineering philosophy for limited beings: piecewise approximations to reality*. Harvard University Press, Cambridge, MA, in press.

Winfree AT. *Biological rhythms and the behavior of populations of coupled oscillators*. Journal of Theoretical Biology: 16, 15–42, 1967.

Winfree AT. *Some principles and paradoxes about the phase control of biological oscillations*. Journal of Interdisciplinary Cycle Research: 8, 1–14, 1977.

Wouters A. *The function debate in philosophy*. Acta Biotheoretica: 53, 123–151, 2005.

SECTION IV

Organization in biological systems

10

The biochemical factory that autonomously fabricates itself: A systems biological view of the living cell

Jan-Hendrik S. Hofmeyr

SUMMARY

The essence of life must lie somewhere between molecule and autonomously living unicellular organism. Modern biology generally views organisms as beads along the necklace of lineage; it attempts to explain life from an evolutionary viewpoint, with reproduction (of cells) and replication (of DNA) as defining phenomena. Systems biology, however, studies each bead per se as an autonomous entity. I suggest that, for systems biology, the defining difference between a living organism and any nonliving object should be that an organism is a system of material components that are organised in such a way that the system can autonomously and continuously fabricate itself, i.e. it can live longer than the lifetimes of all its individual components. Systems biology, therefore, goes beyond the properties of individual biomolecules, taking seriously their organisation into a living whole.

The concept of autonomous self-fabrication of systems is of course not new; it has a distinguished history. Although Maturana and Varela's concept of autopoietic systems is perhaps most prominent in this history, I find that for the purpose of formalisation it is less useful than either Rosen's theory of replicative metabolism-repair systems or Von Neumann's theory of self-reproducing automata based on the concept of a universal constructor. Rosen in particular has shown, using category theory, how to describe such organisations in terms of relational models, although he never realised his metabolism-repair systems

Systems Biology
F.C. Boogerd, F.J. Bruggeman, J.-H.S. Hofmeyr and H.V. Westerhoff (Editors)
Isbn: 978-0-444-52085-2

in terms of biochemistry as we know it. I shall show how it is possible to combine these two strands of thought into a relational model that commutes with our current knowledge of cellular biochemical processes. This model, which I call a metabolism-construction-assembly system, also makes explicit the role of information, and identifies unassisted self-assembly as the process that ultimately makes the system self-fabricating (or, using Rosen's words: 'closes the system with respect to efficient causation'). What makes this model even more interesting is that it is consistent with Barbieri's ribotype theory, and, through that, with the body of thought known as biosemiotics.

1. HOW TO BE A SYSTEMS BIOLOGIST

The aim of this book is to explore the possibility that systems biology may need philosophical foundations of its own. I believe it does, and that systems biology should aim to provide a way of thinking about living organisms that will allow us to understand them as autonomous entities. At present the dominant views of biology are through the glasses of evolutionary biology and of molecular biology. Evolutionary biology seeks to understand life in terms of how natural selection has moulded organisms through the millennia. The philosophy of molecular biology is based on the idea that exhaustive knowledge of all the individual molecular components of the cell will afford the best understanding of life. I shall make a case for a philosophy of systems biology that is based on the premise that the living state exists because of a particular organisation of the internal components of cells.

What does systems biology actually entail in practice (Westerhoff & Hofmeyr, 2005)? If one listens to talks at conferences on systems biology (Cornish-Bowden, 2005) or reads editorial introductions to special journal issues devoted to systems biology (Russel & Superti-Furga, 2005), it becomes clear that although many individual scientists who regard themselves as systems biologists have very clear views on what they regard as the gist of their discipline, there is no clear consensus. It also does not help that the phrase 'systems biology' in a grant proposal has become, or is at least perceived as, a means for ensuring funds. Be it as it may, the different views are all compatible and can be consolidated into something along the lines of 'explaining or understanding the emergence of systemic functional properties of the living cell as a result of the interactions of its components'. How this is to be achieved is usually seen to be by means of two approaches, either on their own or in combination (Chapter 2). Both approaches espouse what I would call the 'system-wide' view: the conviction that one cannot understand the cell if one does not consider it as a whole. One approach comes from the age of 'omics' and proposes that, now that the new high-throughput techniques have made it possible, we should measure the amount of everything there is to be measured inside a cell under different conditions (DNA, RNA, proteins, metabolites), and then 'data-mining' will do the rest. The other approach has as its aim the 'silicon cell' (Snoep et al., 2005), a computer simulation of the complete cellular network of reactions and interactions based on the experimentally measured properties of the 'agents' of the cell (enzymes, pumps, receptors, etc.). What is not always so clear is exactly how the results of these two approaches – exhaustive cell-wide data-sets and complete cell models – necessarily lead to deeper understanding. In the rest of this introduction, I would like to propose that what needs to be added first is a clear view of what the 'systemic approach' entails.

The geneticist Theodosius Dobzhansky summed up the dominant explanatory modality for biology of the last century in the mantra 'Nothing in biology makes sense except in the light of evolution' (Dobzhansky, 1973). Ernst Mayr, however, pointed out that there are two types of explanation in biology, ultimate and proximate, which would respectively follow from evolutionary and functional considerations (Mayr, 1988). I suggest that systems biology seeks to expose general principles that underlie proximate explanations of what governs life. My mantra for systems biology would therefore be 'Nothing in an organism makes sense except in the light of context'. There are three words here that need elaboration. First, 'sense' emphasises that what is sought is explanation and understanding, not just description. As an example, an indispensable part of the system-wide study of the cell is to make a complete map of all reactions and interactions that comprise the intracellular network. However, in the words of Count Alfred Korzybski (1994) 'a map is not the territory'; making the map does not in itself afford understanding; neither does measuring the concentrations of all the nodes on the cellular map. Second, 'organism' emphasises that systems biology studies a particular cell or organism as a material system that is to be explained in terms of itself and its interactions with its environment; in contrast, an evolutionary explanation would be in terms of its history. Third, and most important, 'context' captures the essence of the systems approach. Always taking context into account amounts to using a 'macroscope' (de Rosnay, 1979), a tool for studying the very complex (in contrast to using a microscope for the very small and a telescope for the very large).[1] The macroscope is a 'symbolic instrument' that collects a number of techniques and methods into what De Rosnay calls the 'systemic approach', which, in contrast to the analytical approach, takes into account not only all the elements in the system under study but also all their interactions. A 'system' itself is, in de Rosnay's words 'a set of interacting elements that form an integrated whole'. The living cell is prototypic of such systems.

[1] A web-edition of this book is freely available from Principia Cybernetica (http://pespmc1.vub.ac.be/LIBRARY.html). It is recommended reading for all systems biologists.

However, most discussions of the systemic approach remain on a metalevel, and are, for a practising biologist such as myself, ultimately unsatisfactory. I would rather prefer to give an example, first to explain what I think it means to be a systems biologist, and second to demonstrate how taking context into account by using the macroscope has helped me understand deeply and in a radically new way some functional properties of the cell that were considered to have been explained long ago. What I also intend showing is that it is not always necessary to take the whole system into account to understand something; doing systems biology does not necessarily entail doing 'system-wide' biology (and vice versa).

Consider a metabolic enzyme, the systems biologist's favourite molecule (the enzyme marked 1 in Fig. 1a). Consider further a particular kinetic property of this enzyme, say K_p, the Michaelis constant for the product of the enzyme-catalysed reaction, which is an indication of the strength with which the product binds to the active site of the enzyme (the smaller its value, the stronger the binding). This enzyme parameter has a specific value that can be measured experimentally. How can we explain why it has one particular value and not another? There are a number of answers to this question. First, a classical 'enzymological' answer would explain the value as a function of experimentally determined rate constants for the elementary steps in the enzyme mechanism. A second, more modern, 'structural' answer would be based on the three-dimensional conformation of the active site and the complementary structure of the product molecule, which could serve as a point of departure for *ab initio* calculations of the value of the rate constants for binding and dissociation. This type of computational enzymology is fast becoming a reality; it uses the physical and chemical principles of statistical mechanics and quantum mechanics, and they are implemented in computational form using techniques from computational chemistry (Cunningham & Bash, 1997). These two approaches are both reductionistic in that they reduce a property to more elementary properties, here either kinetic or structural; in essence they use the microscope, zooming into the system which here is the

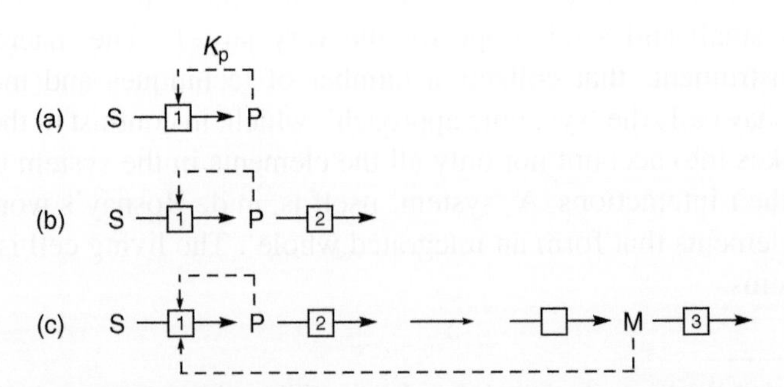

Figure 1 The functional context of an enzyme that converts a substrate S to a product P.

isolated enzyme–product complex. The third answer from molecular biology shows no particular interest in explaining the value K_p, but would rather explain the enzyme as a gene product. A fourth 'historical' answer would explain K_p as an evolved enzyme property that can be followed down the phylogenetic tree and across generations; this is essentially a telescopic approach – but instead of looking far out, it looks far back.

Although all of the above approaches are perfectly valid and necessary, I would argue that none of them truly explain why the enzyme has that particular value for its K_p. The only way to find that out is to study the enzyme in the context of the metabolic reaction network in the cell. Let us therefore use the macroscope to zoom out, taking into account the immediate functional context of the enzyme. We find that the product of this enzyme is a substrate for another enzyme (the one marked 2 in Fig. 1b) and therefore couples the two enzymatically catalysed reactions. Note that context here does not mean spatial environment, but rather network environment. Now the question of why K_p has a particular value takes on a whole new meaning. Metabolic control analysis (Kacser & Burns, 1973; Heinrich & Rapoport, 1974) teaches us that the role that an enzyme plays in controlling steady-state flux and concentrations is determined by the elasticity coefficients of the enzymes that catalyse the reaction network (an elasticity coefficient describes the sensitivity of reaction rate with respect to a chemical species that directly affects the enzyme activity, such as a substrate, product or modifier; mathematically it is the fractional change in reaction rate divided by the fractional change in concentration of the chemical species in question). If, for example, our enzyme reaction is far from equilibrium in the forward direction (so that the mass-action ratio is much smaller than the equilibrium constant) and the K_p has a large value relative to the steady-state concentration of P, so ensuring that the ratio of product concentration p to K_p is very small, the elasticity coefficient with respect to P approaches zero: under these conditions the enzyme is kinetically and thermodynamically insensitive to anything that happens downstream from that enzyme (assuming that the coupling product is the only way through which the downstream reactions can communicate with enzyme 1). Under these conditions the enzyme has complete control over the steady-state flux.[2] The smaller K_p, the more sensitive the enzyme becomes to changes in its product concentrations, and the less its controls the flux.

Now imagine that lower down in the pathway there is a metabolite M that saturates the enzyme for which it is a substrate (enzyme 3 in Fig. 1c) so that the enzyme is insensitive to changes in the concentration of M. In addition, M also feeds back allosterically onto enzyme 1 higher up in the pathway. Hofmeyr & Cornish-Bowden (2000) have shown that in this situation enzyme 1 has no

[2] This explanation is simplified for the purposes of this discussion; it ignores the elasticity coefficients of the other steps in the system which all play a role in determining the control profile.

control over the overall flux through the full system, but completely determines the degree of homeostasis of M. However, this can only happen if enzyme 1 retains flux control over that part of the system that leads up to M (the supply pathway for M). As before this is partly determined by the ratio p/K_p, which must be small. When K_p becomes smaller and p/K_p increases concomitantly, the system becomes structurally unstable and exhibits multistationarity (Hofmeyr et al., 2000). This behaviour only obtains when enzyme 1 binds S and P cooperatively, which is the norm for allosteric enzymes, and it is independent of the specific mechanism (Hofmeyr & Cornish-Bowden, 1997). It is therefore conceivable that evolution has selected for large values of K_p in order to avoid this type of pathological behaviour. However, most kinetic studies of allosteric enzymes have ignored the K_p and it may turn out that in some cases K_p is small, which could cause the system to exhibit switching behaviour instead of a smooth response to changes in the concentration of M. Whatever the case may be, it is clear that one can only understand why K_p has the value that it has by analysing it in its functional context, which need not be that of the whole cell. Having now made and illustrated the claim that nothing in an organism makes sense except in the light of context, it is time to consider the nature of the overall context that the living cell provides for systems biology. This amounts to asking of systems biology how it defines life.

2. THE SELF-FABRICATING CELL: A CONTEXT FOR SYSTEMS BIOLOGY

> Ironically, biology itself provides a ground upon which epistemology and ontology directly meet. Put simply, organisms are themselves fabricators; they build new things, they make new things, they deploy new things. Hence, an essential part of a theory of organism is precisely a theory of fabrication; a theory of invention and deployment. Thus, a theory of organisms has within itself an ineluctable ontological component; a science of fabrication. Nothing shows more clearly than this the unique character of biology among the sciences, and the unique role that its own theory must play in its own application.
>
> (Robert Rosen, *On Theory in Biology*[3])

Biologists, more than ever before, are living in a golden age. The cell, that unit on which all life is based, no longer seems a mystery; in fact, we apparently feel we know and understand it so well and have such advanced technology that we can manipulate life at the molecular level confidently and responsibly. However, keeping in mind E.F. Schumacher's admonition that 'the greatest

---

[3] http://www.rosen-enterprises.com/RobertRosen/BioTheoryHistoryofBiology.html

danger invariably arises from the ruthless application, on a vast scale, of partial knowledge' (Schumacher, 1973), should we not be asking ourselves seriously whether we can we really explain life? How successfully can we at present answer the questions 'why *Escherichia coli*?', 'why *Homo sapiens*?', 'why any organism?

Organisms as we know them are material systems, and according to Aristotle (1998, 350 B.C.E) there are four different ways of answering 'why'-questions about material objects, questions that he placed at the heart of science. Put differently, there are four fundamentally different explanatory factors that together explain any object fully.[4] These four *aitia*, as he called them, are now commonly described as material, efficient, formal and final causes. However, to avoid confounding Aristotelian explanations with 'causation' in the sense of Hume, Cohen's suggestion[5] to replace the noun 'cause' with the verb 'make' is useful:

(1) What is an organism made out of? (its material cause)
(2) What makes (in the sense of '*what is it to be*') an organism? (its formal cause)
(3) What makes (in the sense of '*what produces*') an organism? (its efficient cause)
(4) What is an organism made for? What is its purpose or function? (its final cause)

Biochemistry, molecular biology and molecular genetics have been spectacularly successful in providing us with answers to the first two questions: (i) a century's worth of research tells what organisms are composed of and what the structure of their molecular constituents are, and (ii) after Watson and Crick biologists generally ascribe, rightly or wrongly, the essence of an organism to its DNA. These two answers explain life statically in terms of matter and form, and seem, for many, to suffice.

However, Aristotle insisted that all four explanations are needed for full understanding. The other two questions are questions of process and transformation; they explain why change occurs and lead to dynamic explanations. Currently, biology's answer to the third question of what produces an organism is essentially: 'its parent(s)'. Rudolf Virchow famously summed up this view as *cellula e cellula* (every cell from a pre-existing cell), a phrase actually coined by

---

[4] In contrast with Humean doctrine in which effects and their causes are *events*, Aristotle typically considered the causes of *substances* or *objects*; this approach is particularly applicable to artifacts, whether artificial or natural. Living organisms are the ultimate natural artifacts (Barbieri, 2005).

[5] Lecture on the four causes (http://faculty.washington.edu/smcohen/320/4causes.htm). To quote Cohen: 'Aristotle's point may be put this way: if we ask 'what makes something so-and-so?' we can give four very different sorts of answer - each appropriate to a different sense of 'makes' '.

François-Vincent Raspail, another founder of cell theory.[6] This is of course also the point of departure for the evolutionary view of organisms as beads along the necklace of lineage. The rest of this chapter will argue that there is another, and for systems biology more productive, answer to this question.

The fourth explanation of purpose is generally considered, especially by those of a mechanistic bent, to be outside the realms of science. Contemporary biology has nothing to offer on the question of the final cause of an organism. However, I am of the opinion that pondering precisely this question will lay a path to a philosophy of systems biology. In fact, for organisms it turns out that the answers to the last two questions are one and the same.

Stafford Beer, cyberneticist and systems thinker, said that 'the purpose of a system is what it does' (an idea now entrenched in the acronym POSIWID).[7] If one asks 'what does an organism do?' the usual reply of biologists since Darwin has been 'an organism evolves through natural selection'; related replies are 'an organism reproduces', or, post-Dawkins (1989), 'an organism replicates its genes'. The view of life that leads to these answers is perhaps most clearly enunciated in Dobzhansky's mantra mentioned in the introductory section. However, since the 1960s another answer to the question 'what does an organism do?' has been given with increasing frequency: 'An organism produces itself', by which is meant that organisms constantly and autonomously rebuild or fabricate themselves during their own lifetimes. In the words of Humberto Maturana and Francisco Varela (Maturana & Varela, 1980) organisms are autopoietic.[8] It is probably fair to say that, together, the 'evolutionary' and the 'autopoietic' answer, either on their own or together, form the basis for most current definitions of life (Ruiz-Mirazo & Moreno, 2004). Note also the convergence of causes, alluded to in the previous paragraph, in this concept of self-fabrication: an organism is its own efficient cause in that it autonomously fabricates itself; but then, the purpose of an organism is to fabricate itself – it is its own final cause.

Although the term autopoiesis is associated with Maturana and Varela, the concept of self-fabrication has a long and venerable history, and seems to have been first formulated explicitly by Immanuel Kant, who conceived of organisms as dynamic, functional wholes in which all components are made by and for

[6] http://en.wikipedia.org/wiki/Rudolf_Virchow

[7] Beer said this many times, but never more forcefully than in his address to the University of Valladolid, Spain in October 2001, a month after 11 September: 'According to the cybernetician, the purpose of a system is what it does. This is a basic dictum. It stands for a bald fact, which makes a better starting point in seeking understanding than the familiar attributions of good intentions, prejudices about expectations, moral judgments, or sheer ignorance of circumstances'. (Beer, 2002).

[8] The term 'fabricate' will be used throughout instead of 'build' or 'make' or 'produce', the last being all to often confused with 'reproduce'. 'Autopoiesis' lacks a verb-form, whereas 'fabrication' has one: a system that fabricates itself is self-fabricating.

each other, in contrast with a machine in which components exist only for each other but cannot make each other:

> In such a product of nature *every part not only exists by means of the other parts*, but is thought as existing for the sake of the others and the whole, that is as an (organic) instrument. Thus, however, it might be an artificial instrument, and so might be represented only as a purpose that is possible in general; but also *its parts are all organs reciprocally producing each other*. This can never be the case with artificial instruments, but only with nature which supplies all the material for instruments (even for those of art). Only a product of such a kind can be called a natural purpose, and this because it is an organised and self-organising being [my italics].
>
> (Kant, 1914, §65)

The philosopher of science George Kampis (1991, p. 345) recently put it this way:

> In a component system [a type of system defined by Kampis which includes living organisms], due to the continual turnover that gives rise to the components and then removes them from the system, no component and no higher structure, organised form of the components, can persist, unless produced and renewed over and over again.

I shall take this view of the cell as the foundation on which systems biology must be built. For this to be possible, we must have a formal, abstract language with which to describe the functional organisations that would make autonomous self-fabrication possible. To my knowledge, only two such formalisations have been developed. In the late 1950s, more than a decade before Maturana and Varela invented the term autopoiesis, the theoretical biologist Robert Rosen put forward a formalised treatment in terms of category theory of what he called metabolism–repair or (M, R)-systems, which become self-fabricating when supplemented with a mechanism that he called 'replication' (an unfortunate choice, as it turns out, because it does not agree with modern biology's use of the term) (Rosen, 1958a,b, 1959b 1972) (for a recent review and exploration of (M, R)-systems see Rosen (1991); Letelier et al. (2006)). Later on he would summarise the central property of such systems as being 'closed to efficient causation' (Rosen, 1991). At the same time John von Neumann (1966) was developing his theory of self-replicating automata which centred around the concept of a universal constructor. In the rest of this chapter, I shall provide a rather informal version of Rosen's formal language, and then use it to show how the main tenets of these two theories can be merged and mapped onto cell biochemistry. However, we first need to explore the central concept of autonomy, because it goes hand in hand with the idea of self-fabrication.

3. AUTONOMY OF MATERIAL SYSTEMS: THE NEED FOR SPECIFIC CATALYSIS

A logic of life, at least of earth-bound life as we know it, can be deduced from two basic postulates:

Postulate 1. Living organisms are material objects.
Postulate 2. Living organisms are autonomous.

The first postulate commits us to a view of life that is inextricably linked to chemistry: the science of spontaneous transformation of matter and therefore the science of creativity and what Stuart Kauffman (2000) calls the 'adjacent possible'. The creative nature of chemistry is captured in the concept of 'component systems' (Kampis, 1991). Whether a chemical transformation will actually occur under specified conditions, and if it does, how fast, is answered from thermodynamic and kinetic considerations. An important generalisation of chemical biology is that covalent chemistry is virtually exclusively enzyme-catalysed, whereas the noncovalent chemistry involved in, for instance, chemical recognition, protein folding and self-assembly of macromolecular complexes is largely uncatalysed (although we now know that at least folding is often assisted by chaperones). This distinction between molecular (covalent) and supramolecular (noncovalent) chemistry, made by Jean-Marie Lehn (1995), will be seen further on to be crucial in understanding the ability of living cells to fabricate themselves. Supramolecular chemistry refers to the formation of ordered molecular aggregates that are held together by noncovalent binding interactions. Because these forces are weak, the formation of supramolecular assemblies is usually thermodynamically controlled and therefore a spontaneous process of self-assembly rather than a sequential bond-forming synthesis.

A recent series of papers (Ruiz-Mirazo & Moreno, 2004, 1998, Ruiz-Mirazo et al., 1998, 2004) provide an excellent analysis of the concept of autonomy, not only as a point of departure for a universal definition of life, but also in relation to autopoietic theory (see also Chapter 11). They make a strong and convincing argument that the concept of autonomy is multifaceted; living systems exemplify all these facets, whereas the autopoietic perspective only considers an abstract organisation that recursively produces itself; real-world autonomy cannot escape the requirements of chemistry, energetics and kinetics, and the necessity for spatial autonomy by self-bounding.[9]

Living systems are open and can never be fully thermodynamically autonomous; as dissipative structures they depend on an externally determined

---

[9] However, Ruiz-Mirazo and Moreno emphasise the thermodynamic aspects of autonomy, and virtually ignore the kinetic aspects, which, in my opinion, are just as, if not more, important.

Gibbs energy gradient (Nicolis & Prigogine, 1977). However, living systems also create internal non-equilibrium conditions that allow them a degree of thermodynamic autonomy. As an example, consider a chemotrophic bacterium that not only grows on glucose by fermenting or oxidising it, but also stores glucose as glycogen. If the external glucose is depleted, i.e. if the external Gibbs energy gradient collapses, this bacterium will still be able to survive due to the internal nonequilibrium condition that it has created. As long as its glycogen store lasts it is thermodynamically autonomous with regard to its carbon source. There is, therefore, a difference between a dissipative system in which a certain range of external conditions create and maintain the system (so that if outside this range the dissipative system no longer exists), and an autonomous dissipative system that also actively creates and maintains internal nonequilibrium conditions. A Bénard cell would be an example of the first type, a living cell an example of the latter.

To be kinetically autonomous, the chemical reactions that comprise the system must operate on a faster timescale than the rest of the underlying network of spontaneous mass-action chemical transformations; the greater the separation on the timescale, the smaller the effects of these spontaneous side-reactions and the greater the degree of kinetic autonomy. This can only be achieved by catalysts that are specific with regard to both reactants/products and reaction; kinetic autonomy therefore absolutely requires the existence of catalysts that specifically recognise their substrates and transform them into specific products. If such catalysts are themselves short-lived, the autonomous system must be able to replace them. In short, such a system must itself also be a catalyst factory. However, to fabricate molecular catalysts requires both building blocks and additional machinery, which itself must be made within the system. The building blocks can of course be supplied by the environment, but even if the system has to fabricate them this is not a problem; all it needs is to be able to make the specific catalysts that will accomplish the synthesis. However, the machinery that constructs the catalysts must itself be replaceable by the system, lest it fails; this implies even more additional machinery. It is clearly here that the linear hierarchy of efficient causes followed up to now seems to wander off into an infinite regress that is incompatible with the existence of real autonomous systems. In some way, this hierarchy of efficient causation must fold back into itself, must close, must become circular.[10]

The possibility, mentioned above, for internal creation and maintenance of nonequilibrium conditions and their dynamic adaptation in the face of

[10] The kinetic autonomy that is ensured by specific catalysts is essentially what is lacking from Tibor Ganti's chemoton (Gánti, 2003). There is nothing in a chemoton that would prevent its chemical intermediates dissipating into side-reactions. Much closer to the kinetic autonomy of living systems is that of the autocatalytic Belousov-Zhabotinsky reaction system (Fiel & Burger, 1985) in which the catalytic species are produced within the system itself.

a continually changing internal and external context also depends on catalysts, in this instance on their capacity for being regulated. Hofmeyr & Cornish-Bowden (2000) developed their theory of metabolic supply–demand to describe this aspect of autonomy.

The above logical analysis of the consequences of materiality and autonomy has led inexorably to the need for specific catalysts that are functionally organised in such a way that they form a closed loop of efficient causation. The rest of this chapter explores what this type of functional organisation may look like and how it is realised in living cells as we know them. First, however, we need a way of formally representing a self-fabricating system as an organisation of catalytic components.

4. FABRICATION AND THE LOGIC OF LIFE

What is fabrication? Are there basic principles underlying fabrication? Must a fabricator be more 'complicated' than that which it fabricates? Can a fabricator fabricate itself? One would suppose that by now there would have been developed, either by engineers, technologists or anybody that designs or makes gadgets, a full-fledged theory of fabrication that answers such basic questions. The theory of self-replicating automata developed by von Neumann (1966, 1951) goes some way towards answering these questions, but other than that I have not been able to find a theory of fabrication.[11]

Seen abstractly, fabrication is a process in which a material object is created either by rearrangement of, or by taking away from, or by adding to an existing object. Usually, one assumes that this process is accomplished by a fabricator, which is itself a material object (and, of course, not necessarily alive). However, one has to leave open the possibility that the fabrication process happens spontaneously without assistance from a fabricator. In keeping with my background as (bio)chemist, I take my cue from chemistry (the epitome of a fabrication world). Consider A, B, C and P to be either (i) single molecules in which all the atoms are bonded covalently or (ii) assemblies of molecules that associate through noncovalent forces (ionic and hydrogen bonds, Van der Waals forces, hydrophobic interactions, etc.). One could consider as an example of the first a single polypeptide and of the second an enzyme consisting of noncovalently associated subunits, each consisting of a single polypeptide. A new molecule can form from existing molecule(s) in a number of ways, shown in Fig. 2. Because it is all too easy to forget that physical laws such as the conservation of mass underlie all fabrication processes I depict them using both symbols and schematic

[11] Besides biology, nano-engineering is also a field from which such a theory could arise (Drexler, 1992; Freitas Jr & Merkle, 2004).

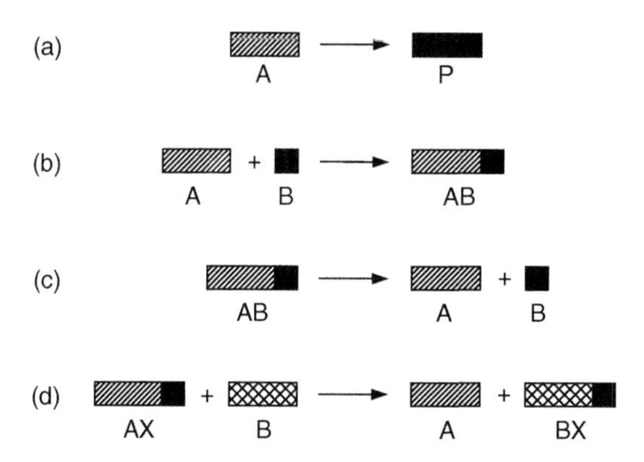

Figure 2 Basic fabrication processes.

(a) The atoms of A are rearranged into a new configuration P. A and P have the same atomic composition and are either structural or conformational isomers of each other; (b) Synthesis of a new molecule AB from A and B, or an association of A and B to form a non-covalently bound complex AB; (c) Degradation of AB to two fragments A and B, or the dissociation of a complex AB into components A and B; (d) The synthesis of two molecules A and BX through the transfer of a part of a donor molecule AX to acceptor molecule B.

representations. For example, I could have written the process $\mathrm{A+B \rightarrow AB}$ as $\mathrm{A+B \rightarrow C}$, but that obscures the fact that C must contain exactly the atoms of A and B.

In general, therefore, I consider both the input and output to a fabrication process to be a material object, which can be considered a unit with a fixed internal arrangement of components ('atoms'). The fabrication process itself involves either an internal rearrangement within an input object, the combining of objects, the splitting of an object, or the transfer of part of one object to another. Following Rosen (1991), the fabrication process can in the abstract be regarded as a mapping f from a domain (a set A of input objects) to a codomain (a set B of output objects). Such a mapping is usually depicted as

$$f: A \rightarrow B \quad \text{or, equivalently,} \quad A \xrightarrow{f} B \tag{1}$$

Any specific conversion of $a \in A$ to $b \in B$ can be depicted with the 'mapsto' notation

$$a \mapsto f(a) \quad \text{or, equivalently,} \quad a \mapsto b \tag{2}$$

This mapping is the fundamental relationship on which Rosen (1991) builds his relational theory of biology.

Whereas, the nature of A and B is reasonably clear, that of f is not. Is it just a process or is it itself a physical object? We shall see below that f can be either.

Rosen (1991) provided the mapping in Eqn. (1) with a natural interpretation in terms of Aristotelean causes: the effect B has material cause A and efficient cause f. In this particular system A and f have only final cause, namely B; the function of A is to serve as material from which B is made, while the function of f is to fabricate B.

What about formal cause? In the above mapping $f: A \longrightarrow B$ there is nothing that can explicitly be interpreted as formal cause. Here, we would have to assume that formal and efficient cause are inseparably part of f (think of sculptor f carving a sculpture according to a vision which exists in her mind only). However, there are clearly situations where formal cause is, at least partly, associated with a separate object (think of an electronic engineer building a circuit board according a design on paper, or a polypeptide being synthesised according to the nucleotide sequence in a particular mRNA).

To account for objects that serve as formal causes of, for instance, macromolecular synthesis, the mapping in Eqn. (1) clearly needs an additional entity. Rosen (1989) suggested the more general formulation

$$\begin{aligned} f &: A \times I \to B \\ (a, i) &\mapsto b = f(a, i) \end{aligned} \tag{3}$$

where I is a set of templates or blueprints. In this formulation f is the efficient and I the formal cause, although the separation need not be absolute; part of the formal cause can remain associated with f itself.

There are two problems with this formulation. The first is that it leads to a logical paradox when an $i \in I$ is the blueprint for f itself, in the sense that f is an element of its own range (Rosen, 1959a, 1962). No mapping can be defined before its domain and range are stipulated; however, if the range contains the mapping itself as an element, it cannot be stipulated before the mapping is given. Thus, in the words of Rosen (1959a), 'neither the mapping f nor its range can be specified until the other is given'.

The second problem is that I appears in the mapping with the same status as A, namely as a material cause. However, the role of I is purely informational; logically, any particular $i \in I$ should be associated with f as the pair (f, i). We should rather consider (f, i) as the efficient cause in which the formal part has been made explicit: f is an agent acting on the information contained in i. In these terms the mapping would be

$$\begin{aligned} (f, i) &: A \to B \\ a &\mapsto b = (f, i)(a) \end{aligned} \tag{4}$$

with (f, i) is an element of the Cartesian product $f \times I$.

Readers interested in how these mappings can be formally composed (combined) into fabrication networks are referred to Rosen (1991). In the following section, I use a more informal approach to develop an understanding of what it would entail for a collection of material components to become self-fabricating. As mentioned above, the interpretation of objects in the diagrams in terms of Aristotelean causes is also due to Rosen.

5. HOW TO CONSTRUCT A SELF-FABRICATING FACTORY

Just as Shakespeare surely found it more profitable to compare his love to a summer's day than to a rock or a whale, we, in order to understand the nature and logical requirements of a self-fabricating system, need to find a useful image to compare it with. From the long line of machine metaphors that have since Descartes been used to describe organisms – through hydraulic automata, clockworks, steam engines, servomechanisms and computers to the vending machine (stick in a gene, pull out a product) – the image of a chemical factory is most useful for the purpose. It embodies the essence of a system that not only consists of fabricators, but, as a whole, is also a fabricator, though, in the case of all factories thus far made by man, not of self.

Let us, therefore, begin by considering a man-made factory as an generalised example of a fabricator. A bird's-eye view of this factory could be Fig. 3a: the factory L is a black box that transforms raw materials P (its input) into products Q (its output). P and Q can be single objects or collections of objects. Fig. 3a) is also a graphical equivalent of the mappings described in Eqn. (1). Q can of course be used as input for another 'downstream' factory M that transforms it to R (Fig. 3b). Zooming our view out even further, L and M may even be viewed as a single factory LM that produces R from P. However, instead of zooming out, let us zoom into the details of our factory. Now, instead of representing the factory as such, L could represent an agent or machine (a simple fabricator) inside the factory; L performs the elementary task of transforming an intermediate widget P somewhere in the production process into the next widget Q. In fact, the production process as a whole can be visualised as a network in which simple fabrication processes are linked as in Fig. 3b. It need not be a linear process: there could be branches where one intermediate widget is used as input for two different processes. Branches can converge; cyclically organised processes can occur. The details of such a transformation network are, however,

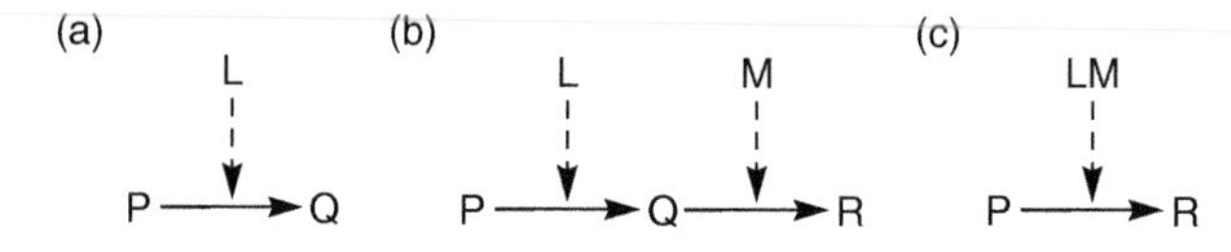

Figure 3 single (a), linked (b) and lumped (c) fabricators.

not important for this discussion. What is important is that, in the picture as painted above, whether it is applied to the factory as a whole or to a machine inside the factory, there is a clear conceptual difference between fabricators on the one hand and their inputs and outputs on the other.

In a computer analogy, P and Q would represent the software which runs on the hardware L and M. However, further on, when our factory becomes more complicated, we shall see that this analogy becomes so ambiguous as to be useless. On the other hand, Aristotelean causal descriptions will prove to be robust. Let us use such a description to explain how the answer to the question 'why L or M?' differs from that to the question 'why P or Q or R?'. For example, the question 'why Q?' is explained by considering Q to be the 'effect' of material cause P and efficient cause L. However, Q can also be considered to have final cause R (the purpose or function of Q is to serve as material cause for R). There is no explicit formal cause for Q in the diagram – it could be considered to be embedded in the properties of L or it could be added to the diagram as information needed by L to fabricate Q. Note that material, efficient and formal cause 'flow forward' to Q, whereas final cause 'flows backwards' to Q. This is always the case. R can be similarly analysed as effect of material cause Q and efficient cause M; unlike Q, R is only effect and plays no functional role within the system. Note that, whereas 'why Q?' and 'why R?' can be answered from within the system (they both have material and efficient causes), 'why P?', 'why L?', and 'why M?' do not have answers within the system. They can only be explained in terms of their final cause: P functions as material cause for Q, while L and M function as efficient causes (for Q and R).

To emphasise that from now on we only consider fabricators inside our factory, i.e. the components that comprise the factory, I switch to different symbols (Fig. 4a). In a perfect world where machines do not deteriorate, the factory will run forever as long as enough input material is available. Consider, however, that the 'hardware' of the factory, i.e. its fabricators C, have a limited

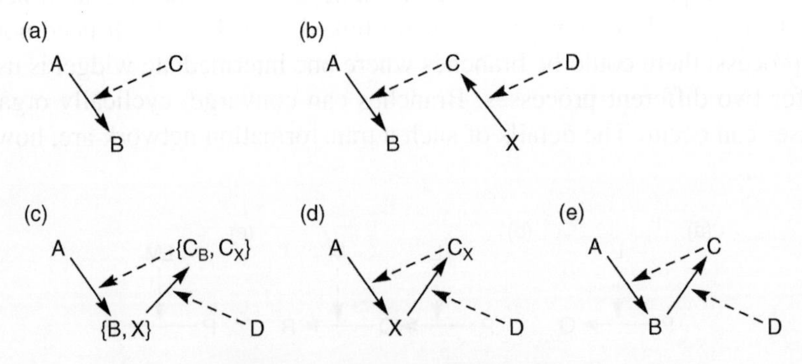

Figure 4 How to build a factory.

lifetime; after a while they malfunction and have to be either repaired or replaced in order for the factory to outlive the lifetimes of its machines. Let C in Fig. 4a malfunction. As shown in Fig. 4b, one possibility for overcoming this problem would be to expand the scope of the factory by acquiring a new fabricator D that builds replacement C from material X (or repairs C using spare parts X). It is clear that we have now started to create a 'fabrication hierarchy' in that C, which acts as fabricator for the lower transformation level, now also is the target of a fabrication process at a higher level. Furthermore, in this expanded description the factory is less autonomous in that it now depends on not only an external supply of A but also of X. What if the supplier of X becomes unreliable? The most effective measure to counter this would be to incorporate additional machinery C_X that can fabricate X from A into the factory, alongside the original machinery that fabricates B (now distinguished as C_B), thereby increasing the degree of autonomy of the factory at the expense of more machinery (Fig. 4c).

However, there is a problem in Fig. 4c. D is required not only to fabricate C_B from X, but apparently also C_X from something. But from what? We could consider new material Y, but in order to become independent of a possibly unreliable supply of Y that would mean incorporating into the factory even more new machinery C_Y to fabricate Y from A. An infinite regress looms. This regress can be sidestepped by ensuring that both C_B and C_X can be fabricated by D from X (in which case Fig. 4c would be a valid diagram). Furthermore, nothing now prevents us from discarding C_B, so forcing the factory to become a producer of primarily C_X instead of B, X now being an intermediate in the process (Fig. 4d). When, further on, we analyse the living cell as we know it in terms of its fabrication hierarchy, we shall see that it has adopted both these strategies. Now that we understand the subtleties involved we shall, in the interest of readability, continue to use the diagram in Fig. 4e.

Just as C, fabricator D is also subject to failure and we can imagine safeguarding the factory (so increasing its relative independence even further) by adding another level to the fabrication hierarchy in which D is fabricated by a new fabricator E from either C (Fig. 5a) or B (Fig. 5b). As before, adding this level to the fabrication hierarchy makes all the lower levels more complicated in that new machines must be fabricated as needed. However, and this is important, the factory still depends on a supply of its input A and the one fabricator, here E, that is not manufactured inside the factory. In fact, as the proportion of endogenously produced fabricators to external fabricator increases

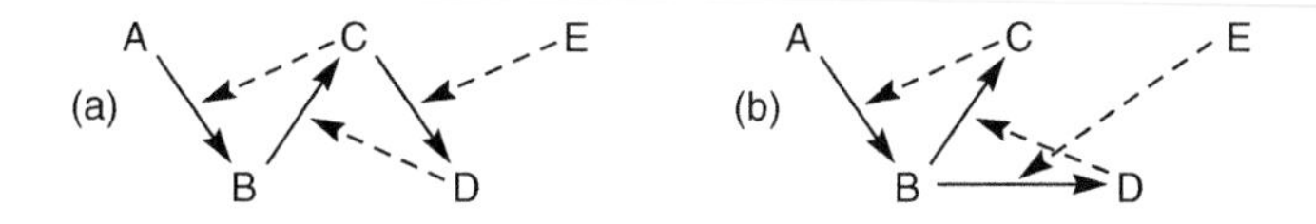

Figure 5 Adding another level to the fabrication hierarchy.

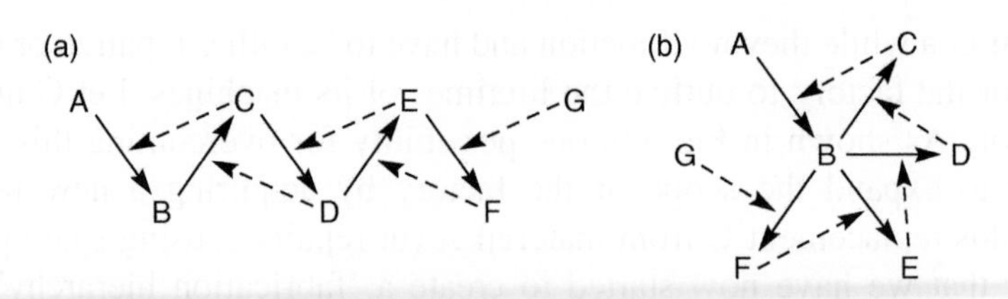

Figure 6 Linear (a) and wheel (b) fabrication hierarchies.

the factory becomes more autonomous. We can easily extend the hierarchies in Fig. 5 indefinitely (Fig. 6 shows examples of such extensions). It is left to the reader to imagine hierarchies that are mixtures of these two motifs.

Starting with Fig. 4e, and even more so with Figs. 5 and 6, the clean differentiation between hardware and software that obtained in Fig. 3 has become obscured. For example, C, which in Fig. 4 was unambiguously the hardware for software A, has itself become in Fig. 5a the software for E, a higher level of hardware. In addition, hardware C is now seemingly fabricated from its own 'software'. Clearly in Fig. 5a the distinction between hardware and software has been lost and it has become meaningless to discuss the factory in these terms. However, Aristotelean language comes to the rescue, as follows: Consider in turn each object in the complicated situation of Fig. 5 a:

- Why A? A functions as material cause for B (A's final cause).
- Why B? B is made from A (its material cause) by C (its efficient cause), and it functions as material cause for C (B's final cause).
- Why C? C is made from B (its material cause) by D (its efficient cause), and it functions as efficient cause for B and as material cause for D (C's two final causes).
- Why D? D is made from C (its material cause) by E (its efficient cause), and it functions as efficient cause for C (D's final cause).
- Why E? E functions as the efficient cause for D (E's final cause).

The beauty of this analysis is that it provides a way of soothing the bugbear of final cause. Consider, for instance, 'why C?' In Fig. 5a, there are ostensibly four answers to this question, but upon closer inspection there are only two: 'because B' and 'because D'; in both cases a final cause of C has become identified with either material or efficient cause of C: on the one hand, C functions as fabricator of B, which itself is the material cause for C; on the other hand, C functions as the material cause for its own fabricator, D. The final causes for C have been absorbed into the system.

Thus far, however, there has always still been one fabricator which cannot be replaced or repaired from within the system; one final cause that has not been internalised. If this fabricator fails, it would cause a domino effect down the hierarchy which eventually would bring the whole factory to a standstill. Clearly the problem cannot be solved by adding extra levels of fabrication. Is it possible to internalise this fabrication process so as to make the system completely autonomous (closed) with respect to fabrication, i.e., self-fabricating? (Note that the factory will always remain open to material cause through its dependence on input A from outside.) Consider the arrangement in Fig. 4e. One conceivable way in which the system can rid itself from the necessity for level E and thereby become self-fabricating is if a fabricator lower down in the fabrication hierarchy is able to manufacture D: In Fig. 7a, C manufactures D from B, while in Fig. 7b B manufactures D from C. Another possibility is that D be able to fabricate itself from either B (Fig. 7c) or C (Fig. 7d).

However, as the following argument shows, there is another incipient regress hidden in our factory: The problem of insufficient numbers of machines. If all the individual steps in all the levels of the fabrication hierarchy are to be performed by a dedicated specific machine, then it is impossible to make the factory self-fabricating using the organisations in Fig. 7. A very simple numerical example demonstrates the problem: Let the set of building blocks B have two members, each fabricated from A in two steps. The set of specific fabricators C therefore needs four members. Let each member of C be made from B in two steps, each step being facilitated by a specific member of the set of D, which therefore needs eight members. If each member of D again needs two specific fabricators then the next level E (Fig. 5) would have contained 16 fabricators. However, in Fig. 7a and b these 16 functions must be performed by members of either B or C. On could conceive of adding extra members to C, but then they would also need to be made from B, which implies that D must be even larger than before, which in turn implies even more C, and so on. The same holds for Fig. 7b, but here we also require the members of B to also be fabricators. The organisations in Fig. 7c and d lead to the same problem. Clearly this option is a logical impossibility. Another way out would be to require members of B or C to become multifunctional. However, even in this simple case it means that

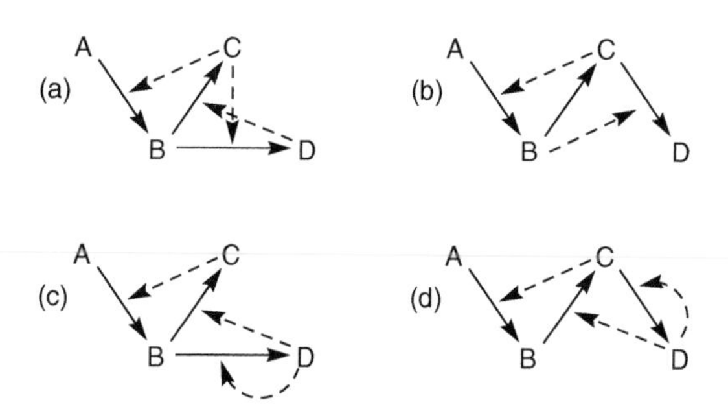

Figure 7 Potential self-fabricating organisations.

the four members of C in Fig. 7a must share 20 functions between them (four to make B and 16 to make D), or that the two members of B in Fig. 7b share 16 functions between them. Whereas this option is not logically impossible, it confronts us directly with an number of crucial questions: Is there by necessity an increase in the degree of complication of fabricators as one goes up the fabrication hierarchy? Intuitively, one would think so. Therefore, if yes, it is possible for a 'simpler' fabricator to make a more complicated fabricator? In fact, can a fabricator conceivably make itself?

It seems to be generally accepted that Von Neumann (1966, fifth Lecture) showed that self-fabrication of a machine (autonomous turnover of self on the basis of a supplied blueprint) and self-reproduction (making a copy of self, including the blueprint) is in principle possible. Von Neumann's so-called kinematic self-reproducing machine consists of a general purpose fabricator $P+\phi(X)$, which is an automaton consisting of two parts: A constructor P that fabricates a machine X from spare parts according to $\phi(X)$, the blueprint for X. When supplied with its own blueprint $\phi(P)$ the constructor makes itself.[12]

The incorporation of a general purpose fabricator such as $P+\phi(X)$ into our factory solves the 'insufficient number of machines' dilemma sketched above; in fact, I suspect this to be the only way to circumvent the problem. In Fig. 8, I sketch a self-fabricating factory that shows how incorporating the Von Neumann architecture makes Fig. 4e self-fabricating. The symbolism used in this figure derives from Eqn. (4); the Von Neumann constructor P is symbolised by r, while the blueprint $\phi(X)$ referred to above is now an element i of the set of blueprints I; any element (r, i) of the Cartesian product $r \times I$ is

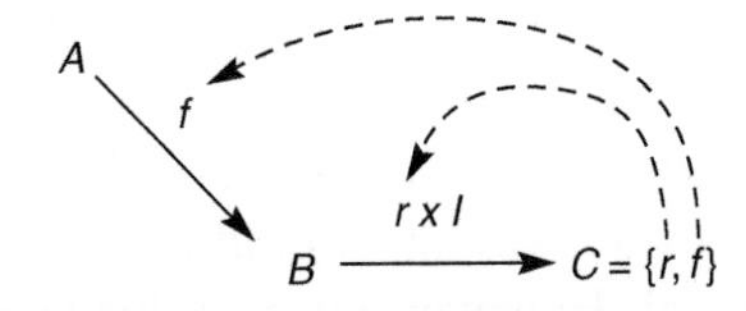

Figure 8 An abstract self-fabricating factory that incorporates the Von Neumann architecture.

Raw materials A are converted into building blocks B for the fabricators in the factory. A Von Neumann constructor r uses the information in the set of blueprints I to fabricate the set of machines C. The fabricator (r, i_k) with $i_k \in I$ can make itself directly when supplied with its own blueprint $i_k = i_r$. With $i_k \neq i_r$ it makes all the other machines f required by the factory.

[12] To give the entire fabricator $P+\phi(X)$ the ability to make a copy of itself, von Neumann added a blueprint copier Q and a controller R so that the fabricator becomes $(P+Q+R)+\phi(X)$, which can make not only X but also a copy of $\phi(X)$. When supplied with its own blueprint, $\phi(P+Q+R)$, it can make a copy $(P+Q+R)+\phi(P+Q+R)$ of itself and of its blueprint, thereby ensuring self-fabrication of the full system. However, for the purpose of analysing self-fabrication we do need to concern ourselves with replication of the blueprint.

a fabricator of a particular X. Following Rosen (1991) we call f a 'metabolic' mapping; r will be called the 'construction' mapping; this is therefore a metabolism construction system.

It should be clear that without the ability of constructor r to make itself with the help of its own blueprint, the factory cannot become self-fabricating. Although it seems to be generally accepted that constructors can in principle do this, what if it turns out not to be so? It is now time to turn to life-as-we-know-it and ask how organisms manage to fabricate themselves. Does what we know about metabolism and protein synthesis match the organisation sketched in Fig. 8 or do organisms do it differently?

6. SELF-FABRICATION IN LIVING SYSTEMS

For the purpose of matching known biochemistry to our abstract representation of a self-fabricating factory, consider the diagram in Fig. 9. It has been pointed out many times in the literature that ribosomes are really the only known examples of Von Neumann constructors. They fit the description perfectly: on its own a ribosome can do nothing, but in conjunction with the information embedded in a messenger RNA molecule that has been transcribed from DNA it can (with the help of a plethora of auxiliary enzymes, cofactors and an energy source, GTP) string amino acids together in the specified sequence. However, and this seems to have been universally ignored, the genetic blueprint for a ribosome is made up of

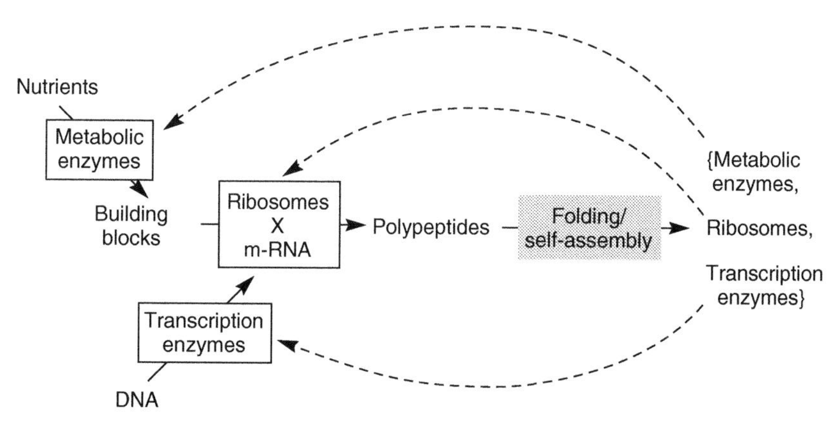

Figure 9 A summary of the biochemistry relevant to self-fabrication.

As explained in the text, the assembly of supramolecular complexes (the shaded box) must be a spontaneous, unassisted process if the diagram is to depict a self-fabricating organisation. For the sake of clarity the diagram has been kept simple; for example, ribosomal RNA and its synthesis by transcription enzymes has been omitted. The chaperones that assist in the folding of some polypeptides are also absent on the diagram, but are discussed in the text.

a set of individual blueprints for the myriad of protein and ribonucleic acid components that make up a ribosome; there is no contiguous genetic blueprint for a complete ribosome. Therefore, a ribosome never directly makes a ribosome, only the protein bits from which it is made up (the ribosomal RNAs are of course made by ribosomally synthesised enzymes). Note that the problem of whether a Von Neumann constructor can fabricate itself directly therefore does not arise in the cell. Nevertheless, we still need to explain how the ribosomal components assemble into a fully functional entity. The fabrication of all ribosomes entails two processes: the construction of the parts (here the polypeptide chains and ribosomal RNA), and their subsequent assembly into a fully functional entity. In fact, there is another process wedged in between, namely that of the folding of newly synthesised polypeptide chains into a functional, three-dimensional conformation.

Above I argued, following Rosen, that for a system to be self-fabricating it must be closed to efficient causation. The existence of the noncovalent, supramolecular processes of folding and assembly therefore forces us to search for their efficient causes inside the system, which immediately confronts us with the 'insufficient number of machines' dilemma described above. Adding extra blueprints does not solve the problem; each addition implies a new polypeptide that has to be accounted for in terms of internal efficient cause for folding and possible association with other proteins. The recently discovered existence of chaperones that assist the folding of some polypeptides also cannot fully solve that part of the supramolecular problem: chaperones are themselves proteins that need to fold in order to become active. There may be chaperones that assist the folding of other chaperones, but somewhere along the line there must then be chaperones that either fold spontaneously or assist in their own folding (or there must be a group that form a closed autocatalytic system). However, as far as we know chaperones fold spontaneously on their own. Similarly, with regard to assembly we are reasonably certain that supramolecular complexes such as ribosomes, spliceosomes, proteasomes, multimeric and oligomeric enzymes self-assemble spontaneously – the efficient and formal causes of self-assembly are embedded in the properties of the subunits of these complexes and in the properties of the environment. It is possible to dissociate these complexes *in vitro* and then have them reassemble themselves spontaneously. There appears to be no need for a physical agent to assist in the assembly process. It therefore turns out that, at least for life as we know it, unassisted self-assembly is the process that makes self-fabrication, and therefore life, possible. It is interesting to note that not one of the myriad of definitions of life listed in Barbieri (2003) and Popa (2004), nor the two regularly quoted sets of criteria for life – the Seven Pillars (de Duve, 1991) or PICERAS (Koshland Jr, 2002) – mention self-assembly as a necessary condition for life. In fact, I conjecture that if we discover life elsewhere in the universe, we shall recognise it by two properties: being autonomously self-fabricating by having learnt how to harness supramolecular chemistry and self-assembly, and having the ability to adapt and evolve by means of near-perfect replication and natural selection.

The abstract diagram in Fig. 8 can be extended to one that matches Fig. 9. This abstract representation of the self-fabricating metabolism-construction-assembly (M, C, A) organisation of living cells is given in Fig. 10. I propose this as an alternative to the replicative (M, R)-systems described by Rosen. Both are closed to efficient causation, but the (M, C, A) description has a number of distinct advantages. First, it maps onto the known biochemistry of the cell, whereas neither Rosen nor anybody else has been able to map the replication aspect of (M, R)-systems (which closes these systems to efficient causation) onto biochemical processes. In the language of category theory the replication component of (M, R)-systems is equivalent to an inverse evaluation map, and nobody seems to have been able to interpret this in terms of a physical process or object. The second advantage is that the (M, C, A) organisation reconciles Rosen's and Von Neumann's treatments. As mentioned in Section 2, Rosen's recasting of the Von Neumann architecture led to a logical paradox which has since apparently served to isolate these two views from each other (Rosen, 1959a, 1962). In a separate paper we shall show how to formally avoid this Rosen-Von Neumann paradox by using the formulation in Eqn. (4) instead of Eqn. (3). A third advantage is that the unassisted self-assembly component of (M, C, A)-systems obviates the need to postulate an agent that directly fabricates itself, a notion that I still find problematical. A fourth advantage is that the (M, C, A) architecture matches the triadic relationship between genotype, phenotype and ribotype suggested by Barbieri (1981, 2003) (see Fig. 11). Barbieri suggested

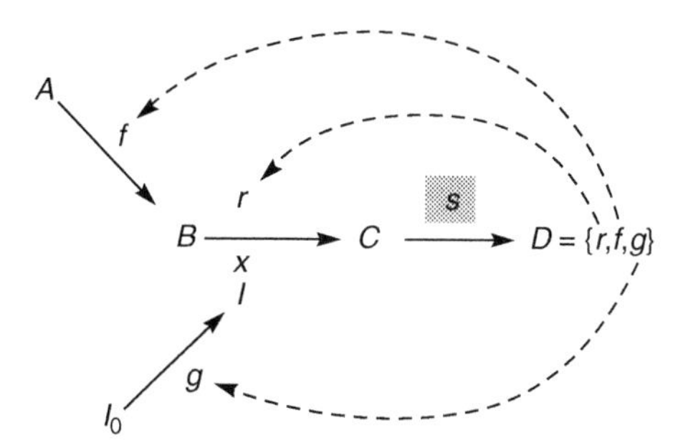

Figure 10 Adding self-assembly and information processing to the metabolism-construction system in Fig. 8.

Here the fabricator r cannot make itself directly, but it can make all its own components and of course those of all the other machines; together they form set C. These machine components then self-assemble spontaneously through mapping s to form the set D, which then contains the constructor r and all other machines. This factory also makes a set of information processors g that translate archival information I_0 into blueprints I.

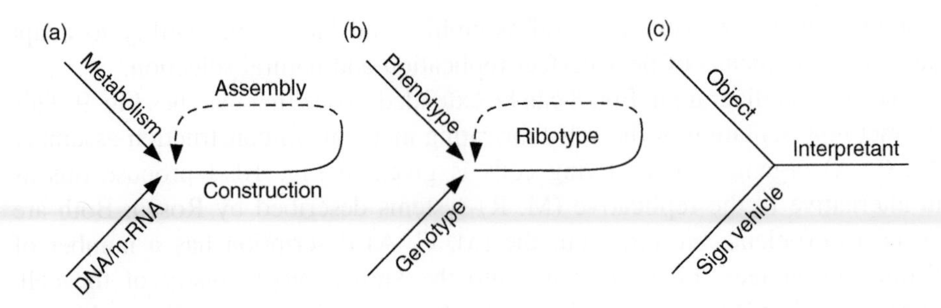

Figure 11 Triadic relationships in (a) metabolism-construction-assembly (M, C, A) systems, (b) Barbieri's genotype-phenotype-ribotype triad of ontologically distinct types, and (c) the logical distinctions between sign vehicle, object and interpretant in biosemiotics.

that the machinery for protein synthesis (ribosomes, associated enzymes, tRNA adaptors) forms a logically distinct ontological type besides the phenotype and genotype. As suggested by Fig. 11, this idea fits perfectly with both (M, C, A)-systems and the triadic logic of biosemiotics as discussed by, for example, Hoffmeyer (1996).

7. CONCLUSION

The model of self-fabrication developed above is at this stage rather simple, informal and certainly not yet mathematically rigourous. However, a formal, mathematical exposition of the theory has been worked out and will be presented elsewhere. I have concentrated solely on the functional organisation of processes that make self-fabrication possible, and have purposefully ignored important aspects such as energy requirements, control and regulation, self-bounding, and communication with the environment to name but a few.

In conclusion, I therefore argue for an epistemology for systems biology that is essentially relational and views everything that happens inside a living cell in the context of a functional organisation that makes self-fabrication possible. Working out all the implications this has for how we study, how we model and how we attempt to manipulate the cell is one of the tasks that systems biology must tackle if we want to lay claim to a deep understanding of life as we know it.

ACKNOWLEDGEMENTS

This project was funded by the Ernest Oppenheimer Foundation in the form of the Harry Oppenheimer Fellowship Award and by the National Research Foundation of South Africa.

REFERENCES

Aristotle, *The Metaphysics.* Penguin Books, London, 1998, (translated by Hugh Lawson-Tancred) A.3.

Aristotle, *Physics.* The Internet Classics Archive (classics.mit.edu//Aristotle/physics.html), 350 B.C.E, (Translated by R. P. Hardie and R. K. Gaye) II.3.

Barbieri M, *The ribotype theory of the origin of life.* J theor Biol, 91, 545–601, 1981.

Barbieri M, *The Organic Codes: An Introduction to Semantic Biology.* Cambridge University Press, Cambridge, 2003.

Barbieri M, *Life is 'artifact-making'.* J Biosemiotics, 1, 107–134, 2005.

Beer S, *What is cybernetics?* Kybernetes, 31, 209–219, 2002.

Cornish-Bowden A, *Making systems biology work in the 21st century (a report on the Biochemicial Society meeting 'Systems Biology: will it work?').* Genome Biol, 6, 317, 2005.

Cunningham MA, Bash PA, *Computational enzymology.* Biochimie, 79, 687–689, 1997.

Dawkins R, *The Selfish Gene.* Oxford University Press, Oxford, New edition, 1989.

Dobzhansky T, *Nothing in Biology Makes Sense Except in the Light of Evolution.* The American Biology Teacher, 35, 125–129, 1973.

Drexler KE, *Nanosystems: molecular machinery, manufacturing and computation.* John Wiley and Sons, New York, 1992.

de Duve C, *Blueprint for a Cell: The Nature and Origin of Life.* Neil Patterson Publishers, Carolina Biological Supply Company, Burlington, 1991.

Field RJ, Burger M, *Oscillations and Traveling Waves in Chemical Systems.* John Wiley & Sons, New York, 1985.

Freitas Jr RA, Merkle RC, *Kinematic self-replicating machines.* Landes Bioscience, Georgetown, 2004.

Gánti T, *The Principles of Life.* Oxford University Press, Oxford, 2003.

Heinrich R, Rapoport TA, *A linear steady-state treatment of enzymatic chains: General properties, control and effector strength.* Eur J Biochem, 42, 89–95, 1974.

Hoffmeyer J, *Signs of meaning in the universe.* Indiana University Press, Bloomington, 1996.

Hofmeyr JHS, Cornish-Bowden A, *The Reversible Hill-Equation: How to Incorporate Cooperative Enzymes Into Metabolic Models.* Comp Appl Biosci, 13, 377–385, 1997.

Hofmeyr JHS, Cornish-Bowden A, *Regulating the cellular economy of supply and demand.* FEBS Lett, 476, 46–51, 2000.

Hofmeyr JHS, et al., *From mushrooms to isolas: surprising behaviour in a simple biosynthetic system subject to feedback inhibition.* In: *Animating the Cellular Map,* (Edited by Hofmeyr JHS, et al.), Stellenbosch University Press, Stellenbosch, 2000, 213–219.

Kacser H, Burns JA, *The control of flux.* Symp Soc Exp Biol, 32, 65–104, 1973.

Kampis G, *Self-Modifying Systems in Biology and Cognitive Science: A New Framework for Dynamics, Information and Complexity.* Pergamon Press, Oxford, 1991.

Kant I, *Critique of Judgment (1790).* Macmillan and Company, London, 2nd edition, 1914, (Translated by J. H. Bernard). Available as e-text at oll.libertyfund.org/ToC/0318.php.

Kauffman SA, *Investigations.* Oxford University Press, New York, 2000.

Korzybski A, *Science and sanity: an introduction to non-Aristotelean systems and general semantics.* Institute of General Semantics, San Francisco, California, 5th ed. edition, 1994.

Koshland Jr DE, *The seven pillars of life.* Science, 295, 2215–2216, 2002.

Lehn JM, *Supramolecular Chemistry : Concepts and Perspectives.* Wiley-VCH, Weinheim, 1995.

Letelier JC, et al., *Organizational invariance and metabolic closure: Analysis in terms of (M,R) systems.* J theor Biol, 238, 949–961, 2006.

Maturana HR, Varela FJ, *Autopoiesis and Cognition: The Realisation of the Living.* D. Reidel Publishing Company, Dordrecht, Holland, 1980.

Mayr E, *Toward a New Philosophy of Biology: Observations of an Evolutionist*. Harvard University Press, Cambridge, Massachusetts, 1988.

von Neumann J, *The General and Logical Theory of Automata*. In: Taub AH (ed.), *John von Neumann: Collected Works. Volume V: Design of Computers, Theory of Automata and Numerical Analysis*. Pergamon Press, Oxford, 1961. Chapter 9, 288–328, delivered at the Hixon Symposium, September 1948; first published 1951 as pages 1–41 of: L. Jeffress, A. (ed), *Cerebral Mechanisms in Behavior*, New York: John Wiley.

von Neumann J, *Theory of Self-Reproducing Automata*. University of Illinois Press, Urbana, Illinois, 1966.

Nicolis G, Prigogine I, *Self-organisation in non-equilibrium systems*. Wiley, New York, 1977.

Popa R, *Between necessity and probability: searching for the definition of life*. Springer, Berlin, 2004.

Rosen R, *A relational theory of biological systems*. Bull Math Biophys, 20, 245–260, 1958a.

Rosen R, *The representation of biological systems from the standpoint of the theory of categories*. Bull Math Biophys, 20, 317–341, 1958b.

Rosen R, *On a logical paradox implicit in the notion of a self-reproducing automaton*. Bull Math Biophys, 21, 387–394, 1959a.

Rosen R, *A relational theory of biological systems II*. Bull Math Biophys, 21, 109–128, 1959b.

Rosen R, *Self-reproducing automaton*. Bull Math Biophys, 24, 243–245, 1962.

Rosen R, *Some relational cell models: The Metabolism-Repair systems*. In: *Foundations of Mathematical Biology*, (Edited by Rosen R), Academic Press, New York, volume II (Cellular Systems), chapter 4, 217–253, 1972.

Rosen R, *The Roles of Necessity in Biology*. In: *Newton to Aristotle*, (Edited by Casti JL, Karlqvist A), Birkhäuser, New York, 11–37, 1989.

Rosen R, *Life Itself: A Comprehensive Inquiry into the Nature, Origin, and Fabrication of Life*. Columbia University Press, New York, 1991.

de Rosnay J, *The Macroscope: a new world scientific system*. Harper and Row, New York, 1979, (Translated by Robert Edwards).

Ruiz-Mirazo K, Moreno A, *Autonomy and emergence: how systems become agents through the generation of functional constraints*. Acta Polytechnica Scandinavica, Ma91 (Complexity, Hierarchy, Organization. Special Issue. Farre, G.L. & Oksala, T. (eds) The Finnish Academy of Technology), 273–282, 1998.

Ruiz-Mirazo K, Moreno A, *Basic autonomy as a fundamental step in the synthesis of life*. Artificial Life, 10, 235–259, 2004.

Ruiz-Mirazo K, et al., *Merging the Energetic and the Relational-Constructive Logic of Life*. In: *Artificial Life VI: Proceedings of the Sixth International Conference on Artificial Life*, (Edited by Adami C, et al.), MIT Press, 1998, 448–451.

Ruiz-Mirazo K, et al., *A universal definition of life: autonomy and open-ended evolution*. Orig Life Evol Biosph, 34, 323–346, 2004.

Russel R, Superti-Furga G, *Systems Biology: understanding the biological mosaic*. FEBS Lett, 579, 1771, 2005.

Schumacher EF, *Small is beautiful*. Vintage, London, 1973.

Snoep JL, et al., *From isolation to integration: a systems biology approach for building the Silicon Cell*. In: Alberghina L, Westerhoff H (eds.), *Systems Biology: Definitions and Perspectives*, volume 13 of *Topics in Current Genetics*. Springer, Berlin, 2005.

Westerhoff HV, Hofmeyr JHS, *What's Systems Biology? From Genes to Function and Back*. In: Alberghina L, Westerhoff H (eds.), *Systems Biology: Definitions and Perspectives*, volume 13 of *Topics in Current Genetics*. Springer, Berlin, 2005.

11

A systemic approach to the origin of biological organization

Alvaro Moreno

SUMMARY

I present here an analysis of the core of biological organization from a genealogical perspective, trying to show which could be the driving forces or principles of organization leading from the physico-chemical world to the biological one. From this perspective the essential issue is to understand how new principles of generation and preservation of complexity could appear. At the beginning, the driving force towards complexity was nothing but the confluence of several principles of ordering, such as self-assembly, template replication, or self-organization, merged in the framework of what I have called a nontrivial self-maintaining organization. The key of this process is functional recursivity, namely, the fact that every novelty capable of contributing to a more efficient form of maintenance will be recruited. This leads us to the central concept of autonomy, defined as a form of self-constructing organization, which maintains its identity through its interactions with its environment. As such, autonomy grasps the idea of (minimal) metabolic organization, which, in turn, is at the basis of what we mean by (minimal) organism. Finally, from the concept of autonomy, I try to show how it has generated a new and more encompassing system in which evolution by natural selection takes over, generating in turn a new form of individual organization (genetically instructed metabolism) erasing the previous ones.

Systems Biology
F.C. Boogerd, F.J. Bruggeman, J.-H.S. Hofmeyr and H.V. Westerhoff (Editors)
Isbn: 978-0-444-52085-2

1. INTRODUCTION

Living systems appear as highly complex integrated units formed out of many different and complex chemical aggregates. Nothing similar in degree of complexity exists in the inorganic world, neither among human-made artifacts, where systems lack the deep integration and autonomy characteristic of living organisms. This organization is an intricate web of chemical reactions organized spatiotemporally. Organisms share the property that the functional molecules of their material make-up (DNA, proteins, fatty acids, etc.) are fabricated by internal processes. One could say that the functional hardware of living systems is continuously changing as new proteins are constructed and other molecules synthesized. The first conclusion, then, is that those components that make up the system as a whole are, in their turn, generated through the web of interactions of the whole system (in the sense that each process depends on several others within the system). This apparent causal circularity – along with the deep integration and high complexity of their components – makes the understanding of biological systems extremely difficult. Thus, the explanation of biological systems requires many different principles: physical and chemical laws, self-organization, and even informational constraints.

Already, certain philosophers acknowledged the problem of deep integration and holism, which are so characteristic of biological systems. More than 200 years ago, I. Kant (1790:1987) remarked that living systems are beings that organize themselves; they are systems whose parts depend on each other so that, taken as a whole, a living being is both the cause and effect of itself. Scientists, however, have begun to see living organisms from a holistic perspective only more recently.[1] In the 1950s Rashevsky (1954) defended the idea that living organization can only be explained in relational terms, i.e., every component in a biological organism will have an explanation in which the other components of the system are involved. R. Rosen, a disciple of Rashevsky, has held a similar view, defining organisms as 'systems closed under efficient causation' (1991). In contrast to any man-made machine, living organisms are self-made machines, in the sense that all the complex components are made within the system. In other words, the complex organization of living systems is a consequence of itself. To consider the whole network as a result of a former (lower) level made up of simpler isolable components whose properties determine their interactions would be partial, and ultimately, useless. The reason is that many of these components can only exist as such, as a consequence of the recursive maintenance of the whole network. In other words, complex components depend on the whole system.

[1] With the important exception of Developmental Biology at the end of nineteenth century and the beginning of the twentieth century.

However, this holistic view of living organization is at odds with the standard evolutionary perspective, which sees organisms as the result of changes in genes that have taken place at the populational level. Modern molecular biology has grounded this view, presenting a highly hierarchical approach of cellular organization. According to this view, the extreme precision of cellular organization relies on the fact that practically all biological reactions are controlled by one kind of macromolecule: catalytic proteins, i.e., enzymes. If we inquire about the cause that produces a given protein we will find a rather peculiar situation. In the first place, we can answer that, as any other component of the cell, a protein is the result of the action of a set of components (which will constitute, in Aristotelian terms, the efficient cause of this particular protein). Concretely, the synthesis of any protein is a direct consequence of the action of tRNAs and peptidyltransferase protein molecules, both involved in the formation of the string of amino acids. Another answer to the question is that the cause of a protein is the smaller molecular aggregates of which it is made of (this constitutes in Aristotelian terms its material cause). Then, amino acids would be considered to be the material cause of proteins.

Unlike other components, proteins are highly specific and complex. As is well known, the specific sequences of amino acids that constitute the proteins of a particular organism are related ultimately to the specific sequences of the nucleotides of DNA molecules. DNA (and RNA) acts, then, as an 'informational' template for the synthesis of proteins, because it contains the necessary instructions for guiding the construction of primary sequence of proteins. Hence, we can say, in Aristotelian terms, that DNA molecules are the formal cause of proteins in biological cells, because their specific sequence of nucleotides conveys the 'form' of the latter.

If we finally ask which is the cause of the specific form that DNA molecules carry, the answer would send us to a more encompassing framework than that of the individual cell. There is a fundamental difference between DNA and the remaining components of the cell. Although materially speaking, DNA is made up of building blocks as any other component, the specific order of the nucleotide sequence of a given DNA is ultimately a consequence of the evolutionary process, which leads us beyond the level of the cell. DNA represents a material connection between the evolutionary and the organismic levels, the collective/historical dimension and the individual organization. This is why DNA 'escapes' in a certain sense the causal closure that characterizes the organization of individual living beings.

This seems to give the final word to (Darwinian) evolution. But what is in fact evolution? Biological evolution, as we know it, requires the existence of certain types of systems: self-reproducing discrete entities based on a full Geno-Phenotype separation (namely, a code-based organization), which compete among each other in a noninterbreeding way. Then, if Darwinian evolution does not exist without organisms (or something almost likewise complex) how to explain the appearance of organisms by evolutionary mechanisms?

The solution to this dilemma leads us to the question of the origins. If evolution requires a certain threshold of organizational complexity, we have to search what kind of principles would explain the appearance of such organization. Which kind of properties/features that could have appeared in a purely physico-chemical scenario (and that, despite such initial simplicity, could be capable of generating a sequence of steps of increasing complexity) should the early – necessarily simple – prebiotic organizations possess? How can such an intricately holistic and complex organization appear from the much more 'simple' and understandable physico-chemical world?

As has been mentioned at the beginning, the understanding of biological organization will require many different principles. I am convinced that, unless we understand how all these organizational principles have become entangled together, we will not achieve a full understanding of life.

2. THE ORGANIZATIONAL PERSPECTIVE

The process by which living organization originated may probably be a sequence of different forms of organization, progressing from relative simple to more complex stages. In a general sense, the term organization means the activity (or result) of distributing or disposing a set of elements properly or methodically. Namely, in an organization there is a nonrandom arrangement of parts, generally serving a purpose or function. However, for obvious reasons, in our case, we have to discard any form of external design and/or purpose. That is why, in a prebiotic context, the idea of organization is rather associated with the formation of dynamical systems in which the random interaction of their parts generates the 'emergence' of a global order.

The spontaneous emergence of order could take two different forms: self-assembly (SA) and self-organization (SO). Unlike evolution, both SA and SO are widespread phenomena, which do not require very complex elements or systems. Self-assembly is a process in which a set of (randomly distributed) elements group together, forming a stable structure (an order), e.g., a crystal. This process is due to the material properties of the elements, namely, to the forces acting among them. Thermodynamically, self-assembly can be described as a process towards equilibrium, ending in a stable structure.

In both SA and SO there is the emergence of order from a set of randomly interrelated elements. However, in the case of SO this order is not a consequence of the structure of the constitutive elements,[2] but of certain boundary

[2] However, as we shall see, complex forms of SO will require the introduction of a variety of specific constraints in the constitutive elements.

conditions in far-from-equilibrium (FFE) conditions: the emergent order is essentially dissipative. Given certain specific boundary conditions in FFE a set of local interactions becomes nonlinear, and a collective behavior – a macroscopic pattern – emerges. Now, the maintenance of this pattern is not only a consequence of the given boundary conditions, but also a result of the inherent recursivity of the process: once it appears, the pattern constraints the dynamics of the system components so that the produced pattern in turn produces itself. For instance, in the case of Bénard convection cells, the emergent pattern (the creation of hexagonal cells) contributes to the self-maintenance, because the fact of belonging to a certain cell is what makes a molecule turn to the left or the right. Thus, recursivity and removal from thermodynamic equilibrium is the key feature of this concept of organization.

Though very different in nature, both SA and SO are important sources of order. In fact, many systems show both forms. However, only SO really holds the dynamical and functional senses of the idea of organization. The term 'organization' implies not only order, but also the usefulness of this order that it effectively does something. And for this 'doing', a continuous process is implicitly necessary. SO is therefore the ground of any organization as it is a dynamical form of order that contributes to the creation and *maintenance* of itself. The minimal (because self-sustaining) meaning of the terms *task* and *function* is that something is contributing to the maintenance of the very organization in which it appears. As we shall see, this internal sense of usefulness that appears in SO is the key for allowing a process of increase in complexity.

3. THE STARTING POINT: NONTRIVIAL SELF-MAINTENANCE

Now, within this idea of organization, what is of interest for our purpose? As the process by which life originated was probably a sequence of different forms of *organization* going from relative simple to progressively more complex stages, we have to look for a kind of organization fulfilling two basic conditions: on the one hand, such an organization has to be, in principle, simple enough for it to be likely to appear from sets of material aggregates formed spontaneously; on the other hand, it has to have the capacity, al least in principle, to further generate other, more complex forms of organization[3] (and so on, until new organizational

[3] Here I will use the term complexity in an organizational frame. This means that by an increase of organizational complexity I am not considering a mere increase in the 'complicatedness' (i. e., more number and variety of components) of the system, but rather a functional re-arrangement of this complicatedness. In this sense, an increase in complexity is linked to the generation within a system of new functional levels of organization. Thus, an increase of organizational complexity can take the paradoxical form of an apparent 'simplification' of the underlying complicatedness when a system creates a new hierarchical level (Simon, 1969) through a functional loss of details (Pattee, 1973).

principles would appear). Accordingly, the fundamental problem can be stated this way: 'How can a system be organized such that (given certain specific but probable conditions[4]) it can generate higher degrees of complexity, and, at the same time, is capable of retaining this new complexity?'

As for the first requirement, we should conceive a set of prebiotically plausible boundary conditions capable of driving a variety of chemical (see next point) systems where the maintenance of the process is (at least in a very simple sense) a consequence of the very organization of the system, namely, a self-maintaining system (SMS). Although such systems had to be probably preceded by many other systems whose maintenance was essentially driven by boundary conditions much more complex than themselves, the appearance of SMS is the fundamental starting point for a sustainable process of increase in complexity. For, if the retention of new complexity in a SMS were mainly dependent on an increasingly complex set of external conditions, we would be only transferring our problem (namely, the natural, prebiotic origin and maintenance of increasingly complex forms of organization) to the external environment. Accordingly, the starting point should be a self-maintaining (SM[5]) organization, namely, a system in which it is the organization itself (rather than external conditions) that explains the maintenance of the process. This leads us to the second requirement.

As for the second condition, the way in which the system realizes its own maintenance has to be nontrivial. By a nontrivial form of self-maintenance (henceforth, NTSM), I mean those systems where there are many simple interactions involved in its realization and (also many) ways to achieve it. In other words, the system has to be capable of increasing the number of different functional relations (within an integrated whole). In the next section we will see how this can be done.

These two requirements should be entangled, namely, the very organization plays some role on itself, such that it creates new internal differences; and some of these organizational differences may later play a new functional role. For instance, a new form of organization (say, a self-enclosing autocatalytic network) might be preserved because it allows a more stable form of maintenance.

But what kind of self-maintaining organization is at the same time minimal (i.e., the simplest) and capable to increase in complexity? This is the kind of organization that I call 'nontrivial SM'. Let us examine this question in several steps. First, I shall characterize what is a trivial form of SM. Second, I will consider the necessary (but not sufficient) condition for a NTSM. And finally, I will discuss the (minimal) organizational requirements for NTSM.

---

[4] In terms of what the physical and chemical evolution of the universe can create in certain places during reasonable temporal periods.

[5] The term SM is usually referred to in studies of the organization of chemical systems (because of the constructive character of these systems), whereas in physical systems the usual term is self-organization. But here, I am using the term SM generically, as equivalent to SO.

3.1. From self-organization to NTSM

We now want to address the relation between SO and complexity. Ordinary forms of SO – those which can appear in relatively simple conditions, be they in natural ones, like thunderstorms, or in artificial ones, like Belousov–Zhabotinsky reactions – may show in certain cases significant degrees of robustness, but they seem unable to achieve further increases in complexity. Let us take the example of the so-called dissipative structures. A Dissipative Structure (DS) is a phenomenon by which a set of nonlinear microscopic processes generates a macroscopic-collective pattern in a situation of distance from thermodynamic equilibrium[6] maintained by the continuous action of a group of constraints, one of which is the very pattern generated by the global dynamics. Now, although in classical DSs (either physical or chemical) the emergent pattern may be very complicated, its contribution to the maintenance of the process is always 'simple', in the sense that such global pattern cannot exert a variety of selective local constraints on the microscopic dynamics. In Bénard's Convection Cells, for example, the pattern only makes a molecule turn to the left or the right. Thus, the generation of new complexity is only possible if the dissipative organization of the system develops a remarkable degree of internal plasticity, so that certain elements (and/or patterns) will be recruited to serve different roles in the system, thus producing a new form of SM. But how?

The answer is a SM organization that produces local and selective constraints (instead of only one or few global patterns). In other words, it has to be a chemical organization, for physical systems in general do not have the capacity to create a wide enough variety of dynamical constraints. On the contrary, the chemical domain is based on relations among elements that generate new elements, and these new elements may in turn give rise to different interactions, which may produce new elements, and so on, bringing about a potentially unlimited set. From the thermodynamic perspective, chemical systems are a special kind of organization in which the construction of new molecular variety through dissipative processes creates new conservative constraints (molecular shapes), which in turn can modify the whole organization, and so on. A chemical organization creates many local, selective constraints (new components) and its global maintenance relies on many of them. Interestingly, some elements – called catalysts – can modify the relations among elements, and therefore a chemical system is potentially a domain where components can become rules and vice versa.

In other words, the jump from physics to chemistry seems necessary for material systems to reach a diverse enough spectrum of dynamic, constructive, and

[6] Unlike the case of self-assembling structures, which keep their order in thermodynamic equilibrium.

emergent behavior.[7] Chemistry allows accumulative construction, a process of interactive feedback between the organization of components and the accumulative assembly of increasingly complex components, which is a consequence of the combinatory – chemical – nature of molecules. This way, chemical networks can reach a special plasticity or potential for diversification. This does not mean that chemistry is enough for this process of increase in complexity. As we see in biological systems self-assembly is also an important factor of the construction of their complex organization, and once a high level of molecular complexity is reached, there are even constructive processes of mechanical nature. But what I want to stress here is that all these processes are possible due to the establishment of a NTSM chemical organization.[8]

Now, what does SM mean in this context? The basic idea is that of autocatalysis: molecule A could catalyze the formation of B, B of C, and C of A (but of course many more reactions and side reactions can occur). Given a set of specific initial conditions, a process of production of components may be triggered, creating new components and catalysts, which in turn will produce new ones, and so on, until the initial set of components is produced and the whole process becomes recursively regenerated.[9] The key feature of a SM organization capable of increasing its complexity is that the circular loop involves the generation of many (some of them stable) components that may act as local and selective constraints (molecules and supramolecular structures). In other words, an increasing part of the dynamical order is recorded in complex stable structures/components (which in turn will allow more complex forms of organization). However, the maintenance of the system depends on the whole set of relations, and is therefore a distributed, holistic phenomenon. The interesting point here is that, once a certain threshold of complexity is crossed, the maintenance of an autocatalytic organization lies in the entire network, instead of in single reactions (see footnote 12). Thus, the main characteristic of a system of

[7] It is necessary but not sufficient. In fact, a chemical DS like a BZ system is a type of organization in which the emergent pattern depends on dynamical factors, such as (the behavior of) concentrations or diffusion of the components of the system, but not on changes of the relations of production (catalytic reaction rates) among components. Thus, the emergent pattern is (like in purely physical DSs) nothing more than a global correlation of billions of molecules allowing a different (according to some authors, more efficient) way to dissipate energy but not a selective re-organization of the component–production relations that may change the identity of the system. For example, a burning candle is a self-maintaining chemical system, in which the macroscopic pattern–the flame itself–contributes to maintaining the organization of the system (by vaporizing wax). However, the flame as a macroscopic, global constraint only plays this action, and therefore it is a trivial form of SM.

[8] This process, however, would be similar to what Evelyn Fox Keller (this issue) refers to as 'the iterative processes of self-organization that occur in heterogeneous systems over time', because she considers the formation of composite systems rather than composite structures (components).

[9] Autocatalysis constitutes a simple example of the entangled relation between construction and organization maintenance. A dissipative process is maintained through the construction of certain aggregates (catalysts) that in turn modify the interactions in such a way that the organization will again produce such components.

this kind is that some components are products of reactions catalyzed by other components whose formation depends recursively on the action of the whole network.

A NTSM organization constitutes a special case of the general principle of self-maintenance. Whereas the common feature of all forms of SM is a more or less robust but simple form of recursivity, what is specific in our case is that the whole organization is a potentially self-modifying form of SM.[10] This organization is potentially capable of generating open forms of recursivity, as a change in the initial conditions can bring about new molecular structures, which in turn may produce new forms of organization indefinitely. Such an organization can explore an unlimited[11] variety of new molecular combinations and transformation processes leading to new, more complex forms of organization in the context of reaction webs where a reaction can directly or indirectly influence another, until a certain recursivity (closure) property is reached. The system is self-maintaining because of the recursive and holistic form of its organization: the global dynamics is necessary for the maintenance of (many of) the reactions of the system and these reactions are in turn necessary for the maintenance of the whole system ('metabolic' closure[12]). This is the basic idea behind Rosen's (1971, 1973) 'M, R-systems' or Kampis's (1991) 'component production systems'. A NTSM organization is in fact the core of a metabolic organization.

Along with this capacity to generate internal variety, a sustainable process of increase in complexity requires ways of preserving it. Only those novel forms of complexity contributing to the maintenance of–or being functional in – the organization that generated them can be preserved. Typically, the way to retain complexity is by stabilizing it. There are several ways to create stability within a dissipative organization: from the production of certain structurally complex components that are thermodynamically stable and 'localize' degrees of complexity to functional reorganization in hierarchical levels.[13] To sum up, a SM organization has the capacity to preserve the generated complexity if it is able to recruit it for a further, more efficient (ultimately, more stable), form of SM. In other words, by 'jumping' to another more efficient form of SM, a system can preserve the generated complexity.

---

[10] M. Bickhard (2000) uses the term 'recursive SM' with a very similar meaning.

[11] This limitless is only potential, because, as we shall see, as the complexity of the organization increases, its brittleness also increases.

[12] As has recently been pointed out by Letelier et al. (2006) a nontrivial idea of 'metabolic closure' is more demanding than that of autocatalytic closure, such that in the former the circularity is a property of the global connectivity in the entire network, not a property of a single reaction.

[13] Fast and reliable reproduction is also an indirect way to increase the stability of a given organization. We shall further discuss the origin and consequences of this mechanism.

As a consequence, a NTSM organization is capable of a true complexity 'bootstrapping', in the sense that (some of) the new organizational varieties (that can be generated) can in turn create new levels of functionality and SM. The creation of increasingly larger and richer chemical networks can even bring forth functionally hierarchical organizations, as for instance, cellular organizations, which in turn generate new forms of SM, increasingly autonomous of the environmental conditions.

3.2. The problem of the origin of NTSMSs

So far the conditions of possibility for the appearance of a prebiotic evolution were analyzed, concluding that the key step is the appearance of certain kind of chemical systems based on what we have called a NTSM organization.

We do not know how these kinds of systems could have appeared, neither what the threshold is of complexity of the chemical aggregates necessary to generate them, nor under which set of boundary conditions. These are questions whose answers depend largely on empirical research. Unlike ordinary examples of DS that appear spontaneously, present-day life has created a wide organizational 'no-man's-land' (because life has changed primitive conditions and also because of its capacity to eliminate less efficient competitors). This space has probably been occupied by a long sequence of increasingly complex systems (which must in turn have eliminated their precursors). Thus, the appearance of a NTSM organization is nowadays only possible in lab experimental conditions.

We may imagine that a great variety of chemical systems appeared on special places of the Earth (or other similar planets) during the period of chemical prebiotic evolution that took place when the planet cooled down. Among these systems, some would constitute autocatalytic cycles, leading to increasingly large self-maintaining networks, namely, systems where all components and component aggregates (directly involved in their organizational dynamics) must be products of a reaction network that constructs itself. As Kauffman (1993) has pointed out, 'The origin of life was a quite probable consequence of the collective properties of catalytic polymers. More probably (...) many properties of organisms may be probably emergent collective properties of their constituents. The origins of such properties then find their explanation in principles of self-organization rather than sufficiency of time.' In support of this claim, Kauffman (Farmer et al., 1986) and more recently, also other authors (Fontana, 1992, Steel, 2000, Hordijk & Steel, 2004) have presented different computer simulations showing the emergence of this kind of systems from a small set of relatively simple components.[14] However, most of these simulations ignore or neglect the

[14] By now the only evidence of the spontaneous appearance of NTSM systems is computational.

thermodynamic requirements for self-maintenance. From a physically realistic perspective, a NTSMS cannot have appeared unless there exists a set of conditions ensuring both the adequate concentrations of its fundamental components and a specific order among their constitutive reactions (endergonic reactions have to be adequately coupled with exergonic ones). As F. Harold has pointed out (2001, p. 251) 'a credible biopoietic theory will be one that generates mounting levels of complexity naturally, by providing the means to convert the flux of energy into organization'.

Given the deep interdependence between the degree of complexity of an organization and its capacity for developing real self-maintenance, it is hard to imagine the early steps of the process. In the absence of self-generated mechanisms for maintaining the conditions of functioning, the earlier SMSs were probably fragile or/and dependent on highly specific external conditions. A good example is the hypothetical protometabolism proposed by G. Wächstershäuser (1988). This author argues that certain mineral surfaces could have created the adequate conditions for the appearance of a kind of what he considers a protometabolic organization. He defends that the first 'surface metabolic network' was established on the interface of solid/liquid phase (mineral surface/water). Under these conditions the system could 'act' on the environment only establishing a buffering of pH or bringing down the tendency towards hydrolysis in its surroundings. Other authors (Harold, 1986, Morowitz, 1992, Deamer, 1997, Segré et al., 2001), however, consider very unlikely the possibility of a metabolic organization, however primitive it may be, in the absence of any enclosure. Such scenario would imply the necessity to elaborate models that integrate metabolism and compartmentation simultaneously (Penny, 2005).

Be it as it may, it is likely that the earlier SM systems were highly dependent on a set of external conditions upon which they did have few (if any) control. Thus, their degree of robustness would be seriously restricted, and the accessible level of complexity would also be low. In consequence, the appearance of NTSM systems is linked to crossing a frontier between those organizations whose identity was more dependent on a set of external conditions (let us call them 'predominantly passive SM systems') and those whose maintenance depends more on their own organization (let us call them 'predominantly active SM systems'). This frontier is precisely what defines the *autonomy* of the system.

4. NTSM ORGANIZATION AND AUTONOMY

The question of whether the earlier NTSMSs appeared in certain mineral surfaces, enclosed in primitive compartments or otherwise, is an empirical issue. However, what in any case is essential is that the organization of the primitive NTSMSs could control the external constraints necessary for their maintenance.

Then, how can an NTSMS reinforce its own self-maintenance? This question is strongly related with the problem of the increase in the complexity. Let us see why.

In a scenario populated by primitive SM chemical systems the appearance of new systems based on new components is very likely, therefore allowing different forms of SM. Some of these new forms of SM may be simpler, but others will probably create larger and more diverse networks. The interesting point is that, within this variety of systems, some of them can generate new, more efficient ways to achieve SM – for instance, increasing both internal diversity and stability. In other words, a more complex organization may be reached and preserved if it increases the capacity of self-maintenance of the whole system. This fact is important because, in general, as the complexity of a FFE organization increases, its maintenance becomes harder.

These considerations point to a key question. We have seen that the condition to preserve the new complexity generated is that it is functional for the maintenance of the system. This can take place if the system becomes less dependent on (the presence of) certain external conditions and, consequently, more dependent on itself. Thus, the possibility of any significant increase in complexity of the primitive SM chemical organizations implies a progressive takeover of the external conditions necessary for their viability; and the only way for this to happen is that the system recruits its own organization for the active reconstruction of the necessary conditions of its maintenance. In other words, by increasing their degree of autonomy with respect to the earlier minimal case. By this, I mean a progressive takeover by the NTSM systems of the environmental conditions necessary for their viability. Accordingly, a system can be considered autonomous if its maintenance is more the consequence of its own organization than of the structure of its external environment.

A crucial step in the evolution towards autonomy was the appearance of systems whose organization includes the construction of a selective and functionally active membrane.[15] Such organizational change allows the components of the system to be produced in much more favorable and stable conditions (regulation of concentrations, selection of kinds of components, etc.). In this way the generation and stability of more complex systems becomes possible. Now the (internal) organization will appear much more integrated and complex in respect to its environment than either the primitive autocatalytic networks without physical border or the primitive micelles or vesicles.[16]

---

[15] The incorporation of a membrane is a process that transcends the chemical frame, as it implies the recruitment of certain independent processes of self-assembly.

[16] It is likely that cellularity integrated with the inner organization was preceded by more primitive cellular systems (like micelles or vesicles enclosing independent SM organizations).

Self-encapsulation will sharply differentiate the organization of the system (the set of relations that constitute it as a distinct unity) from the environment, where different interactions occur. In this way, a clearly distinct inner medium is created: a space where not just concentrations, but even components will be different from those of the "external" medium. However, the most important issue is that the boundary is produced by the very organization of the system (as it is an integral and integrated part of the metabolic network, not a mere 'wall' whose properties are externally defined). This entanglement, between the physical border and the recursive process of production of components constituting the system as an autonomous unit, is basically the idea of an autopoietic system (AS), as has been explained by Maturana and Varela (1973) 30 years ago.

However, the idea of autopoiesis has been formulated in a rather abstract way. If we take into account the thermodynamic requirements, an AS should also manage autonomously the flows of matter and energy necessary for its maintenance. To be autonomous, a primitive SM system should be capable of constraining the flows of energy (and matter) to ensure the TD realization of the processes that constitute it as a system. In other words, the constructive logic (the recursiveness of the relations among the components of the system) should be entangled with the energetic logic (Moreno & Ruiz-Mirazo, 1999, Ruiz Mirazo & Moreno, 2000, 2004). This implies that the membrane is not only a physical border ensuring adequate concentrations, but also a key element in the energetical maintenance of the system[17] (Pereto, 2005). In any known living system a phospholipid membrane and its molecular machines actively control the flows of energy between the inside and the outside. The essence of the energetic self-maintenance of the cells lies in the asymmetric disposition of molecular machines on the cell membrane allowing efficient coupling between primary energy sources and metabolic networks.

This thermodynamic view of autonomy goes far beyond the idea of logical closure of autopoiesis. As a FFE organization, the constitutive processes, the recursive network of production of components, of an autonomous system is essentially entangled with a set of interactive processes. Thus, an autonomous system is an agent, namely, a system that maintains its own identity by performing functional processes on its environment (Moreno & Etxeberria, 2005, Moreno & Barandiaran, 2005) (see Fig. 1).

These interactive processes are different from mere physico-chemical reactions constantly taking place in both directions (and which, in fact, occur in all kinds of systems, from the simplest to the most complex). As the interactive processes are embedded in the self-constructing dynamics of the system, they

[17] Thus, such an interface needs to be a semi-permeable structure where coupling mechanisms (particularly energetic transduction and active transport mechanisms), which are basic for the complete self-construction, are anchored.

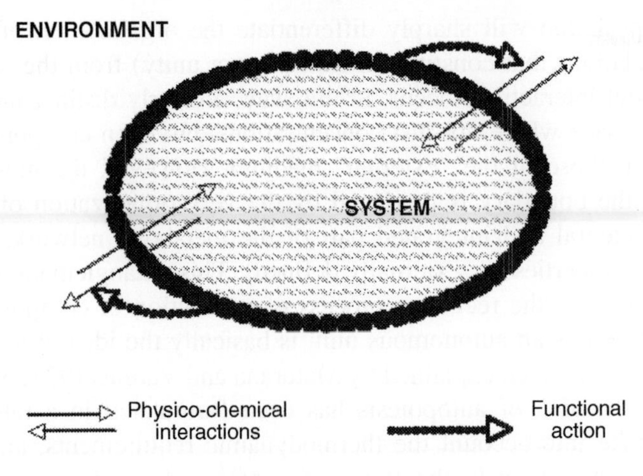

Figure 1 Agency in a basic autonomous system

will constitute a functional loop of the system itself, i.e., agency constitutes an environmentally mediated process, but clearly asymmetric.[18] The interactive processes are driven by the internal organization of the system. An example of a very basic form of agency (universally present in all cellular life) is the mechanism of active transport, which is an in/out flow of matter against a concentration gradient driven by a free energy source.[19] Agency involves a functional action on the environment, modifying (and later controlling) certain environmental conditions for the system's dynamics. In short, seen from a thermodynamic perspective, autonomy is the necessary expression of nontrivial self-maintenance. Without autonomy, either the organization of a system is extremely simple or, if complex, it is dependent on another more complex one (whose origin requires explanation).

Let us summarize. Starting from simple dissipative chemical organizations, with maintenance highly dependent on a set of specific external conditions, a process of exploration of new organizational forms and retention of the functional ones leads to progressively more complex organizations. As this process takes place in a chemical scenario, the increase in complexity of the systems creates an accumulative process of construction of bigger, more complex components, which in turn are the necessary elements for the creation of more complex organizations.

The progress occurs when certain specific form of the former collateral complexity is recruited. Thus, in every step, the new organization achieves SM

[18] Asymmetric in the sense that the environment cannot establish such recursive interaction processes with the agent, unless we are speaking of an interactive coupling between two agents.

[19] This mechanism must have appeared in early forms of cellular organizations because it is required for avoiding an osmotic crisis, which would lead to the bursting of the cell.

in a different, more efficient, and robust way, so that it increases its efficiency (in capturing the material and energetical resources necessary for its maintenance). This way, certain systems became capable of using their own organization for their self-maintenance. In other words, they became autonomous systems.

It is frequently argued that a metabolic-like organization – what we characterize as basic autonomy – is the result of evolution by Natural Selection, which, understood in a more simplified way, would only require a scenario of self-replicating molecules, provided that they are able to introduce and transmit variations and generate themselves (along with its environment) a selective pressure, leading to the survival of the "fittest". However, this argument fails. For, on the one hand, an evolutionary process of hereditary self-replicating molecules is hard to conceive in the absence of a functional organization of the energy necessary to keep the process going (i.e., a rudimentary metabolic organization). On the other hand, an evolutionary process leading to higher forms of complexity would require self-replicating entities that could in principle produce an open space for functional variations. Now, an open space for functional variations requires in turn a scenario capable of translating sequential variations into different functional properties (catalytic, for instance) which ultimately influence the way of realization of self-maintenance: more or less capacity to gather external energetical resources; resistance to perturbations; endurance (through either recursive self-production or self-reproduction). Thus, only an organizational framework allows an open space of ("phenotypical") functions.[20] Last, but not least, without a form of enclosure it is hard to conceive a workable process of selection. The conclusion is therefore that basic autonomy is the precondition for the appearance of a (Darwinian) evolutionary system (and not the contrary).

5. THE EMERGENCE OF A HISTORICAL–COLLECTIVE DIMENSION

5.1. Autonomous systems with memory

The basic form of autonomy that we have analyzed in the previous section implies a primitive form of metabolic organization, but not necessarily an instructed one (Ruiz-Mirazo & Moreno, 2004). We can conceive such systems as capable of a nonreliable form of self-reproduction. Thus, it is likely that the maintenance of populations of these primitive autonomous systems permitted that some of them increased the efficiency of their metabolic machinery, allowing the production of more complex components.

[20] This is the core of the argument presented by Wicken (1987) against the evolutionary possibilities of populations of self-replicating modular templates (like RNA-type molecules) in the absence of metabolic organizations.

However, as complexity grows, so does fragility. The organization of basic autonomy, just by itself, does not solve the problem of preserving (for long-term periods) the new complexity that it can generate, and therefore, neither the problem of how this complexity could grow indefinitely. Of course, the productive and reproductive dynamics of a scenario of basic autonomous systems would allow the maintenance of certain level of complexity. But this – still rudimentary – functional dynamics cannot ensure that the components (together with their way of organization) remain unaltered for much longer than their typical life spans (or the one of the whole organization), and the system faces a bottleneck: as complexity rises, its preservation becomes more and more difficult. Therefore, only those autonomous systems that developed specific mechanisms to stabilize and retain the increasing structural and organizational complexity with a fairly high degree of reliability could begin to unfold new, subsequent levels of complexity and, furthermore, set up the first pillars to ensure their *long-term* maintenance.

Now, how to do that? As Szathmáry and Maynard-Smith (1997) have pointed out, the way to preserve the specificity of an increasingly complex organization is through what they call a mechanism of 'unlimited memory'. This mechanism consists in linking the sequential structure of certain stable and self-replicating components (to be more precise, the specific sequence of modular templates[21]) with the most complex structural–functional properties. This allows a 'storage' or 'recording' of these complex (and highly specific) functions, which in turn permits – if these material records become replicated – a reliable form of reproduction (regardless of how complex the organization of the system is). Thus, the requirement for the start of an unlimited hereditary memory is the generation by (and integration within) the organization of autonomous systems of suitable components for this storage.

There are two important aspects here. The first is that with the introduction of these components the maintenance of the specific structure of the organization changes: Instead of a mechanism in which the specific order of the system lies in the dynamically dissipative maintenance of the whole organization, an important part of such order is now frozen in the linear sequence of some components (as this linear sequence will specify increasingly complex key catalytic functions of the system, and therefore it will help to stabilize it). The second is that, because such components have template capacities, they can transmit the specific

[21] A 'modular' molecule is a big structure made of a sequence (a small set) of subunits. In this component, the global shape is a consequence of the particular order of the subunits. A molecule is considered a (catalytic) template when its structure acts as blueprint, inducing the formation of copies of such structure. Although modular templates are considerably complex molecules that hardly must have appeared before the evolution of autonomous systems simple kinds of templates also probably played an important role in previous stages. For instance, certain mineral surfaces would probably have played a role as catalysts in noncellular protometabolic systems, or the very membrane in encapsulated systems.

order they store by direct replication. This way the renewal of the increasingly complex structure of certain components does not depend any more only on the maintenance of an increasingly complex and holistic process, but on the template-directed replication (which relies in a specific sequential order of their building blocks). The robustness of the system may also significantly increase because of the possibility to reliably produce new highly functional components (despite their increased structural complexity). The so-far distributed dissipative organization becomes this way preserved by means of a local conservative order stored in modular template-like components. In other words, the maintenance and reproduction of the organization becomes instructed.

Interestingly, the introduction of modular templates specifying the most complex part of the functional organization of the system conveys an important change: the formerly highly distributed and holistic organization opens the way to a much more modular organization. This is an advantage, as in a modular system the organization as a whole is less likely to be disrupted by a localized failure. Thus, seen in an evolutionary perspective, modularity increases flexibility and minimizes cascading malfunctions (Harold, 2001, p. 212). In total, the introduction of this mechanism of unlimited memory changes so deeply the organization and capacities of the primitive autonomous systems that it seems convenient to give them a specific name. We have elsewhere (Ruiz-Mirazo et al., 2004) used the term of hereditary autonomous systems (HAS).

Let us go now to the more fundamental question. Given that modular templates now control both the maintenance and reproduction of the system, all random changes in these components (allowing viable reproduction) will lead to a process of exploration of the sequential space linked to a correlative selective retention of the more efficient organizational forms.[22] This allows the recruitment by the individual autonomous systems of the results (end-products: selected patterns) of a slow process of natural selection, which is much more encompassing as it takes place beyond these individual autonomous systems. Thus the evolutionary process in which the whole population and its environment are involved largely determines the changes that take place at the level of the sequences of the templates. These templates become a kind of material memory, which can reliably transfer organizational changes from one system to another. In this way, the systems endowed with modular templates can combine coherently and consistently the individual dimension of their activity (related to the self-construction/self-maintenance of each of them) with a progressively more important temporal and spatial dimension (related to their long-term maintenance and evolution as a whole population). But the most important consequence is the increase in the complexity of the metabolic organization that this insertion in a historical and collective dimension allows.

[22] Provided there exist certain additional constraints, like competition for resources in a limited space.

The beginning of a process of evolution based on the transmission of modular templates probably inaugurated a stage of both vertical and horizontal exchanges of hereditary components (as suggested by Woese[23]). As a consequence, individual organisms crucially depend on the (long-term) selection of the functional hereditary components; and in turn this global process of selection of functional hereditary components from a random process of variation will depend on the performances of the individual organisms they instruct. From now on the maintenance of the autonomous systems will depend on the maintenance of this global historical–collective structure of relations. Furthermore, even the structure of the autonomous systems will progressively depend on these larger and wider relations that we know with the term "biological evolution" (recall Dobzhansky famous dictum: "nothing in biology makes sense except in the light of evolution").

5.2. The origin of an informational organization

Hereditary autonomous systems must have been the immediate precursors of present-day living organization. As has been mentioned in the previous section, the organization of HAS was likely based on one single type of polymer to support at the same time template and catalytic functions.[24] Now, this organization cannot yield to an unlimited increase in complexity, and therefore the evolutionary possibilities of HAS are blocked. The reason is the following: there is a trade-off between the realization of catalytic and storage/replicative functions; the better a given type of polymer is suited for template tasks, the worse it is for exploring the catalytic space, and vice versa.[25] Accordingly, the only way for an unlimited increase in complexity is by introducing two different types of polymers, devoted, respectively, to template and catalytic tasks (Ruiz-Mirazo et al., 2004). This way, the systems that start to produce two different – and

[23] Such modular organization should have helped the exploration of new, more efficient forms of organization. As Woese (2002) has pointed out, the beginning of cellular evolution was necessarily a collective process, where different cellular designs evolved simultaneously, systematically exchanging genetic material (what he calls 'horizontal gene transfer'). So, this early (pre) Darwinian evolution would allow an exploration of different forms of organization, until a 'modern design' was reached.

[24] The current view of the origin of life postulates a stage of prebiotic systems based on certain type of bi-functional polymers (like RNAs) capable of performing both template and catalytic functions, although in a much less suitable way than DNA and proteins. Hence, despite its evident limitation in the exploitation of both template and catalytic functions, this solution is organizationally much more simple (because it allows the direct conversion of a specific sequence into a specific catalytic task) and therefore more likely to have occurred. This is in fact the hypothesis of the so-called RNA world.

[25] This problem has a rather simple chemical interpretation. Template activity requires a stable, uniform morphology, suitable to be linearly copied (i.e., a monotonous spatial arrangement that favors low reactivity and is not altered by sequence changes), whereas catalytic diversity requires precisely the opposite: a very wide range of 3D shapes (configuration of catalytic sites), which are highly sensitive to variations in the sequence (Moreno & Fernández, 1990, Benner, 1999).

complementary – polymers create two new ‘operational subsystems’: one in charge of the reliable recording, storage, and replication of certain sequences (which are crucial for the correct functioning of the system and that of similar systems in future generations) and the other strictly in charge of carrying out – with increasing efficiency – these metabolic tasks required for the continuous realization of the individual autonomous system (mainly contributing to the generation of more and more sophisticated mechanisms of catalysis). Then, these former template–catalytic components (RNA) can be substituted by two others: pure templates now completely free of any catalytic task will become tools for an unlimited memory (as we know them in present-day DNA), and pure catalysts, optimally suited for translating sequential variations into 3D diversity (as happens in present-day proteins).

It must be stressed at this point that the emergence of these two new types of processes cannot occur but from a common metabolic platform (i.e., from the characteristic organization of the previous stage), which now becomes responsible for keeping a constant bond between the two, and enables their complementary development as generations flow. Now, as these two new types of macromolecular components (the new templates and the new catalysts) cannot be made of the same kind of monomers or the same kind of chain bonds, an indirect, mediated connection turns to be a requisite in order to achieve an effective functioning between the two operational levels of the system. Thus, the connection becomes a mechanism of a contingent nature (which does not mean that it will be established randomly or free from any constraint). This mechanism, known as the genetic code, is what underlies the duality between genotype and phenotype.[26]

The scenario that this transition brings about is a new form of organization based on two quite different – but complementary – forms of operation: one involved in the fundamental productive-metabolic processes (i.e., dynamic, rate-dependent processes) and the other (constituted by the genetic processes), partly decoupled from all that muddle of on-line chemical reactions, puts together a group of special processes and components (rate-independent processes), with particular rules of composition and functioning (Pattee, 1977). Now, despite their dynamical decoupling, both forms of operation in the system are so deeply linked that none of them can work without the other, and therefore, the whole system depends on their causal connection. However, as the genetic strings are dynamically decoupled from the metabolic processes, their causal action is of a

[26] The genetic code, rather than a mechanism, is in fact the expression of an organization, which Pattee (1982) has called Semantic Closure. The basic idea is that gene strings are self-interpreting symbols because their action – specific but arbitrary as it is mediated by the recognition of certain functional components – is the synthesis of those components (tRNAs and synthetases) that allow the causal action of the very genes. Thus, by contributing to the maintenance of the whole cellular organization, genes in fact achieve their own interpretation.

quite different nature: In fact the causal action of the DNA is the transmission of a specific order or form. DNA 'selects' certain specific sequences for the different amino acids building up proteins. Therefore, the causal role of DNA is to instruct the synthesis of otherwise highly improbable proteins.[27] On the other hand, the almost inert character of DNA molecules and the dynamical decoupling from the metabolic processes permit to see the changes in the DNA strings as practically independent of the metabolic organization (as if they were 'compositional' changes). This dynamical decoupling along with its causal role expresses the informational nature of the genetic material in living organization. Thus, in this new form of organization, causal action is structured in different levels: within the dynamics of the metabolic process and between the dynamically decoupled operational levels of genes and proteins.

Ultimately, this decoupling of the genetic material from the metabolic dynamics is the expression of the radical insertion of organisms, as autonomous systems, into a historical–collective (meta)system, where the 'slow' processes of creation and modification of informational patterns take place and where an additional circular relation of cause and effect is established between individual organizations and the eco-evolutionary global dimension. The origin of information (of genetic information) takes place precisely when the link between both dimensions is articulated. As a consequence, the appearance of information opens a new scenario incorporating continuously new causal relations in the individual organization. Each time a new genetic component, linked to the production of a new functional protein, enters into the organization of a cell, and if this modification turns out to be viable and advantageous for that cell, a new causal link becomes stabilized.

Thus, in functional terms, the causal action of information allows, on the one hand, the robustness of the processes associated with self-maintenance in the early living systems and, on the other, the increase in the complexity of living systems. The informational components, shaped through a collective and historical process, re-arrange material subsets of structures so that highly organized systems are generated and preserved. One important feature of this new organization is that the specification for the maintenance of the system is hierarchically organized: an important part of these specifications is stored/recorded in the informational sequences of the genetic components. This allows the robust maintenance of much more complex networks (which in turn will support more specifications in their connectivity). So, the informational organization will allow

[27] However, the final result of the DNA – the function of the proteins – is also due to the very materiality of the protein. In fact, one of the key aspects of this process is an 'opportunistic' use of the self-assembling properties of the material properties of components. Protein folding is largely a tacit process where the explicit information contained in the linear sequence is only a short part of the causes that govern the process of folding, and therefore the expression of its function. Thus, the specific materiality of the protein is crucial.

a self-sustained feedback between records and metabolic networks, allowing further increases in complexity.

This organization, which we can now call living organization, will permit a limitless exploration of a potentially huge sequence space (of the modular templates) that can also be matched with an unlimited space of functions.

6. THE OPEN STRUCTURE OF DARWINIAN EVOLUTION

Unlike the limited world of HAS, the appearance of informational organization generates a new form of self-preservation: it generates a process in which (a set of) individual organizations reproduce their basic functional dynamics, bringing about an unlimited variety of equivalent systems, of ways of manifesting that dynamics, which are not subject to any predetermined upper bound of organizational complexity (even if they are, indeed, to the energetic-material restrictions imposed by a finite environment and by the universal physico-chemical laws) (Ruiz-Mirazo, Umerez & Moreno, submitted). Thus, the key of this form of self-preservation is the following: On the one hand, any of these systems can adopt a particular, hereditary, form of organization. On the other hand, all of them, however different they may be, share a basic common organizational form (as has been described) whose long-term preservation depends precisely on this capacity of continuous, unlimited variation.

Let us examine the different elements and relations underlying this new form of preservation through open-ended evolution. First, we need a set[28] of certain individual 'units', which have to be autonomous agents (organisms) capable of template reproduction and whose organization is based in two complementary polymers (which implies a full Geno-Phenotype duality: the genetic level is separated from the organization of the agent). We shall call this organization, as has been described, a basic informational organization (BIO). Second, these units will constitute a web of recursive interactions. Third, this web is a populational and ecological system. Populational: it requires a critical mass of individuals sharing the same specific form of SM (phenotype); a certain degree of variability with buffering (within which selection, namely, differential reproduction with conservation of form, is generated) is also necessary; Ecological: it requires the existence of different kinds (species) of individual components (organisms). However, from an initial unique type a diversity of kinds will be generated, so that the boundary conditions for the SM of a given type start to depend on the interactions with other types (ecological relations). As a consequence, the different reproductive rates will depend on the recursive relations between types (and between individuals of the same type).

[28] The system has to include certain degree of redundancy: A unique individual component is not enough.

Accordingly, the specific way in which the units achieve self-maintenance and self-reproduction is deeply entangled with the recursive relations among them. Given all that, the temporal unfolding (evolution) of this system is such that the kinds (species) change according to the transformations of the boundary conditions of the self-maintenance and self-reproduction of their units (individual organisms), which in turn are the result of the recursive interactions at the populational and ecological level. The variation possibilities of (both) individual components and populations are, in principle, open, but the whole evolutionary system is not.[29]

This set of relations would constitute the nucleus of the form of evolution that we know in current life, namely, Darwinian evolution. Let us summarize the requirements for a Darwinian system.

(1) At the level of the individual components: They have to posses an internal organization such that (a) they can reliably self-reproduce the common basic informational organization (CBIO) that characterizes them as individual agents; (b) this CBIO has to admit an unlimited variety of forms, which (c) they will indefinitely and reliably reproduce.
(2) At the global level, the populationally and ecologically generated boundary conditions will act on a particular variety of forms (but never on the CBIO). The effective variability of these forms (unfolded in time) depends on these boundary conditions, and in turn, these ones ultimately depend on the way the former interact with each other, recursively, so that the whole system achieves long-term SM.
(3) This long-term SM of the whole system is (as a consequence of these two previous requirements) necessarily an open process not subject to any pre-determined upper bound of organizational complexity[30] (except to the energetic-material restrictions imposed by a finite environment and by the universal physico-chemical laws[31]). (See Fig. 2)

Notice that we have used the term "system" instead of "organization" for this set of relations. The reason is the following: Whereas in any of the cases where we have used the term organization there is a form of closure, the evolutionary 'self-maintenance' never closes. It is essentially an open-ended process.

[29] Until new properties of the components giving rise to new forms of interaction capable to generate new forms of SM (societies, technologies, etc . . .) emerge, and this way, the fundamental biological organization would be transcended.

[30] As any more complex form would not be preserved unless it were compatible with this organizational structure.

[31] This implies an endless process of creation of new organizational forms. There are, however, certain restrictions in this huge unlimited space (body plans, internal SO laws, etc.) but, on the other hand, these restrictions can act as a set of new rules that create new primitives and relations allowing new forms of increasingly complex organizations.

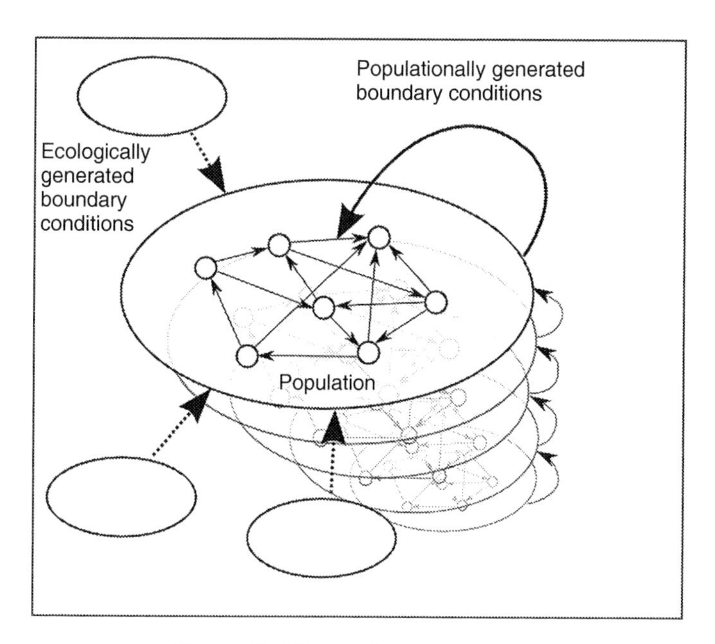

Figure 2 The evolutionary system

This open-endedness constitutes the main difference between the concept of organization based on the principle of nontrivial closure/SM and that of Natural Selection. This latter, by its nature, tends to generate processes that never close,[32] and, with the invention of the informational organization, this potential openness of NS may finally unfold.

Therefore, the invention of the basic living organization was qualitatively different from any other precedent invention in terms of long-term preservation, admitting ever new increases of complexity. Once the first form of living organization appears, not only will it be preserved, but it also becomes the condition of possibility for further and more complex organizational steps. All further forms of organization that have appeared in the course of evolution not only have retained the basic organization of early life, but also crucially depend on it (just think on how bacterial life is the condition of possibility of the emergence of more complex living systems).

7. CONCLUDING REMARKS

The origin of living organization occurred through a long-term process involving different steps. During this process each new form of organization erased these

[32] Although these processes may hit a ceiling, as we have seen for those HAS based on a unique type of polymer.

previous ones that brought them forth. Hence, the origin of life is a set of self-erasing increasingly complex organizational steps leading to a ratchet-like form of organization that is able to increase its complexity only by preserving its basic form. However, once the first form of living organization appeared, it was not only preserved, but also became the condition of possibility for further and more complex organizational steps. Therefore the invention of the basic living organization was qualitatively different from any other precedent invention in terms of long-term preservation, regardless of the eventual increases of complexity.

At the beginning, the 'driving force' was nothing but the confluence of several principles of ordering, like self-assembly, template replication, or self-organization, merged in the framework of what I have called a nontrivial self-maintaining organization. Given this special form of self-maintenance, and provided a long and wide enough scenario, the complementary action of these principles will lead to a process of increase in complexity.

The key to the beginning of this process is functional recursivity. Every novelty capable of contributing to a more efficient form of maintenance will be recruited. This process makes possible the appearance of new, increasingly complex forms of organization. But the key to their stability and capacity of preservation is the achievement of a basic autonomy. Only autonomous systems could attain the threshold of complexity necessary for the appearance of the new driving force of Natural Selection.

Natural selection brings about a mechanism for exploring new causal relations in a much wider dimension, which will progressively make the organization of autonomous systems more complex. But the increase in their complexity will also produce deep changes, integrating them in a historical and collective dimension. Self-maintenance transcends the level of the individual, autonomous organization.

The basic organization of living systems is what generates the mechanism of Darwinian evolution, which is in turn the driving force that ensures their long-term maintenance and increase in complexity. But, ultimately, the open-endedness of the evolutionary process has its roots in that organization.

ACKNOWLEDGMENTS

This work is supported by the Comisión Interministerial de Ciencia y Tecnología (CICYT) of Spain, Grants HUM2005-02449/FISO and BMC2003-06957, and by the University of the Basque Country (UPV/EHU) 9/UPV 00003.230-15840/2004. The author acknowledges the critical comments of previous versions of the manuscript made by J. Umerez and A. Etxeberria.

REFERENCES

Benner SA. *How small can a microorganism be?* In: Size Limits of Very Small Microorganisms. National Academy Press, Washington, D.C., 126–135, 1999.

Bickhard MH. *Autonomy, function and representation.* Communication and Cognition - Artificial Intelligence, 17(3–4): 111–131, 2000.

Deamer D. *The first living systems: a bioenergetic perspective,* Microbiology and Molecular Biology. Review: 61, 239–261, 1997.

Farmer JD, Kauffman S & Packard NH. *Autocatalytic replication of polymers.* Physica, 22, D 50–67, 1986.

Fontana W. *Algorithmic chemistry.* In: Artificial Life II (Eds.: Langton CG Taylor C, Farmer JD & Rasmussen S). Addison-Wesley: Redwood City. 159–209, 1992.

Harold FM. *The Vital Force: A Study of bioenergetics.* Freeman, New York, 1986.

Harold FM. *The way of the Cell.* Oxford University Press. Oxford, 2001.

Hordijk W, & Steel M. *Detecting autocatalyctic, self-sustaining sets in chemical reaction systems.* Journal of Theoretical Biology: 227(4): 451–461, 2004.

Kampis G. *Self-modifying systems in biology and cognitive science: A new framework for dynamics, information and complexity.* Pergamon Press, Oxford, 1991.

Kant I. *Critique of Judgment.* Hackett Publishing Company, Indianapolis-Cambridge, 1987.

Kauffman S. *The Origins of Order: Self-organization and Selection in Evolution.* Oxford University Press, Oxford, 1993.

Letelier JC, Soto-Andrade J, Guíñez Abarzúa, Cornish-Bowden A & Cárdenas ML. *Organizational invariance and metabolic closure: analysis in terms of (M, R) systems.* Journal of Theoretical Biology: 238, 949–961, 2006.

Maturana II & Varcla FJ. *Dc máquinas y seres Vivos -Una teoría sobre la organización biológica.* Editorial Universitaria S.A., Santiago de Chile, 1973.

Moreno A & Fernández J. *Structural limits for evolutive capacities in molecular complex systems.* Biology Forum: 83, 335–347, 1990.

Moreno A & Ruiz Mirazo K. *Metabolism and the problem of its universalization.* BioSystems: 49(1), 45–61, 1999.

Moreno A & Etxeberria A. *Agency in natural and artificial systems.* Artificial Life Journal: 11(1–2), 161–176, 2005.

Moreno A & Barandiaran X. *A Naturalized Account of the Inside-Outside Dichotomy.* Philosophica: 73, 11–26, 2005.

Morowitz HJ. *Beginnings of Cellular Life.* Yale University Press, Binghamton, New Cork, 1992.

Pattee HH. *The physical Basis and Origin of Hierarchical Control.* In: Hierarchy Theory (Ed.: Pattee, H). George Braziller, New York, 71–108, 1973.

Pattee HH. *Dynamic and linguistic modes of complex systems.* International Journal of General Systems: 3, 259–266, 1977.

Pattee HH. *Cell psychology: An evolutionary approach to the symbol-matter problem.* Cognition and Brain Theory: 4, 325–341, 1982.

Penny D. *An interpretive review of the origin of life research.* Biology and Philosophy: 20, 637–671, 2005.

Pereto J. *Controversies on the origin of life.* International Microbiology: 8, 23–31, 2005.

Rashevsky N. *Topology and life: In search of general mathematical principles in biology and sociology.* Bulletin of Mathematical Biophysics: 16, 317–348, 1954.

Rosen R. *Some realizations of (M.R)-systems and their interpretation.* Bulletin of Mathematical Biophysics: 33, 303–319, 1971.

Rosen R. *On the dynamical realizations of (M,R)-systems.* Bulletin of Mathematical Biophysics: 35, 1–9, 1973.

Rosen R. *Life itself: A comprehensive inquiry into the nature, origin and fabrication of life.* Columbia University Press, New York, 1991.

Ruiz-Mirazo K & Moreno A. *Searching for the roots of autonomy: the natural and artificial paradigms revisited.* Communication and Cognition - Artificial Intelligence: 17(3–4): 209–228, 2000.

Ruiz-Mirazo K & Moreno A. *Basic Autonomy as a fundamental step in the synthesis of Life.* Artificial Life Journal: 10, 3 235–260, 2004.

Ruiz-Mirazo K, Pereto J & Moreno A. *A Universal Definition of Life: Autonomy and Open-ended Evolution.* Origins of Life and Evolution of the Biosphere: 34(3) 323–346, 2004.

Ruiz-Mirazo K, Umerez J & Moreno A. *Open-ended evolution requires informationally organized systems* (submitted).

Segré D, Ben-Eli D, Deamer DW & Lancet D. *The lipid world.* Origins of Life and Evolution on the Biosphere, 31, 119–145, 2001.

Steel M. *The emergence of a self-catalysing structure in abstract origin-of-life models.* Applied Mathematics Letters: 13(3), 91–95, 2000.

Simon H. *The Sciences of the Artificial.* MIT Press, Cambridge MA, 1969.

Szathmáry E & Maynard Smith J. *From replicators to reproducers: The first major transitions leading to life.* Journal of Theoretical Biology: 187, 555–571, 1997.

Wächstershäuser W. *Before enzymes and templates: Theory of surface metabolism.* Microbiological Reviews: 52, 452–484, 1988.

Wicken JS. *Evolution, Thermodynamics and Information. Extending the Darwinian Program.* Oxford University Press, Oxford, 1987.

Woese C. *On the evolution of cells.* Proceedings of the National Academy of Sciences: 99(13), 8742–8747. 2002.

12

Biological mechanisms: organized to maintain autonomy[1]

William Bechtel

SUMMARY

Mechanistic explanations in biology have continually confronted the challenge that they are insufficient to account for biological phenomena. This challenge is often justified as accounts of biological mechanisms frequently fail to consider the modes of organization required to explain the phenomena of life. This, however, can be remedied by developing analyses of the modes of organization found in biological systems. In this paper I examine Tibor Gánti's account of a chemoton, which he offers as the simplest chemical system that exhibits characteristics of life, and build from it an account of autonomous systems, characterized following Moreno as active systems that develop and maintain themselves by recruiting energy and raw materials from their environment and deploying it in building and repairing themselves. Although some theorists would construe such self-organizing and self-repairing systems as beyond the mechanistic perspective, I maintain that they can be accommodated within the framework of mechanistic explanation properly construed.

[1] I thank Fred Boogerd, Frank Bruggeman, Andrew Hamilton, Alvaro Moreno, Adam Streed, and Cory Wright for the very useful discussions and helpful comments on earlier drafts of this paper.

Systems Biology
F.C. Boogerd, F.J. Bruggeman, J.-H.S. Hofmeyr and H.V. Westerhoff (Editors)
Isbn: 978-0-444-52085-2

1. INTRODUCTION

The reference to systems in the name *systems biology* points to a holistic emphasis that opposes an extreme reductionistic, mechanistic approach to biology that champions the decomposition of biological systems into their molecular constituents and emphasizes such constituents in explanations of biological phenomenon. For some theorists who adopt the name systems biology (see, for example, van Regenmortel, 2004; Kellenberger, 2004) this entails repudiating the whole tradition of mechanistic biology. On this view, only by maintaining a focus on the whole system in which biological phenomena occur can one hope to understand such phenomena. Many other advocates of systems biology, including the editors of this volume (Boogerd et al., 2002; Boogerd et al., 2005; Bruggeman et al., 2002; Bruggeman, 2005), view the focus on systems as providing an important corrective to overly reductionistic mechanism, but construe the resulting understanding to be compatible with a mechanistic perspective. To evaluate the fate of mechanism within systems biology requires us to examine carefully the commitments of mechanism. Mechanism, I will argue, has the conceptual resources to provide an adequate philosophical account of the explanatory project of systems biology, but it can do so only by placing as much emphasis on understanding the particular ways in which biological mechanisms are organized as it has on discovering the component parts of the mechanisms and their operations.

For philosophy of science, the emergence of antimechanistic voices in biology is ironical as philosophers of science have only recently recognized and appreciated the ubiquity of appeals to mechanism in biological explanations and offered models of explanation in terms of mechanisms (Bechtel & Richardson, 1993; Glennan, 1996; 2002; Machamer et al., 2000; Bechtel & Abrahamsen, 2005).[2] These accounts of mechanistic explanation (which I discuss in Section 2) attempt to capture what biologists themselves provide when they offer explanations of such phenomena as digestion, cell division, and protein synthesis. Like the biological accounts on which they are modeled, the philosophical accounts of mechanisms have tended to focus more on the component parts and operations in mechanisms than on how they are organized. Thus, while these accounts have identified organization as an important aspect of any account of a mechanism, they have not focused on the particular modes of organization that are required in biological systems. As a result, they fail to answer the objections of holist

[2] Until the recent rise of mechanist accounts, most philosophical accounts of explanation viewed universal laws as the key element in an explanation (see, for example, Hempel, 1965, for the canonical presentation of the deductive-nomological model). This has seemed particularly problematic in the context of biology, as biologists infrequently offer laws and, when offered, they seem to describe the phenomena more than to provide explanations of it.

critics (discussed in Section 3) who claim that mechanisms and mechanistic science are inadequate to the phenomena of life.

Part of the challenge of developing an adequate account of mechanism stems from the fact that when thinking about how mechanisms are organized, humans tend to think in terms of linear pathways: the product of the operation of one part of a mechanism is passed to another part of a mechanism, which then performs its operation.[3] But natural systems (and increasingly engineered systems) rely on far more complex, nonlinear modes of organization. Understanding the significance of nonlinear modes of organization is daunting, as the history of the development of the concept of negative feedback exemplifies. Many centuries passed between its first known application by Ktesibios in approximately 270 BCE to ensure a constant flow of water into a water clock, and its recognition as a principle of organization that enabled controlled behavior by complex systems. In the subsequent two millennia it had to be repeatedly rediscovered in different contexts in which control was needed (Mayr, 1970). For example, windmills need to be pointed to the wind, and a British blacksmith E. Lee developed the fantail as a feedback system to keep the windmill properly oriented. When furnaces were developed, temperature regulation became important and Cornelis Drebbel designed such a regulator around 1624. A major turning point in the recognition of negative feedback as a common design principle followed James Watt's introduction in 1788 of a governor for his steam engine (Fig. 1). This became the focus of mathematical analysis by James Clerk Maxwell (1868). The idea of feedback control was further developed and utilized in a variety of fields in the late nineteenth and early twentieth centuries. For example, it was employed for automated ship and airplane navigation; Elmer Ambrose Sperry developed a version of the gyroscope adequate for such functions in 1908 and in 1910 founded the Sperry Gyroscope Company. During World War I he became involved in the design of devices to guide anti-aircraft fire and continued to provide guidance to the US military in the interwar period.

Although the system Sperry developed, the T-6 antiaircraft gun director, used negative feedback in its internal analog computations, it did not use feedback from the target (Mindell, 1995). In the 1930s, Norbert Wiener and Julian Bigelow at MIT tried to apply feedback from the target to control anti-aircraft fire. They soon encountered an obstacle: if the feedback signal was at all noisy and the system responded too quickly, the feedback caused it to go into uncontrollable

[3] This linear focus is highlighted in Machamer, Darden, and Craver's characterization of mechanisms as providing continuous accounts from start up to termination conditions. This emphasizes the role of mechanisms in producing things, but at the cost of downplaying the often cyclic nature of their internal operation. This tendency is exhibited in the biochemists' portrayal of chemical pathways such as fermentation as linear streams from starting substances (glucose) to products (alcohol). Various reactions such as the reduction of NAD^+ are shown to the side of the main linear pathway, but following these reactions often reveals cyclic relations which link different reactions in the main pathway (see Bechtel & Richardson, 1993, Chapter 7).

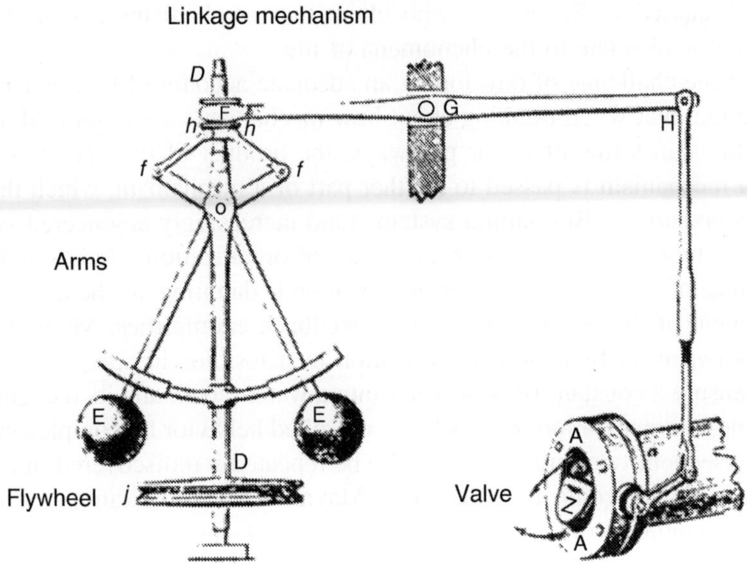

Figure 1 A schematic representation of the governor James Watt designed for his steam engine.

The speed of the flywheel determines how far out the angle arms move by centripetal force. They are in turn linked to the valve in such a way that when the flywheel is turning too quickly, the steam supply is reduced, and when it is turning too slowly, the steam supply is increased. Drawing reproduced from J. Farley (1827), *A treatise on the steam engine: Historical, practical, and descriptive*, London: Longman, Rees, Orme, Brown, and Green.

oscillations. On consulting the Mexican physiologist Arturo Rosenblueth, they learned of a similar behavior in human patients with damage to the cerebellum and recognized the importance of dampening the feedback signal to achieve reliable control. The limitations they found in negative feedback did not dissuade them of its importance; on the contrary, it suggested to them that it was a fundamental principle of design in biological systems and, they proposed, in social and engineered systems as well. In a paper published in *Philosophy of Science*, they argued that negative feedback provided a means of resuscitating notions such as purpose and teleology, enabling these concepts to be applied to both biological and engineered systems without invoking vitalism (Rosenblueth et al., 1943). Their idea was straightforward and powerful – if feedback enabled the system to maintain a given temperature, then maintaining that temperature was that system's goal or telos. Wiener and his collaborators championed the notion of negative feedback as a fundamental principle of design, and with support from the Macy Foundation, they established a series of twice-yearly conferences known as the Conference for Circular Causal and Feedback Mechanisms in Biological and Social Systems. Wiener later coined the term 'cybernetics' from the Greek word for 'steersman' (Wiener, 1948) for feedback control. Thereafter, the conference he and his collaborators had begun was called the *Conference on Cybernetics* and the term 'cybernetics' was applied generally to the movement that attempted to understand control in biological and artificial systems in terms of negative feedback. In section 4 I will show how negative feedback, along with such notions as maintaining a constant internal environment, provided an important step in biologists' attempt to address the concerns of vitalists.

As challenging as it was for humans to master the concept of negative feedback, it is the simplest of nonlinear modes of organization to understand. Once the idea of control via negative feedback is explained, the principle becomes relatively intuitive. The functionality of other nonlinear modes of organization is more difficult to understand intuitively. Organization involving positive feedback systems has often been rejected as leading only to systems that run out of control.[4] But in the twentieth century, theorists gradually recognized that in some cases positive feedback can enable systems to self-organize. It is interesting to note, though, that when Boris Belousov first proposed the reaction now known as the Belousov–Zhabotinsky or B–Z reaction, his paper was rejected on the grounds that such self-organizing reactions are inherently impossible. In the half century since his pioneering investigations, self-organizing systems of reactions have come to be viewed by many theorists as providing the basis for understanding such things as the origins of living systems (Kauffman, 1993; Nicolis & Prigogine, 1977). Nonetheless, in many circles the potency of self-organizing positive feedback to create and maintain organized systems remains underappreciated.

Cyclic organization came to the focus of biology in the twentieth century with the discovery that large numbers of biological systems involve cycles. These include various biochemical metabolic cycles, the cell cycle and cycles of reproduction, and cycles through the biosphere such as the carbon and nitrogen cycle (Smil, 1997). In Section 4, I focus on the discovery of cycles in biochemistry and their significance for understanding mechanisms in living organisms. I draw, in particular, on the work of Tibor Gánti, who, recognizing how cycles differed from ordinary chemical reactions, developed a way to represent stoichiometric

[4] The concern is, in fact, often well-founded. Basic metabolic pathways such as glycolysis involve a positive feedback system: ATP, which is produced by the pathway, is also used to prime the pathway, being consumed in the initial reactions of the pathway to produce hexosemonophosphate and fructose 1,6-biphosphate. In normal yeast, a negative feedback from hexosemonophosphate (through the production of trehalose 6-phosphate, which inhibits hexosekinase, the enzyme responsible for the synthesis of hexosemonophosphate) serves to regulate the priming step so that too much hexosemonophosphate and fructose 1,6-biphosphate do not accumulate. In a mutant in which this negative feedback is removed, however, positive feedback continues unabated as long as ATP is available and the yeast fails to grow despite plentiful glucose and an otherwise intact glycolytic pathway. (See Teusink et al., 1998; I thank Fred Boogerd for pointing me to this example.)

relations within cycles so as to trace the flow of matter through such systems. From this starting point Gánti went on to articulate an account of how cyclic organization could be harnessed to provide the core of a minimal chemical machine that would exhibit many of the fundamental properties of a living organism.

Nonlinear modes of organization provide the tools for understanding the organization of living systems, but further conceptual analysis is required to understand the significance of such organization to account for the features of life. Such a framework can be provided if we focus on living systems and some of their salient features. One of the major claims of the cell theory, as it developed in the nineteenth century through the endeavors of Schleiden, Schwann, Virchow, and others, is that the cell is the fundamental living unit (Bechtel, 1984; 2006). There are two salient features in this claim. First, the cell is a unit – it is an entity whose identity is maintained over time despite exchanges of matter and energy with its environment. Second, as a living entity, a cell is an active agent. Unlike a rock or a crystal, for example, it initiates operations that affect both itself and its environment. To these basic claims recent theorists have added a third claim – many of these internally initiated operations serve to maintain the existence of the system itself. These operations loop back to the cell itself so as to form and repair its own structure to enable it to continue its operations. The term 'autonomy' is often used to describe this capacity of cells, as the most fundamental units of life, to initiate operations that maintain themselves. I will develop the idea of organization that maintains and enhances autonomy in Section 6.

Many of the theorists who have pursued the sorts of themes about nonlinear organization that I discuss in this paper have presented them as undermining mechanistic science. A clear example is Robert Rosen who emphasized the role of nonlinear organization in maintaining metabolic repair. As I discuss in Section 6, he focused on how repair in organisms originated within the system so that the whole system was 'closed to efficient causation'. He presented such organization as a radical departure from the principles of mechanistic Newtonian science and offered them as the basis for a very different, nonmechanistic science (Rosen, 1991; see also Mikulecky, 2000). While I too will be making much of the importance in living systems of cyclic organization, I do not consider this as requiring a rupture with mechanistic science, but as helping to fill out the picture of what mechanisms are capable of doing when they are organized appropriately.[5]

---

[5] A similar integration of the basic mechanistic view with a focus on system organization is advocated by Ruiz-Mirazo and Moreno (2004, p. 238): '*system thinking* does not imply forgetting about the material mechanisms that are crucial to trigger off a biological type of phenomenon/behavior; rather, it means putting the emphasis

2. THE BASIC CONCEPTION OF MECHANISM

I begin with a basic characterization of mechanisms that captures many of the features that have figured in recent philosophical accounts of mechanism. I will then elaborate it into a framework for mechanistic explanation. A mechanism

> 'is a structure performing a function in virtue of its components parts, component operations, and their organization. The orchestrated functioning of the mechanism is responsible for one or more phenomena'.
>
> (Bechtel & Abrahamsen, 2005)

The first thing to note about this characterization of mechanism is that a mechanism is responsible for a phenomenon (Bogen & Woodward, 1988) that is here characterized as the function of the mechanism.[6] The identity conditions for a mechanism are provided by the phenomenon such that what count as parts, operations, and organization are determined by the phenomenon (Kauffman, 1971). By characterizing a mechanism as a structure I mean to emphasize that it consists of an arrangement of parts and has at least some enduring identity. Sequences of causal operations not organized into an enduring system are not mechanisms on this account.

Just as mechanisms themselves are identified in terms of phenomena, mechanistic explanation starts with a characterization of the phenomenon to be explained and seeks to characterize the responsible mechanism. Researchers do not simply hunt for mechanisms, but seek them to explain an already identified phenomenon. Part of identifying a phenomenon involves empirical research that identifies environmental conditions under which the phenomenon will appear. For example, Pasteur determined that yeast perform fermentation when in an oxygen-free environment. This does not mean that the characterization of the phenomenon remains fixed; on the contrary, investigating the responsible mechanism may lead researchers to revise their conception of the phenomenon (in Bechtel & Richardson, 1993, we characterized this as 'reconstituting the phenomenon'). For example, research on metabolic mechanisms in organisms began by construing them as responsible for the generation of heat (Mendelsohn, 1964). Only after Karl Lohmann's (1929) discovery of adenosine triphosphate was it

on the interactive processes that make it up, that is, on the dynamic organization in which biomolecules (or, rather, their precursors) actually get integrated'.

[6] There are two important features of this characterization. First, in characterizing the resulting phenomenon as the function of the mechanism, I am not committing myself to an evolutionary analysis of function in the manner of Wright (1972). Indeed, as I note below, the construal of biological systems as autonomous systems provides the basis for a very different characterization of function. Second, while there is a significant amount of flexibility available to the scientist in demarcating the phenomenon and determining whether the researcher is dealing with one or more phenomena, if the researcher demarcates multiple phenomena, then the researcher will offer multiple, potentially overlapping, mechanisms to account for them.

recognized that the point of metabolism was to convert the energy of foodstuffs into a chemical substance which could then provide energy to other vital processes. The phenomenon was then reconstituted in a very different way.

The second feature of mechanistic explanation to emphasize is that the component parts and operations of a mechanism are within the mechanism and can only be identified by taking the mechanism apart, either literally or conceptually.[7] There is a compositional relation between parts of a mechanism and the mechanism itself, and so it is useful to characterize the parts as at a lower level than the mechanism itself.[8] Appealing to parts and their operations is reductionistic, in a sense familiar to scientists, although not necessarily to philosophers.[9] But it is important to be clear on what the appeal to parts and their operations is designed to do – it explains what resources a given mechanism has that enable it to behave in a particular way when in the context in which it functions. It does not, in any way, supplant the need to identify the manner in which the mechanism as a whole operates under various conditions in its environment. Moreover, and especially important for the purposes of the current paper, it does not mitigate the importance of considerations of how the parts are organized for explaining what the mechanism does or indeed for understanding how a part of the mechanism is operating. It is an important feature of mechanistic explanation to recognize that parts will operate differently under different conditions,[10] and that the organization in which they are incorporated is often a major factor in determining these conditions and hence the operation performed by the part.

---

[7] Although literal decomposition has often been a productive strategy in biological research for identifying the operation associated with a component (e.g., isolating an enzyme through fractionation), it can also disrupt the operation when it is dependent on coordinated interactions with other components. A clear example is that the rate at which an enzyme catalyzes a given reaction is dependent on the concentrations of substrates, products, and effectors. Often the effects of being embedded in a particular organization are only realized after noting the differences in behavior in the original system and the isolated component and determining the role of the organized setting in determining the operation of the component. Hence, in the end, decomposition is often conceptual rather than literal – in a model the theorist specifies an operation performed by the part and how that performance relates to other operations occurring within the mechanism.

[8] Although this conception of levels is compositional in the sense articulated by Wimsatt (1976), the levels that result are only defined locally within the mechanism. Moreover, there is no requirement that the parts which interact with each other are of remotely the same size dimensions – in one mechanism membranes may interact with whole bacterial cells whereas in another they may interact with ions. Accordingly, the conception of reduction that emerges is local because the levels to which a scientist appeals are only identified in the context of the attempt to explain a given phenomenon. Moreover, typically a given investigation goes one or two levels down. There is no goal of reducing all sciences to some most fundamental one (Bechtel, 2006).

[9] In particular, this sense of reduction does not focus on theories and logical relations between them, as in classical philosophical accounts of theory reduction (Nagel, 1961). For discussion, see Bechtel and Hamilton (in press).

[10] Boogerd et al. (2005) emphasize the fact that parts behave differently under different conditions and invoke it as part of their case for the claim that biological systems exhibit emergent properties. Under their analysis, a property is emergent if it cannot be predicted from what is known by studying the part in isolation or in the context of simpler systems.

Identifying parts and operations is a challenging activity. Not every way of carving up a mechanism yields working parts – parts which perform operations that figure in the explanation of the ability of a mechanism to perform its function. Moreover, it is often challenging to figure out what sorts of operations might give rise to the function. Biochemistry only made headway when biochemical groups were identified and biochemical reactions were recognized as operations over such groups (Holmes, 1992). There are many cases in which researchers sought evidence for one sort of operation only to discover later that a very different type of operation was responsible for the phenomenon. A classic example is that biochemists, following the lead of E. C. Slater (1953), assumed that the energetic intermediate in ATP synthesis in oxidative phosphorylation would involve a chemical compound and only gradually recognized, after Mitchell (1961) advanced a very different alternative relying on an ion gradient over a membrane, that they were seeking the wrong kind of operation (Allchin, 1997).

A common frustration in biological research is the inability to reproduce the phenomenon once one has assembled what appear to be all the component parts and operations. In many cases the failing is that there remains yet unknown parts or operations. But in other cases the failing involves the third feature in my characterization of mechanisms, organization. Lip service is often given to the fact that components of a mechanism must be organized, but the importance of organization is often underappreciated. Yet, it is organization that causes parts of the mechanism to behave in ways they do not in isolation and enables the mechanism as a whole to accomplish things that none of the components alone can do. What is possible when components are put together in creative ways is often obscured when one focuses just on the components themselves. What is learned about the part in conditions in which a researcher has removed it from the context of the mechanism may not include how it will operate in the organized structure. One can appreciate this point better by turning one's focus from science to engineering. Engineers do not build new devices by creating matter with distinctive properties *ab initio*. Rather, they start with things that already exist and put them together in novel ways. What can be accomplished when the parts are put together is typically far from obvious. Creativity is required, and accordingly engineers can win patents and fame for developing a new design that enables old parts to perform new operations. The only thing the engineer added to what already existed was organization, but this is what is critical in developing mechanisms that perform tasks that initially seemed impossible to perform with existing components.

3. THE VITALIST CHALLENGE

Although the search for mechanisms within biology has not been particularly controversial since the beginning of the twentieth century, it was heatedly debated in the nineteenth century. The opponents to explaining biological phenomena in terms of mechanisms were often labeled *vitalists*. Although there was no official vitalist doctrine (just as there was no official doctrine of those advancing mechanism), vitalists tended to insist that ordinary physical objects could not generate the phenomena associated with living organisms and to maintain that explaining such phenomena required appeal to additional factors such as vital forces or vital powers. Some vitalists proposed that there was a nonphysical component of living organisms that gave them their distinctive properties. Others downplayed the radical nature of such appeals, arguing that vital forces were not substantially different from the sorts of forces Newton had invoked to explain the behavior of physical objects. The behavior of living organisms could be described in distinctive laws of biology that invoked vital forces, and biologists were no more obliged than Newton to explain them in more fundamental terms.[11]

The historical importance of vitalists lies not in their positive doctrine of vital forces or vital powers, but in their critique of mechanism. They drew attention to phenomena exhibited by living organisms that mechanists seemed incapable of explaining. The natural tendency of mechanists was to focus on the phenomena they could explain, and to divert their attention from the phenomena that proved more difficult. Vitalists thus provided an honesty check on mechanists, keeping in focus the phenomena of life that were recalcitrant to existing mechanist explanatory strategies.

An exemplar of the vitalist challenge to mechanism is offered by the French anatomist Xavier Bichat. His project began in the fashion of a mechanist as he proposed a decomposition of living systems into 21 different types of tissues distinguishable in terms of their sensibility and contractility. He then appealed to these properties to explain the phenomena associated with organs built from these tissues. But with the catalogue of tissues, Bichat contended, this explanatory project reached a limit. He highlighted two reasons for resisting any attempt to explain the phenomena associated with tissues in terms of their material composition. The behavior of living organisms was simply not sufficiently deterministic to be explained mechanistically: ‘The instability of vital forces marks all vital phenomena with an irregularity which distinguishes them from physical phenomena [which are] remarkable for their uniformity’ (Bichat, 1805, p. 81). Moreover, he contended that living organisms operate to resist external factors that threaten their existence, construing life as ‘the sum of all those forces which resist death’. This last characterization is particularly potent. The notion of resistance points to self-initiated action, where that action is directed at maintaining the living organism as a distinct system. As we will see, the construal

[11] For an account of the positions of various vitalists and mechanists in the history of physiology, see Hall (1969).

of biological systems as autonomous systems provides a means of capturing the key insight in Bichat's characterization of life.

As I will develop in the next section, in the nineteenth century Claude Bernard, by focusing on features of the organization of biological systems, took major steps toward answering Bichat. But the spirit of Bichat's objection lives on in such attempts as those of Robert Rosen (1991; for analysis, see Mikulecky, 2000), noted above, to contrast living systems with mechanisms. Attention to the differences between biological systems and humanly engineered mechanisms is not limited to critics of mechanism. I will develop below some of the fundamental insights into the nature of biological mechanisms advanced by Tibor Gánti, but before developing his positive account, it is worth noting how he contrasts biological systems with extant humanly engineered machines:

> First, living beings are soft systems, in contrast with the artificial hard dynamic systems. Furthermore, machines must always be constructed and manufactured, while living beings construct and prepare themselves. Living beings are growing systems, in contrast with technical devices which never grow after their completion; rather, they wear away. Living beings are multiplying systems and automata (at least at present) are not capable of multiplication. Finally, evolution – the adaptive improvement of living organisms – is a spontaneous process occurring of its own accord through innumerable generations, whereas machines, which in some sense may also go through a process of evolution, can only evolve with the aid of active human contribution'.
>
> (Gánti, 2003, pp. 120–121)

Many of these features, such as multiplication and adaptive change through evolution, are salient differences between extant machines and living systems, but I take the most fundamental of the features Gánti lists to be the engagement of living beings in self-construction and growth so that they do not merely wear away or dissipate in the fashion of ordinary physical objects. These capacities must be exhibited by any system that is to be a candidate for reproduction and evolution and are not found in extant machines. Hence, they are critical phenomena for which any viable mechanistic account must offer an explanation.

4. FIRST STEPS: BERNARD, CANNON, AND CYBERNETICS

As I noted above, Bernard (1865) took major pioneering steps toward answering the objections to the mechanistic approach to explaining life advanced by Bichat. Fundamental to Bernard's view of science was that causal processes are deterministic; accordingly, it was critical for him to explain away the apparent indeterminism in the activities of living organisms that Bichat had highlighted.

The key to Bernard's response was to focus on the internal organization of living systems and to argue that the internal parts of a living mechanism resided in an internal environment that is distinct from the external environment in which the organism as a whole dwells. This provided a ready account of the indeterminacy exhibited by living mechanisms. Whereas there might not be strict determinism in the response of a part of an organism to changes in the external environment, he maintained that strict determinism could be found in its response to conditions of the internal environment. For example, whereas fluctuations in the sugar available in food might not lead to changed metabolic activity in somatic tissue, decreased glucose levels in the blood would result in lowered metabolic activity in somatic tissue. The focus on the internal environment also provided Bernard the beginnings of a response to Bichat's contention that organisms are not mechanistic insofar as they operate to resist physical processes in their environment. The internal environment provides a buffer between conditions in the external environment and the reactive components of the mechanism, insulating component parts of the mechanism from conditions in the external environment. Bernard proposed that this buffering is achieved by individual components of the organism, each performing specific operations that served to maintain the constancy of the internal environment.[12] Insofar as some of its mechanisms are designed to maintain a constant internal environment despite changes in the external environment, a living system can appear as an active system doing things that resist its own demolition.[13]

Although emphasizing the role of organs in the body in maintaining the constancy of the internal environment, Bernard did not provide a detailed account of how organs might operate in this way. Walter Cannon (1929) introduced the term homeostasis (from the Greek words for *same* and *state*) for the capacity of living systems to maintain a relatively constant internal environment. He also sketched a taxonomy of strategies through which animals are capable of maintaining homeostasis. The simplest involves storing surplus supplies in time of plenty, either by simple accumulation in selected tissues (e.g., water in muscle or skin) or by conversion to a different form (e.g., glucose into glycogen) from which reconversion in time of need is possible. A second kind of homeostasis involves altering the rate of continuous processes (e.g., changing the rate of blood flow by modifying the size of capillaries to maintain uniform temperature). Cannon noted such control mechanisms are regulated by the autonomic nervous system.

[12] Bernard, for example, says 'all the vital mechanisms, however varied they may be, have only one object, that of preserving constant the conditions of life in the internal environment' (Bernard, 1878, p. 121, translated in Cannon, 1929, p. 400).

[13] Bernard's focus seems to have been more on the constancy of the internal environment than on just what the conditions in the internal environment were. A key feature of living systems is that via control of flow of materials across membranes they create environments different from those found outside them and in these internal environments component parts operate differently than they do in the external world.

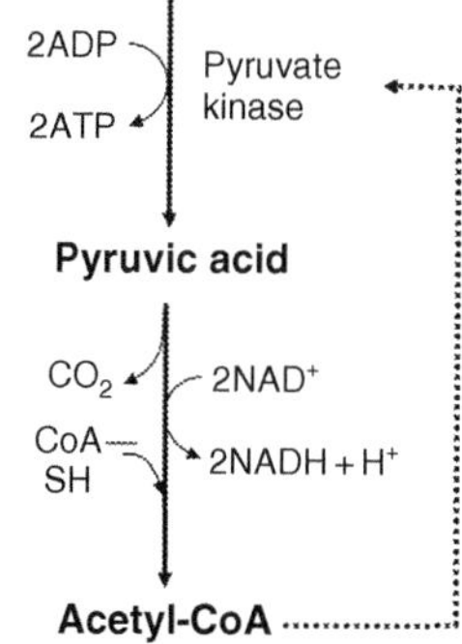

Figure 2 Feedback loop in the linkage between glycolysis and the citric acid cycle.

In the final reaction of glycolysis, phosphoenolpyruvate produces pyruvic acid. Pyruvic acid then produces acetyl-CoA, some amount of which is needed to continuously replenish the citric acid cycle (not shown). If more acetyl-CoA is produced than can be used in the citric acid cycle, it accumulates and feeds back (dotted arrow) to inhibit pyruvate kinase, the enzyme responsible for the first step in the reaction. This in turn will stop glucose from entering the glycolytic pathway.

The crucial idea required to flesh out Bernard's and Cannon's insights into how biological systems are organized to maintain themselves in the face of external challenges is that of negative feedback. As I noted above, the notion of negative feedback was repeatedly rediscovered in different engineering contexts in which it was important to regulate or control a process so as to insure a regular output over 2000 years. With the development of the cybernetics movement and control theory in engineering in the twentieth century, it came to be viewed as a general design principle. Although enriched by a variety of tools, such as the use of off-line emulators and filtering techniques (Grush, 2004), negative feedback remains at the center of the modern field of control theory. It also plays a critical role in the understanding of biological systems. Feedback loops provide a way of insuring that critical processes, such as the consumption of nutrients to generate ATP, only occur when they are required. Figure 2 illustrates an instance of negative feedback at the junction between glycolysis and the Krebs cycle which halts the generation of pyruvate from phosphoenolpyruvate (coupled with the synthesis of ATP) when there is already a plentiful supply of acetyl-CoA waiting to enter the citric acid cycle.

5. CYCLIC ORGANIZATION AND GÁNTI'S CHEMOTON

As important as negative feedback is in providing control mechanisms in living organisms that enable them to maintain homeostasis, it is not sufficient to explain the tendency of organisms to resist death. A major additional form of organization involves cycles. Cycles appear in biological accounts at various levels of organization (many of the best known involve relations between organisms and their environment, such as the nitrogen cycle), but I will focus on cycles at the level of basic metabolic processes. The pioneers in biochemistry at the turn of the twentieth century generally assumed that the chemical reactions in living organisms occurred serially and hence the pathways of reactions would be linear. They sought to identify the intermediates in, for example, the oxidation of fatty acids or the oxidation of glucose to alcohol that could be generated through known chemical reactions (oxidations or reductions of substrates, additions or removals of carboxyl groups, etc.). Often, however, they could go only so far in arranging intermediates in linear sequences; proposed reaction sequences would yield a substance which could not be processed in the same manner. The discovery of perhaps the most famous biochemical cycle, the citric acid cycle, resulted from such a circumstance. After adopting Wieland's (1913) account of oxidation as involving the removal of a pair of hydrogen atoms from a substrate (dehydrogenation), Thorsten Thunberg established that a number of organic acids (lactic acid, succinic acid, fumaric acid, malic acid, citric acid, etc.) could all be oxidized in cellular extracts in such a manner (Thunberg, 1920). He then tried to fit the reactions together in a coherent pathway involving a chain of reactions.[14] Relatively easily he was able to fit the various compounds together into the sequence:

$$\text{Succinic acid} \rightarrow \text{fumaric acid} \rightarrow \text{malic acid} \rightarrow \text{oxaloacetic acid} \rightarrow \text{pyruvic acid} \rightarrow \text{acetic acid}$$

He then faced a problem in specifying what happened next – it was not possible to remove two hydrogen atoms from acetic acid. In response to this problem, Thunberg offered a bold proposal – he proposed 'a reaction in which two acetate molecules are simultaneously each deprived of one hydrogen atom, with the joining of their carbon chains into one. The substance which must therefore form is succinic acid' (Thunberg, 1920, passage translated by Holmes, 1986, p. 69). The reaction Thunberg proposed was the following:

$$2CH_3\text{–}COOH \rightarrow COOH\text{–}CH_2\text{–}CH_2\text{–}COOH + H_2$$

This resulted in the pathway becoming a cycle, as shown in Figure 3a.

---

[14] Thunberg actually had the idea of a sequence of reactions as early as 1913, before he encountered Wieland's conception of dehydrogenation: 'The oxidative processes in living cells must be thought of as forming chain reactions, a series of reactions connected to one another in such a way that, by and large, none of the links in the reaction chain can proceed more rapidly than the others' (Thunberg, 1913, translated in Holmes, 1986, p. 68)

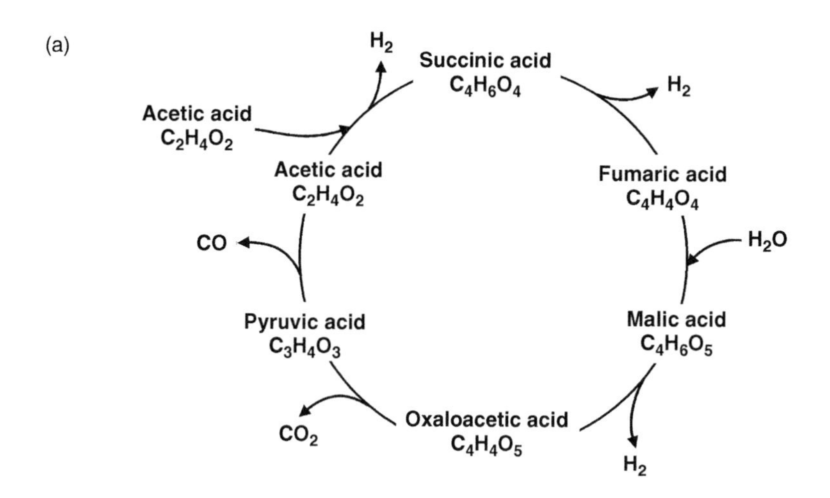

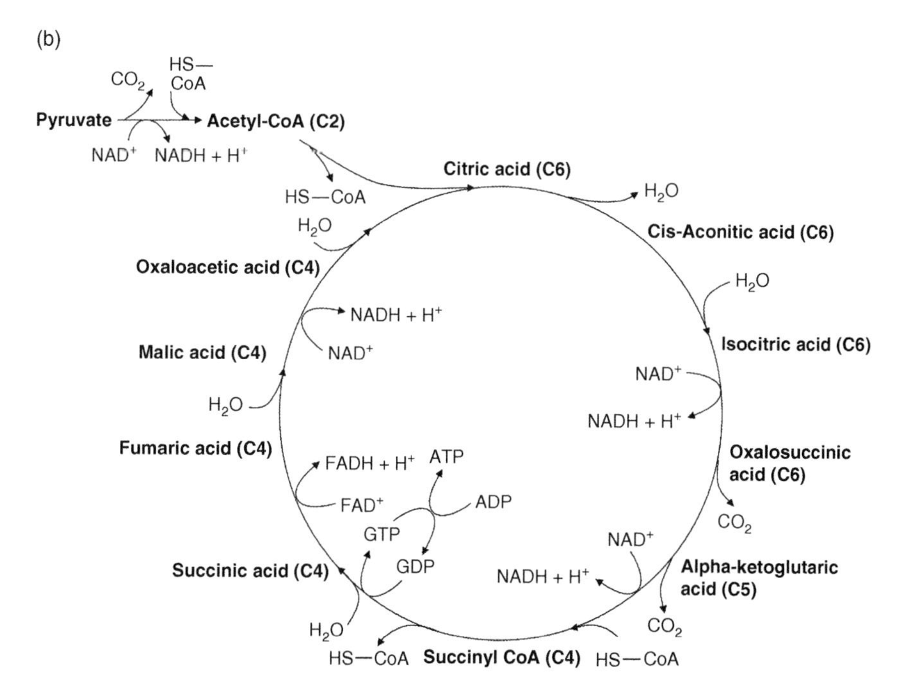

Figure 3 Two versions of he TCA cycle.

(a) Cycle proposed by Thunberg in 1920 to link various three carbon sugars that figure in oxidative metabolism. (b) the citric acid cycle as understood today.

Thunberg's proposed cycle was speculative, and a good deal of research went into trying to determine the actual intermediates in the oxidation of three carbon sugars until Krebs and William Johnson (1937) published their account of the citric acid cycle (also known as the tricarboxylic acid cycle and later as the Krebs cycle). As shown in Fig. 3, instead of two molecules of acetic acid combining to form succinic acid, Krebs proposed that oxaloacetic acid combined with a three-carbon substance, which Krebs designated simply as *triose*, to generate citric acid. He then proposed that citric acid underwent a sequence of reactions resulting in succinic acid. After the discovery of coenzyme A in the 1940s, biochemists recognized that it was acetyl-CoA that entered the cycle by combining with oxaloacetic acid to form citric acid.

What is noteworthy about Thunberg's and Krebs' research, as well as that of many other biochemists who proposed cycles in the early decades of the twentieth century, is that they were guided purely by the desire to articulate plausible pathways of chemical reactions that would provide complete accounts from known starting points to appropriate endpoints. Cyclic organization was not construed as theoretically significant. Subsequently, as the ubiquity of cycles became apparent, some theorists did take up the question of why cycles were so common. One advantage of cyclically organized processes is that they provide a means of effective negative feedback regulation – only when the products of the cycle are generated will new substances be able to enter the cycle. But they play an even more significant role in organizing chemical operations in biological systems.

An illuminating perspective on the function of cycles in biochemistry is offered by Tibor Gánti (see his 2003 for a synthetic statement of an account which he initially advanced in the 1970s). Gánti's project has been to identify the simplest chemical system that exhibits the distinctive features of life, a system he calls the *chemoton*.[15] In pursuing this project, Gánti takes his lead from the minimal biological system exhibiting the properties of life, the cell. He identifies three subsystems as common to all cells: 'the cytoplasm, the membrane, and the genetic substance'. His analysis plays particular attention to the cytoplasm, which he characterizes as 'the chemical motor'.[16] He then notes a general feature

---

[15] It is important to note that Gánti is not trying to provide a detailed account of first organisms or the origins of life but, as Griesemer & Szathmáry (forthcoming) emphasize, a heuristic model to help probe the organization required in living systems: 'the chemoton model is not intended as an accurate representation which, if implemented exactly, could live. It is instead a heuristic guide to the organizational properties of chemical systems that would minimally fulfill the living state'. Although the account is not presented as accurate in detail, in another sense Gánti is seeking to be true to the systems modeled – the general architecture of the model is intended to correspond to the architecture of the actual system. This is a feature of many similar modeling efforts that attempt to demonstrate how, possibly, a given phenomenon might be produced.

[16] Although emphasizing all three sub-systems, Gánti explicitly comments on how cytoplasm is the most complicated and in many respects the most critical:

of motors designed to produce work – they are not one-time causal agents but must be continually able to produce work:

> 'Continuous work performance can only be achieved by means of suitable work-performing systems characterized by changes occurring through a series of constrained motions, such that the inner organizational characteristics of the system remain unchanged'.
>
> (p. 68)

In order to maintain the system unchanged while still performing work, cyclic organization is required, and this he contends is true of both humanly designed mechanical systems and biological systems:

> In an internal combustion engine the explosion moves the piston from its original location, but the engine is so constructed that the displacement occurs on a constrained path and after performing work the piston returns to its starting position. . . . The ability of non-mechanical systems to perform continuous work also depends on cyclic processes or, as they are often call for simplicity, cycles.
>
> (p. 72)

In the case of the internal combustion engine, the component parts are (relatively) permanently fixed. As Gánti puts it, they exhibit a 'geometrical structure of fixed materials'. While there is some material fixity in living systems, for the most part organization involves what he terms a 'soft geometrical structure' (pp. 64–65).[17]

> 'Firstly, it is the chemical motor. The cytoplasm contains the system transforming the chemical energy of nutrients into useful work. Secondly, it is the homeostatic subsystem compensating the influences of the external world by dynamically responding to them. Therefore, the cytoplasm is responsible for the dynamic and organizational responsibility of the cell. However, it is also responsible for sensibility and excitability, since the accomplishment of homeostasis is nothing more than excitability. To achieve all of these it is necessary that processes in the cytoplasm should occur in a regulated order and thus that the cytoplasm carries the property of regulation and is considered a soft system. Finally, the raw materials necessary for the growth and reproduction of all three subsystems (cytoplasm, cytomembrane, and genetic substance) are also delivered by the cytoplasm; thus it is a self-reproducing soft system'.
>
> (pp. 83–84)

In locating the motor in the cytoplasm, Gánti is reflecting a pre-Mitchell conception of the critical chemical reactions. According to Mitchell's chemiosmotic account, a central part of the mechanism transforming chemical energy into ATP is the inner mitochondrial membrane, across which a proton gradient is established that then provides the energy for ATP synthesis.

[17] Soft organization is central, as per Gánti, for cell life: 'Now what makes the cell living? The soft organization of its inner events and occurrences. Thus, if we are looking for the fundamental laws, for the principle of life, we have to establish the connections of this soft organization' (p. 66).

Gánti uses several biochemical cycles to illustrate the ideas in the chemoton. He begins with the citric acid cycle introduced above and initially de-emphasizes the importance of new inputs of acetyl-CoA and instead emphasizes that oxaloacetic acid and all other intermediates are regenerated by the cycle. He represents this by the following formalism (Scheme 1)[18]:

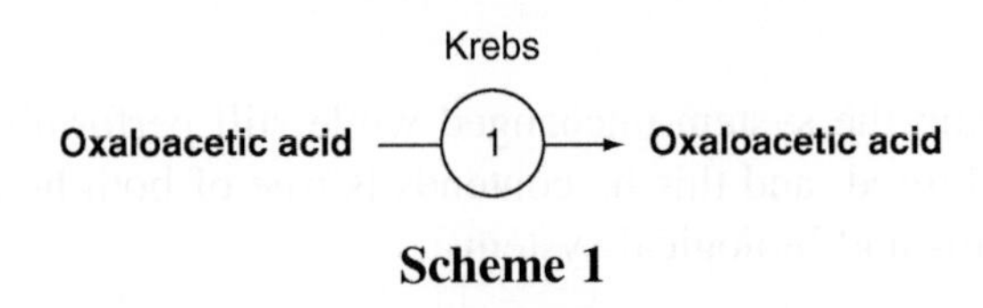

Scheme 1

To show that the nutrient material is entering the cycle and waste products are leaving, he expands the formalism (Scheme 2):

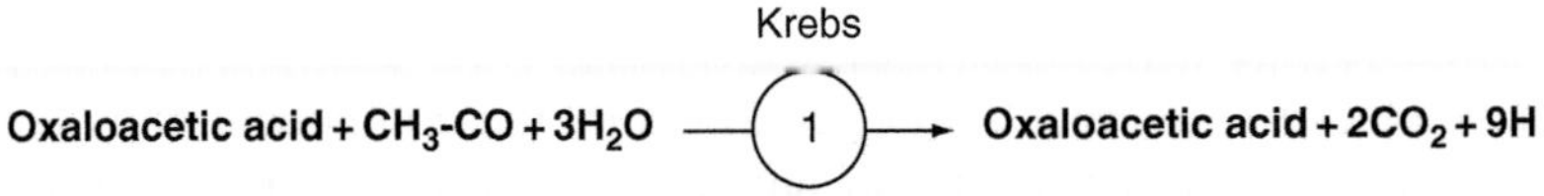

Krebs

Oxaloacetic acid + CH_3-CO + $3H_2O$ —(1)→ Oxaloacetic acid + $2CO_2$ + 9H

Scheme 2

More abstractly, Gánti uses **A** for the components of the cycle, **X** for the reactants entering the cycle, and **Y** for the reaction products (Scheme 3):

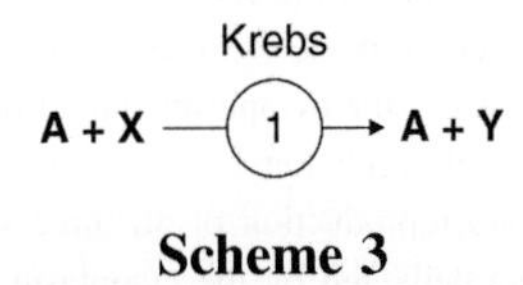

Scheme 3

For Gánti the citric acid cycle is only the starting point. He views living systems as fundamentally growing systems and so turns to autocatalytic cycles

---

[18] This formalism, as Griesemer & Szathmáry (forthcoming) discuss, was introduced by Gánti so as to draw attention to the stoichiometric requirements of catalyzed reactions. This attention to the flow of matter through the system by balancing each reaction is an important feature of Gánti's approach.

which create more of the cycle intermediates on each iteration (Gánti presents the malate cycle as an example of such a cycle) (Scheme 4):

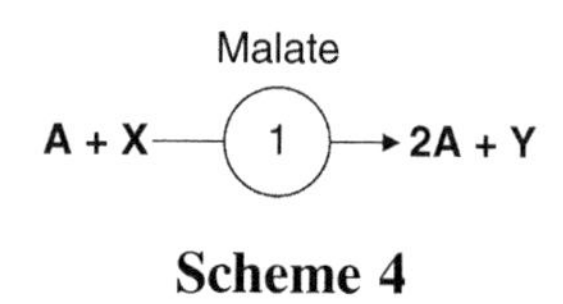

Scheme 4

To appreciate the significance Gánti attaches to such autocatalytic cycles, we need to bring in the second subsystem of a chemoton, the membrane.[19] For Gánti, the membrane not only isolates the autocatalytic system (insuring, for example, that the concentration of intermediates is sufficient that ordinary diffusion will bring reactants together) but also allows for the control of admission and expulsion of materials from the system. (Insofar as it is a selective semi-permeable barrier, the membrane itself is a sophisticated and complex mechanism – a nontrivial component for the system to build and maintain.) It is critical for Gánti's account that the chemoton creates its own membrane, and to explain how it might do this, Gánti further amends his account of the metabolic cycle so that it not only generates more intermediates of the cycle but also components of the membrane, which Gánti represents as **T** (as Gánti has now moved into the realm of a purely theoretical cycle, he designates it simple as A) (Scheme 5):

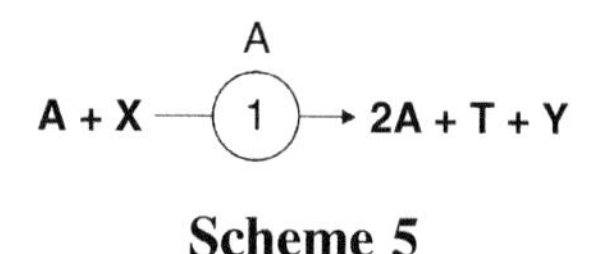

Scheme 5

Assuming that the membrane-bound system naturally takes the shape of a sphere, Gánti notes that such a stoichiometric relation would lead to the membrane increasing more rapidly than the volume of metabolites enclosed. The

[19] The membrane was not part of Gánti's initial account (see Griesemer & Szathmáry, forthcoming) and was introduced only as he recognized a need, when dealing with reactions occurring in a fluid milieu, to keep reaction components together in sufficiently high concentrations. Moreover, Gánti's account underplays the role membranes play in actual living cells – they provide not only a way to create distinct environments, but also a potent tool for energy storage. In oxidative metabolism, for example, a differential concentration of protons across a membrane, as a result of the oxidations along the electron-transport system, results in a proton-motive force that then drives the synthesis of ATP.

solution he proposes is that the system will divide into new spheres when the membrane grows sufficiently to close in on itself and bud.[20]

The metabolic and membrane (including membrane generation) systems together give rise to what Gánti characterizes as a supersystem which exhibits biological features:

> We have combined two systems of a strictly chemical character into a 'super-system' (or, to put it another way, we have combined two chemical subsystems), and we have obtained a system with a surprising new property of expressly biological character. What can this system do? It is separable from the external world and its internal composition differs from that of the environment. It continuously consumes substances that it needs from the environment which are transformed in a regulated chemical manner into its own body constituents. This process leads to the growth of spherule; as a result of this growth, at a critical size the spherule divides into two equal spherules, both of which continue the process. (p. 105)

But, according to Gánti, this system is still not living because it lacks an information-storing or control subsystem.[21] Gánti proposes to provide an information-storing subsystem by having the metabolic system also add a monomer to a polymer that is built along an existing polymer template. The length of the polymer is thereby able to carry information about the number of cycles completed.[22] (See Fig. 4 for Gánti's portrayal of the complete chemoton.)

Gánti seems to have been led to insist on this third subsystem only because such an information storage system, in the form of DNA, has been found in extant organisms: 'This property is not one of the classical life criteria, but on the basis of knowledge gained from molecular biology, it has been selected as an absolute life criterion' (p. 106). Gánti in fact says little about what the information system is to be employed for and one might ask why such an information system is required in a living system. An appreciation of its significance is provided by Gánti's own coupling of the notions of information and control. In the two-component supersystem, the metabolic and membrane systems were strictly

---

[20] Gánti's account of the membrane is overly simplistic. In order to deal with the osmotic crisis that results from concentration differences inside and outside the enclosure and the tendency of water to spontaneously enter the enclosure, resulting in its swelling–bursting, the membrane must from the outset be active in pumping materials in and out (Ruiz-Mirazo & Moreno, 2004, p. 244).

[21] James Griesemer and Eörs Szathmáry include marginal notes accompanying Gánti's text, and Griesemer notes at this point that, had Gánti not been focused on a template-based information system, he could have included an information encoding structure within the membrane system by allowing, for example, the incorporation of a variant molecule into the membrane that will be replicated as the membrane is replicated, resulting in what Jablonka and Lamb (1995) describe as a 'structural inheritance system'.

[22] Griesemer provides some suggestions as to how polymer length carries information. For example, if one molecule is added to the polymer at each turn of the metabolic cycle, it can provide a more reliable indicator of the growth that has already occurred and when the next division should occur.

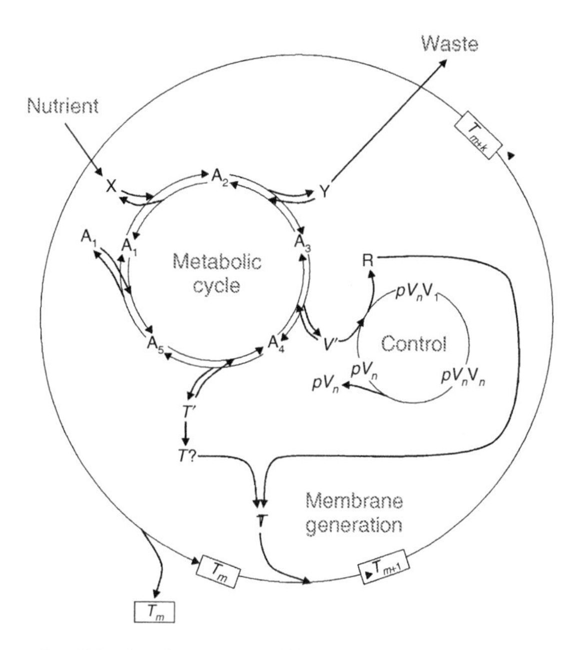

Figure 4 Gánti's chemoton with the three subsystems labeled.

linked with each other such that as the metabolic system produced metabolites it also produced membrane. This will work as long as the environment of the system continues to provide the system exactly what it needs, regularly removes its waste, and does nothing to interfere selectively with either the membrane or the metabolic process.

Even slight variations in the environment may disrupt such a system. Imagine the environment changed so that a new substance entered the system which would react with existing metabolites, either breaking down the structure or building new additional structure. This would disrupt the delicate balance between metabolism and membrane generation that Gánti relies on to enable chemotons to reproduce. What this points to is the desirability of some procedures for controlling operations within the system that are not directly tied to the stoichiometry of the metabolic reactions themselves. Although stoichiometric linkages between reactions are effective for insuring linkages between operations, they do not provide a means for varying the reactions independently. Such independent control can only be achieved by a property not directly linked to the critical stoichiometry of the system.

Griesemer and Szathmáry (forthcoming) provide an account of the stoichiometric freedom made possible with the information subsystem Gánti proposes. If, instead of just one type of molecule being combined into the polymer, two or more constitute the building block, then the polymer will exhibit both a composition of monomers in specific concentrations and a sequence. The concentrations, like other features of the chemoton, will depend on specific stoichiometric relations. The sequential order, however, will not–it is a 'free' property which can then be linked to component operations in the chemoton in other ways so as to control them. Moreover, it can also undergo selection. This indicates to a general point about control systems–if control is to involve more than strict linkage between components, what is required is a property in the system that varies independently of the basic operations. Particular values of this variable property can then be coordinated with responses by other components so that the property can exert control over the operation of the other component.

What is required for an information-based control system that goes beyond direct negative feedback loops, then, is a property that is sufficiently independent of the processes of material and energy flow such that it can be varied without disrupting these basic processes (which may themselves be maintained by negative feedback), but still be able to be linked to parts of the mechanism to enable the modulation of their operations. Such properties need not involve a template matching system but there are reasons to employ such a system. We can appreciate this by considering the role of enzymes in catabolism. On some occasions[23] Gánti tries to characterize the chemoton without invoking catalysts or enzymes to promote the reactions, but it is ultimately unlikely that a system of uncatalyzed reactions could maintain the flow of matter and energy needed in the chemoton. If enzymes are to play a role, however, they must be synthesized by the system. The method for synthesizing proteins, including enzymes, in actual living cells utilizes a common mechanism (a ribosome) that adds amino acids to a developing string. If this mechanism is to produce different enzymes on different occasions (as it must if the result is the set of proteins needed to catalyze a given reaction), a stable information source for specifying the sequence in each protein is required (plus a component for insuring that the right instructions are followed at a given time). In living cells, mRNA plays this role. Here we see a compelling reason for an informational system – it insures that the system is able to make the needed components so that it is able to function. The fact that it also provides the retention component for a process involving variation and selective retention, i.e., natural selection (Lewontin, 1970), is an added plus but not an essential feature of the information-based control system.

Gánti's chemoton proposal is highly speculative, but it offers an intriguing perspective on the centrality of cyclic organization in biological systems. Cycles provide a vehicle through which a set of operations that depend on each other can be maintained even as the overall mechanism performs work. When part of

[23] When he appeals to examples such as the Krebs' cycle, however, Gánti is already invoking an enzyme-catalyzed metabolic cycle. Enzyme-catalyzed systems pose an additional challenge for Gánti's account as they must themselves be made by the system. One way to view this is to differentiate the function of procuring raw materials and energy from that of synthesizing new products, and distinguishing both of these from the functions of differentiating the system from its environment and of storing information in a template.

the work done involves building a semipermeable membrane, the cycle provides the central resource for a mechanism that can obtain a critical autonomy from its environment. Moreover, when cycles are autocatalytic and produce multiple copies of themselves, cycles provide a way to account for growth and reproduction. These features, as we will see, are all critical for providing an adequate answer to vitalists such as Bichat, but to see how they fit into such an account, it will be useful to consider a different theoretical perspective.

6. FROM GÁNTI'S CHEMOTON TO AUTONOMOUS SYSTEMS

During the same period in which Gánti was developing his ideas of a chemoton, other theorists were focusing on efficient causation and ways of closing chains of efficient causation. As Rosen characterized a mechanism, which he contrasted with an organism, there was always some component operation in it that was not caused by other components within the mechanism. It would always be possible to incorporate the efficient cause of that operation within the description of the mechanism, but then one described a new mechanism that had a new uncaused operation. For Rosen the key contrast between mechanisms and organisms was that organisms had no uncaused operations – they are 'closed to efficient causation'.

To see more concretely what Rosen was contemplating, consider a metabolic mechanism in which component f is responsible for metabolizing A into B. Using open arrows to represent efficient causes and closed arrows to represent material causation, Rosen represented the relation as in Scheme 6.

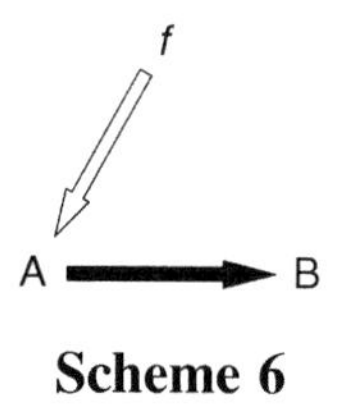

Scheme 6

But f is uncaused in this account. One possibility is to use material in B to make or, in the case Rosen considers, to repair f. But what is the efficient cause of this change? Rosen introduces a new component Φ, yielding the relations shown in Scheme 7.

But now Φ is uncaused. Perhaps materially it was produced from f. But there would have to be an efficient cause of making Φ from f. Rosen initially considers introducing a component β as the efficient cause, but notes that this is a recipe for an infinite regress. The way out, Rosen proposes, is to let B serve as

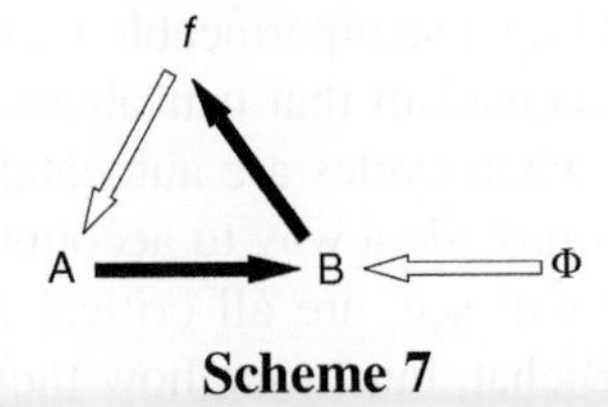

Scheme 7

the efficient cause of making Φ from f. Now each causal transformation (from A to B, from B to f, and from f to Φ) has an efficient cause (f, Φ, and B, respectively) (Scheme 8).

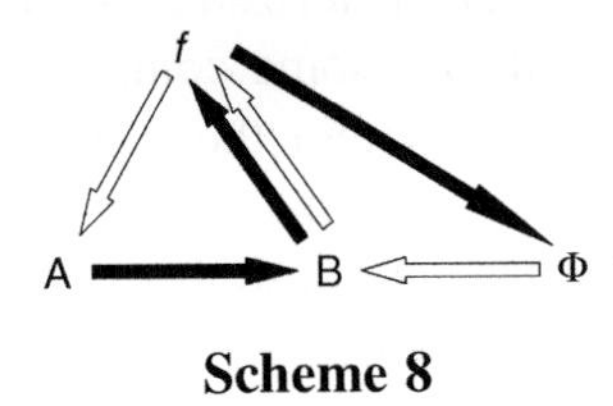

Scheme 8

The result is a material system closed to efficient causation (there are no components which undergo a material change without an open arrow coming in to them), thereby satisfying Rosen's characterization of an organism. (Φ does not have an open arrow coming into it, but it is not changed into anything else in the system, so there is no efficient causation to account for.) Importantly, the system is still materially open because A is not produced from anything within the system. (Although Rosen does not emphasize it, it is also energetically open.)

A similar emphasis on efficient causation lies at the heart of Francesco Varela's characterization of a living system (although Varela, unlike Rosen, begins with the idea that living organisms are machines and focuses his attention on delineating what sort of machine they are). He begins with the idea that a living system has an identity in terms of an organization which it maintains 'through the active compensation of deformations' (Varela, 1979, p. 3). Here Varela invokes Cannon's notion of homeostasis, on which he then relies to develop what he takes to be the key concept of *autopoiesis*. He does this by expanding on homeostasis in two ways: 'by making every reference for homeostasis internal to the system itself through mutual interconnections of processes, and secondly, by positing this interdependence as the very source of the system's identity as a concrete unity which we can distinguish' (pp. 12–13). In other words, all homeostatic operations in organisms are efficiently caused from within the system and it is the continued existence of the set of causally dependent processes that constitutes the continued existence of the system. Varela then provides his canonical characterization of autopoiesis:

> An autopoietic system is organized (defined as a unity) as a network of processes of production (transformation and destruction) of components that produces the components that: (1) through their interactions and transformations continuously regenerate and realize the network of processes (relations) that produce them; and (2) constitute it (the machine) as a concrete unity in the space in which they exist by specifying the topological domain of its realization as such a network.
>
> (Varela, 1979, p. 13; see also Maturana & Varela, 1980)

The crucial idea, as it was for Rosen, is that all the pertinent causal processes needed to maintain the network of causal processes have their efficient cause within the system itself.[24]

Autopoiesis is important according to Varela because autopoietic systems can be *autonomous*, where autonomous systems are those that perform the necessary operations to maintain their own identity.[25] This notion of autonomy provides a powerful way to conceptualize what is special about living systems. It also provides a perspective from which to view any additions to the initially conceived minimal autonomous system – they are ways of extending the autonomy of the system. I will return to this point in the final section of this paper. But before moving on, it is important to qualify the notion of autonomy. We should not view an autonomous system as completely encapsulated. Minimally, on both Rosen's and Varela's conception, such systems are materially and energetically open to their environments. As a consequence, they are also potentially open to efficient causation, but not in a sense that Rosen was seeking to avoid. With respect to autonomy, what is crucial is what a system does when causal processes impinge on it. If the system responds to such impingements through operations it initiates, then the critical autonomy is preserved. If, on the other hand, external efficient causes simply change the system (e.g., a falling rock smashes a cell or wind transports it to a new location), the resulting effects are not due to the system's autonomous action.

---

[24] In his last published paper (Varela & Weber, 2002), Varela traces the roots of the idea of autopoiesis back to Kant.

[25] 'Autopoietic machines are autonomous: that is, they subordinate all changes to the maintenance of their own organization, independently of how profoundly they may be otherwise transformed in the process. Other machines, henceforth called *allopoietic* machines, have as the product of their functioning something different from themselves' (p. 15). Varela later generalizes the notion of autonomy so as to apply beyond autopoietic machines: 'Autonomous systems are mechanistic (dynamical) systems defined as a unity by their organization. We shall say that autonomous systems are organizationally closed. That is, their organization is characterized by processes such that (1) the processes are related as a network, so that they recursively depend on each other in the generation and realization of the processes themselves, and (2) they constitute the system as a unity recognizable in the space (domain) in which the processes exist' (p. 55).

There are a couple of features of both Rosen's and Varela's account that merit comment. First, both of them dissociate the organizational features that are crucial for living systems from their material realization. Varela, for example, comments:

> We are thus saying that what defines a machine organization is relations, and hence that the organization of a machine has no connection with materiality, that is, with the properties of the components that define them as physical entities. In the organization of a machine, materiality is implied but does not enter *per se*.
> (p. 9)

In making this dissociation, both Varela and Rosen endorse the sort of multiple realizability argument that has figured so strongly in functionalist theories of mind and has led numerous philosophers of mind to downplay the significance of the brain to the understanding of cognition. The plausibility of multiple realizability claims, however, often results from adopting overly simplistic accounts of functional organization and from focusing on abstract accounts of machines, such as the Turing machine. In real machines, the relations in which a component can stand are significantly limited by the material out of which they are constituted – change the material, and there will be detectable changes in functionality. When one takes the organizational demands on complex systems very seriously, it is far from clear that there are multiple material realizations that could produce the same functionality (Bechtel & Mundale, 1999).[26] A second, and related point, is that neither focuses on the energy requirements of the systems they consider. In this regard, it is interesting to focus on the fact that the first component that Rosen added to his metabolic operation was a repair operation. Indeed, repair is a crucial feature of living systems, but the reason it is so important is that biological systems, as highly organized systems, exist far from thermodynamic equilibrium. As such, they will dissipate, and in so far as they are chemical, not solid systems, such dissipation will be relatively rapid.[27]

---

[26] A second consideration that leads other theoreticians to attempt a materially independent characterization of living systems is found in the artificial-life community which seeks to understand life in its full generality and not to be 'earth chauvinists'. Seeking universal categories, however, is not the only way of avoiding earth chauvinism. A different strategy is to begin with the concrete case we know – life on earth – and to branch out from it by considering variations that are possible. This fits the common strategy in biology of starting with a model system (e.g., the giant squid axon), develop an account of the mechanisms operative in it, and then investigate the similarities and differences found in other related species. One reason this is likely to succeed often on earth is the conservation of mechanisms that results from natural selection, which of course will not apply to nonearth-based life. Nonetheless, the strategy of starting with mechanisms we know to function successfully on earth and then considering variations is far more likely to succeed than a strategy of seeking generality by discounting what we know of earth-based mechanisms. The latter is a strategy that may only result in vacuous generalities.

[27] It is perhaps because of the problem of dissipation that humanly engineered machines have historically been made of solid, rigid components. Although such machines do experience 'wear and tear' and so need repair,

Thermodynamic considerations apply to all machines, not just biological ones. All must operate within an energy flow between a high-energy source and a low-energy sink and must draw upon the energy available at the source and release its waste products, now in a lower energy state, into the sink. A waterwheel, in which water at a higher elevation stores more free energy than the water at a lower elevation, often provides a fruitful metaphor for this process. What is critical for any mechanism is that the energy liberated in this flow is employed to perform work and this requires that it be channeled in the appropriate way. This typically means that it is transformed into a different format – the waterwheel converts the energy stored in water at a higher elevation into the rotational energy of the wheel's axle, and linkages appended to the axle in turn convert the motion into the form required.

What is distinctive in the case of living organisms is that the thermodynamics must be regulated to enable the organism to maintain itself – to repair itself or to build itself initially. This need is brought into clear focus by considering the autocatalytic networks, to which theorists such as Stuart Kauffman (1995) have turned in their accounts of the origin of life. Such networks, as well as other self-organizing systems such as hurricanes and tornadoes, are extremely fragile and are not able to maintain themselves for long. Part of the reason is that they rely on an energy source which may be quickly expended. But a further part of the reason is that they are not organized to channel the energy they secure to construct themselves so as to extract more energy from the source in the future (Ruiz-Mirazo & Moreno, 2004; see also Bickhard, 1993; Kauffman, 2000). (This would be pointless if there is no further energy source to tap, but of great importance if there is a continued source of energy, but one that can be utilized only if the system is properly configured.)

Whereas Varela did not focus on the thermodynamics and the management of energy in his account of autonomy, Kepa Ruiz-Mirazo and Alvaro Moreno (2004) have made it central to their account. They begin with the recognition that as organized systems, living systems are far from thermodynamic equilibrium and, in order to maintain that organization, must maintain themselves far from equilibrium (cf. Schrödinger, 1944). Many of the chemical reactions required to maintain such a system are endergonic (require Gibbs free energy) and so must be coupled with those that liberate energy from another source (exergonic reactions). In order to maintain themselves far from equilibrium, Moreno focuses on how the system manages the flow of energy so as to provide for its own construction and reconstruction. The membrane presents one point

the bonds that render them into solids make them less subject to dissipation than structures in fluid milieus. On the other hand, it is harder to design a self-repair process for a system made of solids, which may explain why the strategy for dealing with breakdown in human-engineered systems has been to build in redundancy rather than self-repair.

of management, determining what gets in and out of the system. The metabolic pathways that extract energy and raw materials and then synthesize constituents of the organism's own structure are another. Focusing on these management processes, Ruiz-Mirazo and Moreno characterize basic autonomy as

> the capacity of a system to *manage* the flow of matter and energy through it so that it can, at the same time, regulate, modify, and control: (i) internal self-constructive processes and (ii) processes of exchange with the environment. Thus, the system must be able to generate and regenerate all the constraints—including part of its boundary conditions—that define it as such, together with its own particular way of interacting with the environment.
>
> (Ruiz-Mirazo & Moreno, 2004, p. 240; see also Ruiz-Mirazo et al., 2004, p. 330)

Moreno's notion of basic autonomy provides the appropriate complement to Gánti's conception of the chemoton as the simplest chemical system exhibiting the features of life. Construing the chemoton as an autonomous system requires adopting a perspective that is implicit, but not explicit in Gánti's description of the chemoton.[28] The chemoton takes in not only matter, incorporating some of it within itself and expelling other parts as waste, but also energy and utilizes some of it in the various work it performs and expels the rest as waste (generally as heat or substances with too little free energy to be useful). Biochemists were, it is interesting to note, slow to recognize this aspect of metabolism. Until the 1930s they focused principally on the generation of animal heat (in fact, a waste product) and the incorporation of matter into the organism. It was not until the discovery of phosphocreatine and adenosine triphosphate (ATP) and their linkages to the processes of glycolysis that energetic relations became a central focus of metabolism studies (Bechtel, 2006). Although a latecomer, it is a crucial feature of any metabolic system.

7. CONCLUDING THOUGHTS: BEYOND BASIC AUTONOMY

My focus in this paper has been on how organization is far more critical than often recognized in mechanistic science and philosophical accounts of mechanistic explanation. Only by keeping a keen eye on the organization at play

[28] Prior to introducing the chemoton itself, Gánti makes the point about energy: 'The operation of every machine, device, instrument – every continuously operating system – is based on energy flow. Energy enters the system from somewhere and eventually leaves it. While it is within the system it is manipulated so that part of it is forced to make the system operate, whereas the other part leaves the system, mainly in the form of heat' (p. 18).

in living systems is it possible to understand the mechanisms that figure in living organisms. Vitalists and holists play an important function when they remind mechanists of the shortfalls of the mechanistic accounts on offer. Ideas such as negative feedback, self-organizing positive feedback, and cyclic organization are critical for explaining the phenomena exhibited by living organisms. Moreover, the importance of these modes of organization can be appreciated when the relevance of notions such as being closed to efficient causation is taken into account and it is appreciated that as organized systems, living systems are far from equilibrium and require ways of channeling matter and energy extracted from their environment to maintain themselves far from equilibrium. These critical features are nicely captured in Moreno's conception of basic autonomy in which we recognize living systems as so organized to metabolize inputs to extract matter and energy and direct these to building and repairing themselves.

My contention is that recognizing organization does not require a rupture with the tradition of mechanistic science. Mechanism has the resources to identify and incorporate the forms of organization critical in living systems. Moreover, attempts to focus on organization independently of the matter and energy of actual systems are likely to fail, as the organization required to maintain autonomy is an organization that is suited to the matter and energy available to the system. It is in this context that the notion of basic autonomy reveals its importance: it provides a framework for relating organization tightly to the matter and energy of the system as the organization of interest is one which, given the energy and material to be utilized, is able to be built and maintained by the living system.

I have restricted myself in this paper to Moreno's notion of basic autonomy. As the reference to *basic* suggests, there are additional levels of autonomy.[29] These involve functions which can be performed within a system that further enhance the system's ability to maintain itself. Some of these involve ways of interacting with the environment. A basic autonomous system remains highly dependent on the moment-to-moment conditions of its environment as it must continually extract energy and raw materials from it and excrete waste into it. If energy and material resources are not provided in high-enough concentration so that the osmotic or pumping mechanisms in the membrane are able to bring them into the system or if waste accumulates to such a degree that they overwhelm the ability of the mechanisms to expel more waste, the viability of the system is undermined.

[29] Other theorists, such as Collier and Hooker (1999), link the notion of autonomy to more active behaviors of a system such as adapting to varied circumstances and anticipating the response of the environment to its behavior. They maintain that autonomy, adaptivity, anticipation, and reproducibility are all required before one has a living system. While, in fact, most real organisms are adaptive and do anticipate responses of the environment to their actions, and these provide an extremely potent source for learning and hence further development, I think it is conceptually important to focus initially on the basic autonomy and then consider additions to it.

By developing mechanisms to perform operations outside the organism itself and especially to navigate the environment, the organism can take proactive measures to insure the needed conditions of its environment and thereby increase its ability to maintain itself as a functioning system. It makes sense to construe these additional functions as enhancing the system's autonomy. Other ways of enhancing the autonomy focus on the operations internal to the system. For example, by developing ways to perform operations more efficiently, procedures for storing energy and raw materials (or even recycling raw materials), and ways to regulate the internal environment, the system improves its ability to maintain itself.

In considering mechanisms that enhance autonomy, we must bear in mind that an organism must be able to construct all of these additional mechanisms itself. Each structure specialized to perform an additional operation must be constructed from matter and use free energy, the system recruits from its environment and channels into the construction of the structure. Moreover, its operation requires energy that the system has recruited and made available. From the point of view of the autonomous system, each addition only makes sense if the benefit it provides in terms of maintaining the system equals or exceeds the costs of constructing and maintaining the addition. In this way, as Moreno himself emphasizes, the notion of autonomy provides a framework to speak of function not just as something that is done by the system but done for the system: '*Functional* actions in this context are those that ensure the self-maintenance and autonomy of the organization' (p. 241). As it is the organization of the whole that is being maintained, function cannot be assessed locally but only in the context of the whole system.[30]

Evolution via natural selection is a process that, over time, can develop systems with greater autonomy. Although not denying the traditional accounts of evolution (e.g., that evolution requires mechanisms of variation and selective retention[31]), the focus on autonomous systems provides a rather different perspective. First, it places the organism in the central role and emphasizes that an organism needs to be able to maintain itself as an autonomous system. Otherwise,

[30] The sense of function provided by focusing on organisms as autonomous systems is different from the sense of function invoked in purely causal accounts that treat any activity of a system as its function or evolutionary accounts that treat as functional only traits that have been selected in the past (adaptations) or traits that enable the organism to meet current selection forces (adaptive traits).

[31] Ruiz-Mirazo and Moreno argue that before evolution can function to produce systems with greater autonomy, an autocatalytic replication system independent of the catalytic metabolic system is required: 'hereditary autonomous systems have no other possibility but to start producing two types of macromolecular components that will take up different but complementary functions in the organization of those systems. The two types of components (informational records and highly specific catalysts or, equivalently, genotype and phenotype) strongly depend on each other, and their (code-mediated) complex interrelation changes profoundly the organization of autonomous systems, at both the individual (metabolic) and the collective (ecological) level' (Ruiz-Mirazo & Moreno, 2004, p. 251). In Ruiz-Mirazo et al. (2004, p. 330), the authors add evolutionary potential to autonomy in characterizing living beings: '*"a living being" is any autonomous system with open-ended evolutionary capacities*'.

there is nothing to evolve. This does not mean that individual organisms must be totally self-sufficient. Organisms can evolve to rely on features of the environment that are regularly present to them. But they need to create and maintain all the mechanisms upon which they rely in order to use these resources. Second, each addition to the basic system involves a cost in that the system must generate and repair these mechanisms itself. Evolution is not just a matter of introducing and selecting new genes, but requires a system that builds and maintains new traits (i.e., new mechanisms).

I have been emphasizing the value of focusing on organization as it subserves autonomy in understanding biological mechanisms. In concluding, I must acknowledge that in practice a great deal of significant biological research has proceeded without considering how the mechanisms investigated subserve the autonomy of a living system. From this research we have acquired substantial understanding of the mechanisms involved in living systems. This is possible because a researcher can focus on a mechanism as operating in a specified way without considering it as subserving the autonomy of the organism. What, then, is the value of emphasizing autonomy and the demands it imposes on organisms? This can be answered in two ways. First, despite its many successes, when mechanistic science fails to attend to organization it often reaches a point of identifying the parts but not understanding how they succeed in producing the phenomena exhibited in living organisms. Such frustration is part of the explanation of the current appeal of developing a systems approach to biology where tools for mathematically modeling systems complement those for identifying components. Second, discovery of the mechanisms actually operative in nature is often fostered by understanding the constraints under which they work. Lacking constraints, there are often too many possibilities and it is difficult to determine which possibility is the actual one being sought. Maintaining autonomous functioning is a critical constraint on any biological mechanism, and considering the requirements autonomy imposes provides constraint for investigators trying to figure out the mechanisms at work.

REFERENCES

Allchin D. *A Twentieth-Century phlogiston: Constructing Error and Differentiating Domains.* Perspectives on Science: 5(1), 81–127, 1997.

Bechtel W. *The evolution of our understanding of the cell: A study in the dynamics of scientific progress.* Studies in the History and Philosophy of Science: 15, 309–356, 1984.

Bechtel W. *Discovering cell mechanisms: The creation of modern cell biology.* Cambridge University Press, Cambridge, 2006.

Bechtel W & Abrahamsen A. *Explanation: A mechanist alternative.* Studies in History and Philosophy of Biological and Biomedical Sciences: 36, 421–441, 2005.

Bechtel W & Hamilton A. *Reduction, integration, and the unity of science: Natural, behavioral, and social sciences and the humanities.* In: Philosophy of science: Focal issues. (Ed.: Kuipers T), Elsevier, New York, 403–455, 2006.

Bechtel W & Mundale J. *Multiple realizability revisited: Linking cognitive and neural states.* Philosophy of Science: 66, 175–207, 1999.

Bechtel W & Richardson RC. *Discovering complexity: Decomposition and localization as strategies in scientific research.* Princeton University Press, Princeton, 1993.

Bernard C. *An introduction to the study of experimental medicine.* New York: Dover, 1865.

Bernard C. *Leçons sur les phénomènes de la vie communs aux animaux et aux végétaux.* Baillière, Paris, 1878.

Bichat X. *Recherches Physiologiques sur la Vie et la Mort* (3rd ed.). Machant, Paris, 1805.

Bickhard MH. *Representational content in humans and machines.* Journal of Experimental and Theoretical Artificial Intelligence: 5, 285–333, 1993.

Bogen J & Woodward J. *Saving the phenomena.* Philosophical Review: 97, 303–352, 1988.

Boogerd F, Bruggeman F, Jonker C, Looren de Jong H, Tamminga A, Treur J, Westerhoff H & Wijngaards W. *Inter-level relations in computer science, biology, and psychology.* Philosophical Psychology: 15, 463–471, 2002.

Boogerd FC, Bruggeman FJ, Richardson RC, Stephan A & Westerhoff HV. *Emergence and its place in nature: A case study of biochemical networks.* Synthese: 145, 131–164, 2005.

Bruggeman FJ. *Of molecules and cells: Emergent mechanisms.* Vrije Universiteit, Amsterdam, 2005.

Bruggeman FJ, Westerhoff HV & Boogerd FC. *BioComplexity: a pluralist research strategy is necessary for a mechanistic explanation of the 'live' state.* Philosophical Psychology: 15, 411–440, 2002.

Cannon WB. *Organization of physiological homeostasis.* Physiological Reviews: 9, 399–431, 1929.

Collier JD & Hooker CA. *Complexly organised dynamical systems.* Open Systems and Information Dynamics: 6, 241–302, 1999.

Gánti T. *The principles of life.* New York: Oxford, 2003.

Glennan S. *Mechanisms and the nature of causation.* Erkenntnis: 44: 50–71, 1996.

Glennan S. *Rethinking mechanistic explanation.* Philosophy of Science: 69, S342–S353, 2002.

Griesemer JR & Szathmáry E. *Gánti's chemoton model.* forthcoming.

Grush R. *The emulation theory of representation: Motor control, imagery, and perception.* Behavioral and Brain Sciences: 27, 377–396, 2004.

Hall TS. *Ideas of life and matter; studies in the history of general physiology, 600 B.C.-1900 A.D.* University of Chicago Press, Chicago, 1969.

Hempel CG. *Aspects of scientific explanation.* In: Aspects of scientific explanation and other essays in the philosophy of science. (Ed.: Hempel CG), Macmillan, New York, 331–496, 1965.

Holmes FL. *Intermediary metabolism in the early twentieth century.* In: Integrating scientific disciplines. (Ed.: Bechtel W), Dordrecht: Nijhoff, pp. 59–76, 1986.

Holmes FL. *Between biology and medicine: The formation of intermediary metabolism.* Office for History of Science and Technology, University of California at Berkeley, Berkeley, CA, 1992.

Jablonka E & Lamb M. *Epigenetic inheritance and evolution.* Oxford University Press, Oxford, 1995.

Kauffman SA. *Articulation of parts explanation in biology and the rational search for them.* In: PSA 1970 (Eds.: Bluck RC & Cohen RS), Dordrecht: Reidel, 257–272, 1971.

Kauffman SA. *The origins of order: Self-organization and selection in evolution.* Oxford University Press, Oxford, 1993.

Kauffman SA. *At home in the universe.* Viking, New York, 1995.

Kauffman SA. *Investigations.* Oxford University Press, Oxford, 2000.

Kellenberger E. *The evolution of molecular biology: Biology's various affairs with holism and reductionism, and their contribution to understanding life at the molecular level.* EMBO Reports: 5, 547–549, 2004.

Krebs HA & Johnson WA. *The role of citric acid in intermediate metabolism in animal tissues.* Enzymologia: 4, 148–156, 1937.

Lewontin RC. *The units of selection.* Annual Review of Ecology and Systematics: 1, 1–18, 1970.

Lohmann K. *Über die Pyrophosphatfraktion im Muskel.* Naturwissenschaften: 17, 624–625, 1929.

Machamer P, Darden L & Craver C. *Thinking about mechanisms.* Philosophy of Science: 67, 1–25, 2000.

Maturana HR & Varela FJ. *Autopoiesis: The organization of the living.* In: Autopoiesis and Cognition: The Realization of the Living. (Eds.: Maturana HR & Varela FJ), Dordrecht: D. Reidel, 59–138, 1980.

Maxwell JC. *On governors.* Proceedings of the Royal Society of London: 16, 270–283, 1868.

Mayr O. *The origins of feedback control.* MIT Press, Cambridge, MA,1970.

Mendelsohn E. *Heat and life: The development of the theory of animal heat.* Harvard University Press, Cambridge, MA, 1964.

Mikulecky DC. Robert Rosen: *The well-posed question and its answer–Why are organisms different from machines.* Systems Research and Behavioral Science: 17, 419–432, 2000.

Mindell DA. *Anti-aircraft fire control and the development of integrated svstems at Sperry, 1925-1940.* IEEE Control Systems (April): 108–113, 1995.

Mitchell P. *Coupling of phosphorylation to electron and hydrogen transfer by a chemi-osmotic type of mechanism.* Nature: 191, 144–148, 1961.

Nagel E. *The structure of science.* Harcourt, Brace, New York, 1961.

Nicolis G & Prigogine I. *Self-organization in nonequilibrium systems: From dissipative structures to order through fluctuations.* Wiley, New York, 1977.

Rosen R. *Life itself: A comprehensive inquiry into the nature, origin, and fabrication of life.* Columbia, New York, 1991.

Rosenblueth A, Wiener N & Bigelow J. *Behavior, purpose, and teleology.* Philosophy of Science: 10, 18–24, 1943.

Ruiz-Mirazo K & Moreno A. *Basic autonomy as a fundamental step in the synthesis of life.* Artificial Life: 10, 235–259, 2004.

Ruiz-Mirazo K, Peretó J & Moreno A. *A universal definition of life: Autonomy and open-ended evolution.* Origins of Life and Evolution of the Biosphere: 34, 323–346, 2004.

Schrödinger E. *What is life? The physical aspect of the living cell.* Cambridge University press, Cambridge, 1944.

Slater EC. *Mechanism of phosphorylation in the respiratory chain.* Nature: 172, 975–978, 1953.

Smil V. *Cycles of life: Civilization and the biosphere.* Scientific American Library, New York, 1997.

Teusink B, Walsh MC, van Dam K & Westerhoff HV. *The danger of metabolic pathways with turbo design.* Trends in Biochemical Sciences: 23, 162–169, 1998.

Thunberg TL. *Zur Kenntnis einiger autoxydabler Thioverbindungen.* Skandinavisches Archiv für Physiologie: 20, 289–290, 1913.

Thunberg TL. *Zur Kenntnis des intermediären Stoffwechsels und der dabei wirksamen Enzyme.* Skandinavisches Archiv für Physiologie. 40, 9–91, 1920.

van Regenmortel MHV. *Reductionism and complexity in molecular biology.* EMBO Reports: 5, 1016–1020, 2004.

Varela FJ. *Principles of biological autonomy.* Elsevier, New York, 1979.

Varela FJ & Weber A. *Life after Kant: Natural purposes and the autopoietic foundations of biological individuality.* Phenomenology and the cognitive sciences: 1, 97–125, 2002.

Wieland H. *Über den Mechanismus der Oxydationsvorgänge.* Berichte der deutschen chemischen Gesellschaft: 46, 3327–3342, 1913.
Wiener N. *Cybernetics: Or, control and communication in the animal machine.* Wiley, New York, 1948.
Wimsatt WC. Reductionism, levels of organization, and the mind-body problem. In: Consciousness and the brain: A scientific and philosophical inquiry. (Eds.: Globus G, Maxwell G & Savodnik I), Plenum Press, New York, 202–267, 1976.
Wright L. *Explanation and teleology.* Philosophy of Science: 39, 204–218, 1972.

13

The disappearance of function from 'self-organizing systems'

Evelyn Fox Keller

SUMMARY

The term "self-organization" is everywhere; the question is, what does it mean? Immanuel Kant may have been the first to use it, and he used it to characterize the unique properties of living organisms, and the term's subsequent history is inextricably entwined with that of the history of biology. Only in the second half of the twentieth century, however, does it begin to acquire the promise of a physicalist understanding. This it does with two critical transformations in the meaning of the term: first, with the advent of cybernetics and its dissolution of the boundary between organisms and machines, and second, with the mathematical triumphs of nonlinear dynamical systems theory and accompanying claims to have dissolved the boundary between organisms and thunderstorms. But between these two moves a crucial distinction survives, namely, between the emergence and the organization of complexity. I argue that here, in this distinction, we find the questions of function, purpose, and agency returning to haunt us.

For Kant (1790; 1993), the notion of self-organization was crucial to his efforts to address the fundamental question of life sciences, namely, 'What is an organism?'. What is the special property, or feature, that distinguishes a living system from a collection of inanimate matter? In fact, this was the question that first defined Biology as a separate and distinctive science. And by its phrasing (i.e., implicit in the root meaning of the word organism), it also specified at least the form of what would count as an answer. For what led to the common grouping of plants and animals in the first place – i.e., what makes 'the two genres of organized

Systems Biology
F.C. Boogerd, F.J. Bruggeman, J.-H.S. Hofmeyr and H.V. Westerhoff (Editors)
Isbn: 978-0-444-52085-2

beings' (as Buffon referred to them) organisms – was a new focus on just that feature, i.e., on their conspicuous property of being organized and in a particular way. As Francois Jacob observes, by the end of the eighteenth century, it was:

> [b]y its organization [that] the living could be distinguished from the non-living. Organization assembled the parts of the organism into a whole, enabled it to cope with the demands of life and imposed forms throughout the living world.
>
> (p. 74)

Only by that special arrangement and interaction of parts that brings the well-springs of form and behavior of an organism inside itself one could distinguish an organism from its Greek root, *organon*, or tool. A tool requires a tool-user, whereas an organism is a system of organs (or tools) that behaves as if it had a mind of its own, i.e., that governs itself.

Indeed, these two words, organism and organization, acquired their contemporary usage more or less contemporaneously. Kant gave one of the first modern definitions of an organism in 1790 – not as a definition per se, but rather as a principle or 'maxim' which, he wrote, 'serves to define what is meant as an organism' – namely

> an organized natural product is one in which every part is reciprocally both end and means. In such a product nothing is in vain, without an end, or to be ascribed to a blind mechanism of nature.
>
> (p. 558)

Organisms, he wrote, are the beings that

> first afford objective reality to the conception of an *end* that is an end of *nature* and not a practical end. They supply natural science with the basis for a teleology . . . that would otherwise be absolutely unjustifiable to introduce into that science – seeing that we are quite unable to perceive *a priori* the possibility of such a kind of causality.
>
> (p. 558, 66)

As he elaborates:

> In such a natural product as this every part is thought as *owing* its presence to the agency of all the remaining parts, and also as existing *for the sake of the others* and of the whole, that is as an instrument, or organ. . . . [T]he part must be an organ *producing* the other parts – each, consequently, reciprocally producing the others. . . . Only under these conditions and upon these terms can such a product be an *organized* and *self-organized being*, and, as such, be called a *physical end.*
>
> (italics in original; 65, p. 557)

In other words, no external force, no divine architect, is responsible for the organization of nature, only the internal dynamics of the being itself.

The beginnings of biology thus prescribed not only the subject and primary question of the new science, but also its aim. To say what an organism is would be to describe and delineate the particular character of the organization that defined its inner purposiveness, that gave it a mind of its own, that enabled it to organize itself. What is an organism? It is a bounded body capable not only of self-regulation, self-steering, but also, and perhaps most important, of self-formation and self-generation. An organism is a body which, by virtue of its peculiar *and* particular organization, is made into a 'self' that, even though not hermetically sealed (or perhaps because it is not hermetically sealed), achieves autonomy and the capacity for self-generation. 'Strictly speaking,' Kant wrote, 'the organization of nature has nothing analogous to any causality known to us' (65, p. 557).

Admittedly, demarcating organisms in the real world was not always easy. Should we think of ant or termite colonies as organisms? Beehives? Coral communities? Humans too are social organisms – should the societies we form be regarded as organisms in and of themselves? Are they also not 'self-organizing'? And what about natural communities more generally? Is it useful to think of them, as Frederic Clement proposed in 1916, as constituting a 'complex organism', governed by laws of development?

Yet even in the absence of explicit criteria for delineation, these two terms – organism and self-organization – remained tightly linked and clearly set apart from the realm of inanimate objects, especially from those objects, machines, that were designed and built to serve human goals. Machines were designed, and they were designed from without. Of course, organisms too were still seen as designed, but in contrast to machines, biological design – or organization – was internally generated. The burden of the concept of self-organization thus fell on the term *self*, for it was the self as source of the organization that prevents an organism from ever being confused with a machine (see Keller, 2004, for further discussion).

The first major mutation in this tradition came in WW II. 'Out of the wickedness of war', as Warren Weaver put it, emerged not only a new machine, but also a new vision of a science of the inanimate: a science based on principles of feedback and circular causality and aimed at the mechanical implementation of exactly the kind of purposive 'organized complexity' so vividly exemplified by biological organisms. In other words, a science that would repudiate the very distinction between organism and machine on which the concept of self-organization was predicated. Ross Ashby's contributions were crucial. In the 1940s, he built a machine (the 'homeostat') that was to serve as a primitive model for a brain. Ashby's homeostat showed that a properly designed machine could exhibit autonomous, self-organizing, behavior. And this, together with his work on 'Principles of the Self-Organizing Dynamic System' in 1947, served as major stimuli for the rash of conferences on self-organizing systems organized by the Office of Naval Research (ONR) in the 1950s. All of these discussions and investigations were conducted with an explicitly engineering aim, namely, the design and construction of systems that could organize themselves, grow themselves, and perhaps even reproduce themselves.[1] As Ashby and von Foerster formulated it, the crucial property for the realization of these goals lay in the relations between components; out of such relations functionality or purpose would spontaneously emerge. Frank Rosenblatt's *Perceptron* project embodied this vision in its most ambitious form, but despite enthusiastic support in its early stages from the ONR, development of the Perceptron did not continue long enough or far enough to produce anything like a live issue. By the 1960s, support had dried up and the spirit of artificial intelligence, organized around very different principles, had moved elsewhere. Some of Ashby's arguments survived, but in other areas of engineering – most notably, in control theory.

The next important mutation in this history comes with the emergence of studies of nonlinear dynamical systems in physics and mathematics in the late 1970s and 1980s. Here again, the term self-organization surfaces, but now with a further shift in meaning. For Ilya Prigogine, the term referred to the emergence of 'dissipative structures' in systems far from equilibrium and low in entropy. Given a sufficiently large flux of free energy in such systems, one can expect to see the spontaneous emergence of such striking phenomena as eddies, vortices, or Bénard rolls and cells. At roughly the same time, the study of nonlinear dynamical systems by mathematicians was yielding similar insights about these same phenomena, although here they were described in terms of stable attractors and limit cycles. Either way, self-organization now referred to the production of stable patterns observed in physical, and sometimes in biological, systems governed by nonlinear dynamics. Enthusiasm for such analyses ran high, generating some extraordinary expectations. As Paul Davies wrote, 'Mathematically we can now see how nonlinearity in far-from-equilibrium systems can induce matter [here quoting Charles Bennett (1986)] to "transcend the clod-like nature it would manifest at equilibrium, and behave instead in dramatic and unforeseen ways, molding itself for example into thunderstorms, people and umbrellas"'. (1989, pp. 111)

The late 1980s brought a noteworthy (and indeed, much noted) addition to these ideas, namely, Per Bak's notion of 'self-organized criticality' (SOC) (Bak et al., 1987).[2] Bak and Sneppen (1993) developed a model for evolutionary

[1] In a 1958 article entitled 'Electronic "Brain" Teaches Itself', Rosenblatt is quoted as saying, 'In principle, it would be possible to build Perceptrons that could reproduce themselves on an assembly line and which would be "conscious" of their existence.' (New York Times, 13 July 1958, Section 4, p. 9).

[2] In 1996, Bak reported that, since the coining of the phrase SOC in 1987, 'more than 2000 papers have been written on the subject, making' this initial paper 'the most cited in physics . . . ' (Bak, 1995).

dynamics that exhibited all the desired characteristics: An open and dissipative system that organizes itself into a critical state simply by virtue of its intrinsic dynamics, independently of any control parameter. Drawing on an analogy with the physics of phase transitions, the existence of a critical state is said to be signaled by a power law distribution in some variable – to physicists familiar with the behavior of systems at thermodynamic critical points, a seemingly clear indication that long-range interactions have induced a kind of global organization in which details of the particular system get obliterated. The familiar image is that of sand dripping onto a sand pile. Once the sand pile has attained a critical slope, it retains its conical shape as more sand is added; it manages this by setting off small avalanches. The timing and size of individual avalanches are unpredictable, but the distribution of avalanches (in both size and timing) displays the prototypical regularity of the power law.

Self-organized criticality soon became a new buzzword, the latest key to an understanding of emerging structure in complex systems. Within just a couple of years, Laszlo Barabasi and his colleagues extend Bak's idea of SOC to the world of network topology, providing yet another boost to expectations about what the physics of phase transitions might do for biology.

Here is a literature that begins in the world of physics, written by physicists and published in physics journals, but rapidly spreads to other fields, and is soon taken up by a flourishing industry of science popularization. The term self-organization, as used throughout this literature, is severed both from its original biological meaning and from its later engineering sense, stripped of all resonances of design in either of the senses invoked in these disciplines, and instead, appropriated for the categorization of complex phenomena arising out of random ensembles, essentially uniform distributions of simple physical entities. Not only eddies, whirlpools, and Bénard cells are to be understood as arising from homogeneous gases, fluids, and lattices, but also more dramatic eruptions like thunderstorms, earthquakes, and living organisms. Indeed, this is a literature claiming that the emergence of life itself can be seen as a self-organized critical phenomenon.

So what is happening here? In this assimilation of life and familiar physical processes, is biology being reduced to physics, or is physics being revived by the infusion of life? Lee Smolin, one of the more thoughtful writers on this subject, argues that viewing the universe as a whole as a nonequilibrium, self-organizing system has many advantages: in particular, it allows for a world in which 'a variety of improbably structures – and indeed life itself – exist permanently, without need of pilot or other external agent, [and] offers the possibility of constructing a scientific cosmology that is finally liberated from the crippling duality that lies behind Plato's myth'. (Smolin, 1997, p. 160). But what about the costs of such assimilation?

Despite Smolin's caution, and for all his hopes, I claim there is a serious elision here. Analyses of nonlinear dynamical systems clearly demonstrate the ease with which complexity can be generated, but such arguments fail to recognize a crucial distinction – a distinction that Warren Weaver described in 1948 as a distinction between disorganized and organized complexity – a distinction that we might characterize as a difference between the generation (or emergence) of complexity and its organization (Weaver, 1948). In the words of John Mattick (2004), 'The problem is not how to generate complexity – that is easy – but rather how to control it to specify ordered [and, I would add, robust] trajectories that lead to highly organized and complex organisms'. (pp. 317–8) This is an absolutely crucial point for biology, and possibly even for ecology.

Let me start with ecosystems, which Bak viewed as ideal candidates for SOC. Encouraged by the presence of power law distributions in the size of extinctions, Bak argued that SOC offered a way of making sense of ecosystem development without invoking the guiding hand of God, man, or even, for that matter, of natural selection. If we can judge by the frequency of references to SOC in the literature of theoretical ecology today, the influence of his arguments has been enormous.

The difficulty is that power law distributions are everywhere; furthermore, they do not in themselves indicate any particular mechanism of organization (any number of mechanisms can give rise to the same power law), so, in spite of the evident attractiveness of these models, the actual relevance of SOC to ecosystems remains to be determined. I myself am rather skeptical, if only because of their exclusion of so much that seems intuitively crucial. Theoretical ecologist Simon Levin has been particularly active in exploring the role of evolution in ecosystem dynamics. But Levin also – perhaps unwittingly – offers another way of improving on the physicists' notion of self-organization that I believe warrants our attention.

Looking down at his desk, he observes: 'The contents of my desk may not seem to have much to do with ecosystems, but take another look. My office is indeed a self-organized system, with me at the center. It has more the element of design than do ecological systems, yet it still reflects a huge dose of chance and historical influence'. (p. 158). Now, this might be something of a pun, for Levin is the self who is organizing (or not, as the case may be) his desk. However, it is far from customary to include that supremely intentional agent, the human self, in our conception of a self-organized system, so perhaps he should have written 'My office is indeed a myself-organized system'. It should also be said that we tend overwhelmingly to exclude humans from our conception of ecosystem altogether. But why is that? Why is it that, in every conception of self-organization I have discussed, the self that is the source of organization is, in contrast to the human self (or myself), without intentionality or agency? Furthermore, such exclusion seems necessary, for otherwise, what would prevent us from describing the system composed of engineer and his machine as a self-organized system? Or, from describing an urban landscape as an ecosystem? Levin tries to avoid this problem by down-playing his own agency, by emphasizing the randomness of the processes underlying the design of his office (he is not really in charge). But this is just to sidestep the issue of whether or not the agency, or intentional design, can be brought into our concept of a self-organized system, in whatever form that concept takes. To bring the engineer into the system, to put Levin in his office with his agency and intentionality intact, would, I suggest, be to confound the entire tradition which takes human agency and intentionality as a priori unnatural and, accordingly, which pits natural against artificial design. Yet to put this exclusion of agency and intentionality so baldly is to make its absurdity self-evident and to invite us to explicitly challenge the entire tradition on which it rests.

So here is my first proposal: Let us drop intentionality (after all, this too, like free will, may turn out to be an illusion) and focus instead on agency – an attribute we clearly share with many, if not all, other organisms, and that is, both scientifically and philosophically, surely problem enough. And let us think of the machines we create simultaneously as extensions of ourselves (in so far as they enhance our functionality) and, at the same time, as effects our agency has on our environment. In other words, let us try thinking of ourselves in the same terms as we think of other organisms – organisms that shape their environment by their activities and, also, that build entities extending their functionality and can accordingly be thought of as extensions of themselves. Consider, e.g., beaver dams, bird nests, or any of Scott Turner's wonderful examples of the tunnels built by earthworms that serve as accessory kidneys, the bubble gills built by aquatic beetles, the horn-shaped burrows of crickets that amplify song, or homeostatic termite mounds. And now think of all these systems as self-organizing systems. In other words, think of a system that is shaped by the combined activities of all the individual components – activities that are generated inside individual components, but with effects manifested externally to themselves, but all the while remaining inside the composite self that defines the larger system. The modifications engendered by these activities need not be random (and in general, will not be random), nor be aimed at the coherence or survival of the system. It is just that those modifications that do enhance the survival of the system will, by definition, persist. This may lead us to think of everything as a self-organizing system, but doing so need not be bad – as long as it comes with the understanding that the most interesting kinds of self-organizing systems are those that require the participation and interaction of many different kinds of selves.

It seems to me that including the kinds of organization resulting from the complex activities and interactivities of collectivities of different kinds of agents in the concept of self-organization does in fact offer some hope of elucidating some ways by which emerging complexity might come to be organized. Indeed, there is a rapidly developing industry attempting to analyze the dynamics of such collectivities, and it goes by the name, 'agent based modelling' (ABM). True, the agents of ABM have rather less agency than I had in mind, but it is definitely a start, and I am fully persuaded that such models can help us think about certain features of ecosystems, and even about certain aspects of human societies. But what it will not help us think about is the emergence of the agency that appears here simply as posited. Therefore I want to now turn to the question of where it could have come from.

This question of course brings us back to the problem for which the term self-organization was originally introduced, namely, how are we to make sense of such peculiar properties of organisms as agency, function, or purpose – properties that, even now, continue to mark organisms (even if not machines) apart from thunderstorms – indeed, from all the emergent phenomena of nonlinear dynamical systems? For the fact remains that agency, function, and purpose are all conspicuously absent from the kinds of systems with which physics deals, and no one has succeeded in offering an account of how they might emerge from the dynamics of effectively homogeneous systems of simple elements, however, complex the dynamics of their interaction might be. Rather, they seem to require an order of complexity that goes beyond that which spontaneously emerges from complex interactions – a form of complexity that Weaver, Herbert Simon, and now John Mattick and John Doyle have dubbed organized complexity.

For Weaver, the domain of problems characterized by organized complexity – ranging from 'What makes an evening primrose open when it does?' to 'On what does the price of wheat depend?' – lay in sharp contrast to the problems of statistical mechanics that made up the domain of disorganized complexity. For Simon, organized complexity was complexity with an architecture, and in particular the architecture of hierarchical composition (or modularity) whereby a system 'is composed of interrelated subsystems, each of the latter being in turn hierarchic in structure until we reach some lowest level of elementary subsystem' (p. 184). For Mattick, the organization of complexity is mandated by the meaninglessness of the structures generated by sheer combinatorics of complex interactions: '[B]oth development and evolution,' he writes, 'have to navigate a course through these possibilities to find those that are sensible and competitive.' (2004, p. 317). Yet none of these authors quite grapple with the question of just what kind of organization would warrant the attribution of the properties of agency or function, would turn a structure or pattern into a self.

The problem of function seems like a good place to start. Over the last decade or so, something of a consensus has emerged among philosophers of science about how to treat this problem. Proper function, as first argued by Ruth Millikan and as now widely asserted, should be understood solely in the context of natural selection, i.e., the function of X is that 'which caused the genotype, of which X is the phenotypic expression to be selected by natural selection' (Neander, 1998, p. 319). But this is not how I want to proceed. My main difficulty with this formulation, at least for now, is the problem it raises for thinking about how function, understood as a property internal to a biological structure, might have first arisen, particularly in the light of recent arguments that it almost certainly emerged prior to the onset of natural selection. In other words, natural selection, as conventionally understood, requires the prior existence of stable, autonomous, and self-reproducing entities, e.g., single-celled organisms or, simply, stable, autonomous cells capable of dividing. But these first cells were, of necessity, already endowed with numerous subcellular entities (or modules) endowing the primitive cell with the functions minimally required for the cell to sustain itself and reproduce. In other words, even if the first cells lacked many features of the modern cell, they had to have had primitive mechanisms to support metabolism, cell division, etc.; there needed to have come into being primitive embodiments of function that would work keep the cell going and protect it from insult.

What do I mean by function? Let me try to clarify my meaning by taking off from Michael Ruse's argument against the sufficiency of circular causality (Ruse, 2003). Ruse offers the familiar example of the cyclical process by which rain falls on mountains, is carried by rivers to the sea, evaporated by the sun, whereby it forms new rain clouds, which in turn discharge their content as rain. The river is there because it produces or conveys water to form new rain clouds. The rain clouds are a result of the river being there. But Ruse argues that we would not want to say the function of the river is to produce rain clouds, and he is right. What is missing, Ruse claims, is the means by which 'Things are judged useful.' I will not follow Ruse in his deployment of such worrisomely adaptationist notions as 'value' and 'desire'. Instead, I want to salvage Ruse's observation by redescribing his 'judgment' as a measurement of some parameter or, if you like, as an evaluation that is performed by a mechanical sensor and, when exceeding some preset limit, is fed back into a controller which is able to restore the proper range of parameter. In other words, I use the term function in the sense that Nagel and Beckner did, i.e., in the sense of the first of Bill Wimsatt's enumerations: a simple feedback mechanism. Like a thermostat. As Wimsatt (2002, p. 177) puts it, 'To say that an entity is functional is to say that its presence contributes to the self-regulation of some entity of which it is a part.' Once such a mechanism is added to the rain–cloud–river cycle (say, a mechanism that triggers a change in evaporation rates when the water level falls too low) we can, in this sense of the term, legitimately speak of function and say, e.g., that the function of such a mechanism is to maintain the water level within a certain range of parameters. Furthermore, I argue, we can similarly refer to the many different cellular mechanisms (proofreading and repair, chaperones, cell-cycle regulation) that function to maintain various aspects of cellular dynamics.

The scientific question we are then left with (and this is the question in which I am interested) is twofold: how did these early machines first come about, and by what processes (or dynamics) did they come together, combine, and, ultimately, constitute a primitive cell? We take it for granted that all this came about through evolution, but by what kind of evolution? Clearly, not through evolution by natural selection as natural selection itself is a product of this early evolution. What, then, are the alternatives? What other than natural selection, or intelligent design, can provide directionality or sheer accumulation to the random processes of change to which entities in the physical world are subject? In other words, is it possible to account for the emergence of natural design, of a 'self' that can be said to organize – indeed, for the emergence of natural selection itself – from purely physical and chemical principles?

The main scientific alternative that has been promoted in recent years – and for many people, the only possible scientific alternative to evolution by natural selection – is the spontaneous emergence of order associated with the third meaning of self-organization I discussed earlier, i.e., the kind of self-organization seen in nonlinear dynamical systems that can 'mold itself,' as Davies put it, 'into thunderstorms, people and umbrellas'. Or as Stuart Kauffman writes, 'metabolic networks need not be built one component at a time; they can spring full-grown from a primordial soup. Order for free, I call it.' (1993, p. 45). Kauffman is correct. Many complex structures – including networks – can and do arise spontaneously. Indeed, we can find examples of order-for-free all around us. The problem is that such structures do not yet have function, agency, or purpose. They are not yet alive. Self-organization, as mathematicians use the term, may be necessary for the emergence of organized complexity, but as I have already emphasized, and as Stuart Kauffman now acknowledges,[3] it is not sufficient. Something else, something engineers have been struggling to characterize ever since the 1940s, is also required.

Cybernetics, and its emphasis on the relation between feedback and function characteristic of homeostatic devices, offered one clue; I believe that Herbert Simon offered us another. In fact, it is sobering to go back and read Simon's 1962 essay on 'The Architecture of Complexity'. Here Simon introduces a crucial if much neglected argument for a form of evolution that is alternative both to natural selection and to emergent self-organization: evolution by composition. The idea is this: If stable heterogeneous systems, initially quite simple, merge into composite systems that are themselves (mechanically, thermodynamically,

[3] In his more recent work, Kauffman concludes that 'We have no theory of organization', and asks: 'What must a physical system be to constitute an autonomous agent?' (2003: 1089). His answer is that it must couple a motor with self-reproduction/autocatalysis in a nonequilibrium energy flow, and offers as a primitive example a DNA hexamer that can catalyze the linking of two trimers into a hexamer that is a copy of the initial hexamer, provided it is coupled with a PP (pyrophosphate) motor, in a system, far from equilibrium, with an excess of trimers and photons.

chemically) stable, such composite systems in turn can provide the building blocks for further construction. Through repetition, the process gives rise to a hierarchical[4] and modular structure that Simon claims to be the signature of systems with organized complexity. 'Direction,' he explains, 'is provided to the scheme by the stability of the complex forms, once these come into existence. But this is nothing more than survival of the fittest – that is, of the stable'. (p. 191)

In these few words, Simon shows that even pre-Darwinian evolution can be described in terms of a principle of selection, provided, i.e., that selection be reinterpreted in terms of stability or persistence, rather than of survival and reproduction. In doing so he anticipates the recent arguments of Alex Rosenberg, Frederic Bouchard, and others[5] for reformulating (and thereby extending) Darwinian theory by replacing 'fitness' with 'persistence'. In this view, biological survival is just a variant of persistence, and reproduction a way of ensuring persistence through increasing the population (like autocatalysis). But one needs to be a bit careful here in our definition of stability – we are not interested in the stability of rocks, and perhaps not even of the limit cycles of dynamical systems closed to informational or material input. Rather, we are interested in the stability of nonequilibrium systems that are by definition open to the outside world, not only thermodynamically but also materially. Furthermore, we are particularly interested in stability with respect to likely perturbations. Perhaps a better word would be robustness. The systems that endure are those that are robust with respect to the kinds of perturbations that are likely to be encountered. The critical questions then become, first, how do new ways of persisting arise? And second, how are they integrated into existing forms?

In neo-Darwinian theory, novelty arises through chance mutations in the genetic material and integrated into existing population by selection for the increased relative fitness such mutations might provide. In Simon's account, especially as Wimsatt (1974) subsequently elaborated it, novelty arises through composition (or combination), is further elaborated by the new interactions that the proximity of parts bring into play, and, finally, integrated into the changing population by selection for increased relative stability. Of particular importance in Wimsatt's account is the stability of the composite acquired with the passage of sufficient time 'to undergo a process of mutual coadaptive changes under the optimizing forces of selection' (i.e., selection of the stable) (1974, p. 76).

---

[4] Simon's use of the term hierarchy has led to a certain amount of confusion in the literature, for it is often read as referring to relations of authority. Simon attempted to clarify his own use of the term as referring to 'all complex systems analyzable into successive sets of subsystems' (p. 185), and as distinct from authoritarian structures (which he called 'formal hierarchies'), but a certain political valence continues to be read into his arguments (see, e.g., Agre, 2003).

[5] See, e.g., Bouchard (2004); Rosenberg & Kaplan (2005)

Symbiosis provides a good example of all three aspects of the process, and perhaps especially of the ways in which the net effect is to bring into being entirely new kinds of entities.

Such an account would seem to be especially pertinent to the evolution of cellularity: biological cells are replete with devices for ensuring survival, stability, robustness. Consider, for example., the devices that have arisen to regulate cell division, ensuring that cell division is not triggered too early (when the cell is too small) or does not wait too long (when the cell has gotten too big). Or consider the vastly complex kinds of machinery for guaranteeing fidelity in DNA replication, the accuracy of translation, or the proper folding of proteins. Each of these mechanisms, taken by itself, can be considered to embody a function. Much as the function of a thermostat is to maintain the temperature of a room, so too the function of a proofreading mechanism in DNA replication is to maintain the identity of the nucleic acid molecule. The complex of DNA cum proofreading device constitutes a composite with significantly greater stability (or persistence) than either would have by itself. Furthermore, each of these mechanisms has a long evolutionary history, much, if not most, of which preceded the existence of the modern cell. Originally appearing as simple devices, arising out of the fortuitous combination of already existing molecular complexes, it is not hard to imagine their subsequent refinement, elaboration, and integration into every more complex structure, all by virtue of the enhanced stability they would provide. As Simon wrote, 'this is nothing more than survival of the fittest – that is, of the stable.'

Inevitably, different mechanisms for ensuring robustness marked off different evolutionary epochs. Nucleic acid molecules, for example, appearing on the scene long before the advent of anything like a primitive cell, introduced a significant advance over mechanisms of autocatalysis for making more because it made possible the replication of molecules with arbitrary sequences. The subsequent arrival of a translation mechanism between nucleic acid sequences and peptide chains required the combination of already existing nucleic acid molecules and already existing protein structures, but the innovation of a translation mechanism – in effect, the advent of genes – ushered in an entirely new order of evolutionary dynamics. During the subsequent period, the few hundred million years over which cellularity evolved, change seems to have depended primarily on the horizontal flow of genetic bits between porous entities not yet sufficiently sealed off to qualify as candidates for natural selection. Carl Woese argues that cellular evolution, precisely because it needed so much componentry, 'can occur only in a context wherein a variety of other cell designs are simultaneously evolving . . . [and] globally disseminated'. He writes, 'The componentry of primitive cells needs to be cosmopolitan in nature, for only by passing through a number of diverse cellular environments can it be significantly altered and refined.' Similarly, he also concludes, 'Early cellular organization was necessarily modular and malleable.' (2002, p. 8746).

Only with the sealing off of these composite structures and the maintenance of their identity through growth and replication – i.e., after a few hundred million years of extremely rapid evolution – did individual lineages become possible. And with individual lineages (and the predominance of vertical gene transfer), the operation of the entirely new, albeit far slower,[6] kind of selection that we call Natural Selection. [As Freeman Dyson puts it, 'one evil day, a cell resembling a primitive bacterium happened to find itself one jump ahead of its neighbors in efficiency. That cell separated itself from the community and refused to share. Its offspring became the first species. With its superior efficiency, it continued to prosper and to evolve separately'. (2005).] Woese calls this the Darwinian threshold. Much of the variety we see around us is a product of natural selection, but most of the enzymatic reactions on which these forms are based depend on proteins and protein machinery that were already in place with the arrival of single-cell organisms. Furthermore, many of the molecules and machinery responsible for more recent innovations have been formed by cutting and pasting earlier structures. In other words, even with the advent of natural selection, composition (or, to use a currently more popular term, tinkering) seems to have continued to play a critical role in the creation of novelty.

Probably the most important of such an innovation was the invention of development, the process by which higher, multicellular organisms are produced. Development stands in sharp contrast to evolution in a number of ways, but the contrast between Simon's model of evolution of complex structures by assembly or composition and the development of multicellular organisms by differentiation and morphogenesis is particularly sharp. Nevertheless, Simon claimed that the end result is the same – that multicellular organisms have the same hierarchical (or modular) structure as that of systems which had formed by composition. It is the structure, and not its method of production, he argued, 'that provides the potential for rapid evolution'. (p. 193)

For many people, hierarchy and self-organization constitute opposing explanatory schemas, with opposing political valences. Despite Simon's demurral, hierarchical construction is often associated with social hierarchy, while self-organization is understood as organization from below. In fact, however, self-assembly by composition is itself a form of self-organization, and the hierarchy that Simon envisioned is simply the consequence of the iteration of self-organization over time. Where a difference really does need to be marked is between the iterative processes of self-organization that occur in

[6] I have not discussed the rates of evolution here, but one of Simon's most important arguments was that the time required for the evolution of a complex system by assembly is a minute fraction of that required for evolution by sequential accumulation of novelty (see, esp., pp. 1188–189).

heterogeneous systems over time and the one-shot, order-for-free, kind of self-organization associated with the kinds of uniform nonlinear dynamical systems that mathematicians usually study. This, I believe, is the crucial point distinguishing organized complexity – the kind of complexity found in organisms and machines – from the emergent complexity characteristic of thunderstorms. As to what separates organisms from machines is the fact that the organized complexity of the former, unlike the latter, arose and evolved spontaneously. Machines may be self-steering, but even after all these years of mechanical ingenuity, it remains the case that only organisms can be said to be self-organizing in Kant's sense of the term.

REFERENCES

Agre PE. *Hierarchy and history in Simon's 'Architecture of Complexity'.* Journal of the Learning Sciences: 12(3), 413–426, 2003.

Ashby R. *Principles of the Self-Organizing Dynamic System.* Journal of General Psychology: 37, 125–128, 1947.

Bak P. *How Nature Works: The Science of Self-Organized Criticality.* Oxford University Press, Oxford, 1995.

Bak P, Tang C & Wiesenfeld K. *Self-Organized Criticality: An Explanation of 1/f Noise.* Physical Review Letters: 59, 381–384, 1987.

Bak P & Sneppen K.*Punctuated equilibrium and criticality in a simple model of evolution.* Physical Review Letters: 71, 4083–4086, 1993.

Bennett CH. *On the Nature and Origin of Complexity in Discrete, Homogeneous, Locally-Interacting Systems.* Foundations of Physics: 16, 585–592, 1986.

Bouchard F. *Evolution, Fitness and the Struggle for Persistence.* Ph.D. Dissertation, Duke University, Durham, NC, 2004.

Davies PCW. *The Physics of Complex Organisation.* In: Theoretical Biology: Epigenetic and Evolutionary Order from Complex Systems (Eds.: Goodwin B & Saunders P), Edinburgh University Press, 101–111, 1989.

Dyson F. *The Darwinian Interlude.* Technology Review: March, 2005.

Kant I. *Critique of Judgement.* In: Great Books 39: 461, Chicago: Encyclopedia. Brittanica, 1993.

Kauffman S. *The Origins of Order.* Oxford University Press, New York, 1993.

Kauffman S. *Molecular autonomous agents.* Philosophical Transactions of the Royal Society London A: 361, 1089–1099, 2003.

Keller EF. *Marrying the Pre-Modern to the Post-Modern: Computers and Organisms after WWII.* In: Growing Explanations. (Ed.: Norton Wise M), Duke University Press, Durham, NC, 181–198, 2004.

Mattick J. *RNA regulation: a new genetics?* Nature Reviews Genetics: 5, 316–323, 2004.

Neander K. *Functions as Selected Effects.* In: Nature's Purposes: Analyses of Function and Design in Biology. (Eds.: Allen C, Bekoff M & Lauder G), MIT Press, Cambridge MA, 313–333, 1998.

Rosenberg A & and Kaplan DM. 2005. *How to Reconcile Physicalism and Antireductionism about Biology.* Philosophy of Science: 72, 43–68, 2005.

Ruse M. *Darwin and Design: Does evolution have a purpose?* Harvard University Press, Cambridge, MA, 2003.

Simon H. *The Sciences of the Artificial*, 3rd edition. MIT Press, Cambridge, MA, 1996.
Smolin, L. *The Life of the Cosmos.* Oxford University Press, New York, 1997.
Turner S. *The Extended Organism.* Harvard University Press, Cambridge, MA, 2000.
Weaver W. *Science and Complexity.* American Scientist: 36, 536–544, 1948.
Wimsatt WC. *Complexity And Organization.* In: PSA-1972 (Boston Studies in the Philosophy of Science, volume 20). (Eds.: Schaffner KF & Cohen RS), Dordrecht: Reidel, 67–86, 1974.
Woese C. *On the evolution of cells.* Proceedings of the National Academy of Sciences: 99 (13), 8742–8747, 2002.

SECTION V

Conclusion

14

Afterthoughts as foundations for systems biology

Fred C. Boogerd, Frank J. Bruggeman, Jan-Hendrik S. Hofmeyr and Hans V. Westerhoff

This book on the Philosophical Foundations of Systems Biology is the first of its kind. Several books on systems biology have been published or will be published soon. Some of these are text books aimed at teaching and as introductory texts (Kitano, 2001; Klipp et al., 2005; Alon 2006; Palsson, 2006), others were multiauthored volumes reporting on systems biology progress (Kitano, 2001; Kholodenko & Westerhoff, 2004; Kriete & Eils, 2006; Szallasi et al., 2006), and one focused on the definition of what systems biology is (Alberghina & Westerhoff, 2005). None of them deals extensively with the philosophical issues related to and underlying systems biology. For a long time, there have been a number of treatises on the philosophy of biology (Schaffner, 1994; Hull & Ruse, 1998; Sterelny & Griffith, 1999; Sober, 1999; Weber, 2005). Many of these emphasized that biology differs from physics and chemistry because of its perspectives of evolution and function. Earlier the importance of systems and the whole were emphasized (Wiener, 1965; Von Bertalanffy, 1976; Savageau, 1976; Atkinson, 1977; Reich & Selkov, 1981).

This book we have in front of us is special because it came after the genomics revolution, at a time when frustrations are accumulating on the lack of delivery of promises by molecular biology and gene sequencing alone. Perhaps this is a time where the thrust of one paradigm runs out, and where a new science paradigm is taking over, in a process that is not always smooth and accompanied by strong reactions by proponents of the earlier paradigms (Kuhn, 1996). Traditionally, treatises on the philosophy of science have come in the aftermath of the development of new sciences. An early treatise of the structure (cf. Nagel, 1979)

Systems Biology
F.C. Boogerd, F.J. Bruggeman, J.-H.S. Hofmeyr and H.V. Westerhoff (Editors)
Isbn: 978-0-444-52085-2

of a new science should be welcome for a number of reasons. At present many systems biologists are confronted in response to their papers or grant applications that the science they are doing or proposing is not quite proper, being insufficiently driven by minimal hypotheses or being too mathematical for molecular cell biology. Such responses may well derive from a lack of appreciation that there is a paradigm shift to a new science which combines many of the earlier scientific disciplines in a nonlinear way: some but not all the norms and values of the earlier disciplines persist and new standards of how to do science emerge. Establishment of those norms and values for the new science, or at least recognition that the new science will develop new standards on how to do science, may accelerate the development of the new science by removing obstructions deriving from misunderstanding or conservatism. By being relatively early, this book may play a unique role as compared to earlier books on the philosophy of science.

When a new science paradigm comes into existence, response tends to evolve according to the dictum: 'it is nothing new', through 'it is wrong' to 'I knew it all along', where indeed there is an element of truth in all of these, but also an element of conservatism and forgetfulness about the history of science. The same words would apply to statistical thermodynamics or molecular biology. What is clear from the above chapters is that the nature of systems biology is well defined by now and that is does differ from pre existing disciplines. By synergistically combining theory, modeling and experiment, this discipline aims to obtain a molecular mechanistic picture of cells to the extent that cellular behavior can be understood in terms of the behavior of its constituent molecules and higher order consortia thereof. This is possible only now and not, say, 15 years ago because of the recent development of technologies critical to obtaining the required high-quality and experimental data concerning virtually all components of the living cell. Not all of these technologies are perfect yet, but it is anticipated that they reach the required accuracy in not too long a time. These technologies offer two complementary approaches to systems biology: top-down and bottom-up systems biology. Top-down approaches start with experimental data concerning the changes in abundancies of the molecules in the entire system to extract knowledge about the molecular mechanisms that generated the data. Nowadays, this can be done on an organism-wide scale using high-throughput measurements of organisms perturbed by various means, such as changes in nutrient levels or physicochemical environment. Bottom-up approaches start from the knowledge about molecular mechanisms and determine whether our knowledge suffices by comparing the behavior of molecular mechanisms in detailed computer models to their *in vivo* behavior. Discrepancies then pinpoint gaps in our knowledge and offer opportunities for new findings. Such an approach may also lead to the discovery of new regulatory mechanisms.

This book may help clarify many issues concerning the status of systems biology and its methodology. Chapters of this book have described research programs in systems biology and have looked at these from the perspective that they are new ways of doing science (e.g. Westerhoff & Kell, this volume). Because of their ability to capture the whole, genome-wide experimentation should enable a data-driven hypothesis generation that might bring one to a much more complete, perhaps phenomenological, understanding than in other, more unbounded sciences such as physics. Reconstruction *in silico* of living cells or pathway functions thereof, on the basis of quantitative molecular biochemistry and biophysics focusing on interactive properties, should be able to establish a complete description for the other side. The integrative systems biology in which the two approaches meet should merge phenomenological and mechanistic descriptions and enable a validated complete description from molecules to system and vice versa (Krohs & Callebaut, this volume). Some of the discoveries made by the new systems biology have been referenced in this book and therefore, it may now be possible to judge whether these new ways of doing science may lead somewhere and whether the science produced has any scientific value. If so, then this may contribute to a relaxation of the antipathies against the new systems biology, as well define what is systems biology and should be promoted to become even more so, and what is just a linear composition of classical scientific activities. It should put an end to the tendency of some mathematicians and physicists to play down the discipline as it is not sufficiently scientific from the point of view of mathematics and physics. And just as well, it should help reduce the symmetrical effect of molecular and cell biologists suggesting that the modeling and theory efforts in systems biology are inappropriate for their disciplines. If they wish, all these suggestions are right, but this book showed that systems biology is a discipline in its own right, which has its own criteria for scientific endeavor, which differ from and may well be more than the sum of the criteria of the above-mentioned disciplines. This outcome of this book may therewith help the scientists who need to evaluate grant proposals. Perhaps surprisingly, this is quite a practical application of this book on the methodological and philosophical foundations of systems biology, and again this is a unique function of this book.

The ability to describe completely and explain all system behaviors mechanistically reminds us of a few physical chemical systems for which this is also the case, but none of the latter have a complexity that is anywhere near the minimum complexity of living organisms. In both types of systems, the question arises whether such an ability to describe a system completely and mechanistically implies that it is understood completely. Human understanding tends to have an element of simplification, of moving from the detail-laden specific to the general and indeed, as discussed in this book (e.g. Schaffner; Westerhoff & Kell; Wolkenhauer, this volume), another aspect of systems biology is the discovery of general principles of biological systems. Indeed, systems biology is not just specific. It is not about a single organism. It theorizes about organisms in general. The power of its approaches is evident from the successes of bioinformatics and phylogenetics. Yet, realizing that life, as we know it on this planet, is just a subset of all possible living systems may compromise the generality of these principles. Thereby, these 'general' principles or 'laws' of systems biology may not classify as 'laws' according to the definition of 'laws' in physics.

These types of issues are of great interest for the methodology and philosophy of science in general, as biology is only one, but now the best elaborated example of where 'generality' is limited to a large number of related specific cases. Given that human beings can only engage in certain types of behaviors and chimps in others but not all others, sociology struggles when trying to follow scientific paradigms that aim for complete generality, as defined by physics. Here, this book may aid philosophers of science to venture onto new avenues.

And then there is the hyper-interface among philosophy, biology, theology, and ethics. After all, the simplest form of life is the unicellular microorganism. Even the complete understanding of how microorganisms function was once thought to be far outside the scope of natural sciences. Therewith life itself remained outside the domain of complete scientific explanation. This made phenomenology-based biology, theology, the philosophy of being, and ethics, autonomous domains of academic activity. With systems biology, life, first at the simplest level of unicellular organisms, then at the level of all organisms except primates, and perhaps ultimately at the level of intelligent human beings, will become calculable. Herewith the domain of 'metaphysics' that relates to life may be shrunken. The philosophy of what life is may be assisted by computer models that enable them to calculate the behavior of living organisms. Either this leads to a definition of life in terms of minimum behavior that is calculated by those *in silico* replica of living organisms or a new definition of life is achieved, which should then be much clearer than the present in terms of distinguishing it from the one that could be calculated.

What life is has been discussed amply, in the context of systems biology, in a number of chapters of this book. What life is, is perhaps the main issue of biology. Therewith, the philosophy of systems biology is really the philosophy of biology, and this book is the first that highlights this issue. It appears to take the stance that limiting the philosophy of biology to that of the theory of evolution is beside the key issue. The essence of biology is life and not just the way it came about, notwithstanding the importance of evolution for the specific forms that life on this planet has assumed.

In all these respects, this book may spark a reappraisal of the role philosophy can play in biology in general and in systems biology in particular. Although the book's title refers to the philosophical foundations of systems biology, it is much more than that. It provides biology, biochemistry, biophysics, and in fact much of the life sciences with methodological underpinnings. After all, systems biology is not just a subdiscipline of biology; it constitutes much of what twentieth century life sciences are developing into.

But what then did we learn from this book? And how can all the aspects discussed be brought into a single, perhaps multidimensional, perspective? Below we shall highlight what we learnt from the perspective of the scientific revolution in biology leading to the new paradigm, systems biology.

1. SYSTEMS BIOLOGY IS FUNCTIONAL AND MECHANISTIC RATHER THAN EVOLUTIONARY BIOLOGY

Systems biology, especially bottom-up systems biology, puts a strong emphasis on biological function, even though it starts somewhat indifferently from the molecules that do not function themselves. Molecular biology sometimes uses the word function for an activity catalyzed by an individual macromolecule outside the context of the living organism. Only when the context of the whole organism is added, the word function can relate to fitness and become a component in a legitimate teleological explanation. We note that systems biology is still remote from this, because it is often unclear what the true habitat of the organisms harboring the molecule under study has been and what has been or is important for its fitness. Yet, there are well-known activities in this direction, such as flux-balance analysis (Price et al., 2004).

It is important to realize that systems biology tries to understand life as it is now, while it does not focus on evolutionary biology. It may use reasoning derived from evolutionary biology, such as reasoning based on homologies, but it does not yet aim at explaining the evolution of biological systems. This preference reflects the conviction that life should be understandable without reference to the histories of all life forms. Systems biology aims at acquiring a molecular mechanistic understanding of biological systems subject to challenges to their existence from the outside world. For systems biology, the properties of molecules are considered as important as the systemic behavior, and understanding is considered to have been achieved when it can be shown how the latter emerges from the former when their nonlinear interactions and arrangements are taken into account. Because of the complicated organization of cells and nonlinearity, hence condition dependence of the interactions, detailed models are required to 'calculate the emergence of life from its dead molecular constitution'. According to Mayr's analysis of biology in terms of functional and evolutionary biology (Mayr, 1961), where the former would be dealing with how-questions and the latter with why-questions, systems biology deals with the former.

This is not to say that there will be no bridge between systems biology and evolutionary biology (Wimsatt, this volume). Typically, once it is understood why a system is capable of living, one may also begin to note that other constellations of molecules or atoms could also be alive. An example is life where all molecules are the mirror-symmetry isomers of the molecules that do constitute life on this planet. If one were to ask why this planet (probably) harbors only the one form of life we know and not its mirrored analogue (life based on D-enantiomers rather than L-enantiomers), the answer is likely to reside in evolutionary biology. Systems biology does not need to ask this question for it to be a complete science. Many areas of physics could discuss systems that are not actually observed (such as a system identical to ours with all electric charges inversed).

2. SYSTEMS BIOLOGICAL EXPLANATIONS ARE OFTEN MECHANISTIC EXPLANATIONS

An important part of functional biology is about answering 'how' questions. Answers to 'how' questions are usually phrased in causal mechanistic language. One issue, therefore, that ought to be addressed by a philosophy of systems biology is that it should analyze the types of explanations offered by systems biologists and how their structure differs from explanations offered in other sciences and biological disciplines. It can be argued that a philosophy of systems biology should aim for a hybrid form of the unificationist and causal/mechanical type of explanation. Some authors (Westerhoff & Kell; Schaffner, this volume) have discussed the former kind of explanation, whereas several authors (Bechtel; Fell; Richardson & Stephan; Schaffner; Shulman, this volume) have stressed the importance of mechanistic explanations in systems biology. In fact, Westerhoff & Kell engage in both.

In this book, several authors have further argued that answers to 'how' questions are frequently given by discovering and articulating the organization and inner workings of molecular mechanisms underlying behavior. The articulation amounts to showing how the behavior is brought about by the parts when engaging in interactions; this used to be done in terms of a graphical representation of the mechanism (a network depiction of the molecular interactions); more recently, detailed computer models for systems that have been characterized molecularly to a sufficient level of detail have been used for this purpose. This kind of explanation belongs to the causal/mechanical type of explanations in the philosophy of science (Bechtel & Richardson, 1993; Schaffner, 1994; Woodward, 2005; Salmon, 2006) for the way individual macromolecules work, but it rarely extends to the cellular level. An early example of a mechanistic explanation covering systems phenomena is the analysis by Jacob and Monod on how the molecular organization of the *lac* operon underlies diauxic growth in *Escherichia coli*. This example is discussed in depth by Stephan and Richardson in this volume. A second early example was the chemiosmotic coupling hypothesis of free-energy transduction (Mitchell, 1961). When molecular biology and molecular genetics became the mainstream approaches to cellular biology the number of biologists using a more mechanistic approach to systems of molecules was small. However, there always remained a group of mathematical biologists that continued to relate system properties to molecular properties through quantitative methods and in fact this has become one of the roots of contemporary systems biology (Westerhoff & Palsson, 2004). The early examples include: Turing, 1952; Hodgkin & Huxley, 1952; Chance et al., 1960; Goodwin, 1965; Glansdorff & Prigogine, 1971; Kauffman, 1971; Goldbeter & Lefever, 1972; Kacser & Burns, 1973; Heinrich & Rapoport, 1974; Rapoport et al., 1974; Savageau, 1976; Atkinson, 1977; Reich & Selkov, 1981; Westerhoff & Van Dam, 1987.

The unificationist and the mechanistic explanations are not unique to systems biology; they are common to physics and chemistry as well. Yet, they are suspicious, for the unificationist type of explanation is alien to many biologists, who see their discipline as just a large number of special cases. Of course with the sequences of various genomes becoming known, the principle of unity of biochemistry, discovered by Kluyver, is reinforced, and indeed in DNA sequences the communalities are so strong that function can often be predicted on the basis of sequence homology. Clearly there is much space for unificationist explanations in systems biology. The mechanistic explanation is special as it is such an enormous challenge vis-à-vis the enormous complexity of living cells.

3. OTHER TYPES OF EXPLANATION ARE ALSO IMPORTANT FOR SYSTEMS BIOLOGY

A type of explanation that is hardly, if at all, accepted by molecular biologist is the nomological explanation, where from a clearly defined set of premises certain properties are derived. Examples abound in mathematics; the derivation of the second law of thermodynamics through statistical mechanics is an example. In systems biology (and one of its predecessors, metabolic control analysis (MCA)), the summation and connectivity laws are examples (Kacser & Burns, 1973; Heinrich & Rapoport, 1974; Westerhoff & Chen, 1984).

A type of explanation that is symmetrically unaccepted in physics but common to biology is the design explanation (e.g. Wouters, 1999). Here phenomena are explained by specifically referring to the advantage the presence of a certain item or a certain kind of behavior has for the organism in comparison to its absence or to another kind of behavior, respectively. In such explanations, answers to 'why'-questions are given within the realm of functional biology. So, we disagree with Mayr who claimed that giving answers to 'why'-questions is about evolutionary biology (Mayr, 1961). For proper examples of design explanations see Teusink (Teusink et al., 1998), Bakker (Bakker et al., 2000), and Wouters (1999).

Having read about these various types of explanation and their potential for systems biology in this book, we note that the precise conditions under which they can be used are incompletely defined. The philosophical foundations of systems biology may be more diverse than those of other sciences; we believe that they should still be well specified, in order for the quality of this new science to become outstanding and its results reliable.

4. DESCRIPTION OF MOLECULAR MECHANISMS USING MODELS

The detailed description of the organization and inner workings of molecular mechanisms in terms of mathematical models can take many forms, often depending on the purpose of the description. In more applied settings, where models are supposed to have interpolative power, the exact details of the molecular mechanisms may not matter too much, as long as the mechanisms are described to such an extent that the model is a successful predictor. When constructing such models, the structure of the network, description, and the parameterization of its processes can be phenomenological. On the other end of the scale one finds detailed models – 'silicon cell' models – that are careful reconstructions of molecular mechanisms characterized mostly *in vitro*. Such models aim at being or becoming 'replicas' of cells. In between these extreme approaches various models can be distinguished that combine the advantages of both approaches. These models often attempt to mimic the network structure and process characteristics precisely, while parameterization is more phenomenological and derives from fitting the model parameters to experimental system data. This 'hybrid' approach also exemplifies the present lack of knowledge of the kinetic description of many intracellular processes, except for processes occurring in the central metabolism in yeast, *Trypanosoma brucei*, and *E. coli*.

Because of its dependence on complex nonlinear systems, systems biology depends on modeling. The above modeling approaches have therefore been discussed from an epistemological perspective in some of the chapters in this book (Hofmeyr; Schaffner; Westerhoff & Kell; Wolkenhauer & Ullah). Where some authors emphasize that some of these approaches are less valuable than others, we conclude from these chapters that systems biology will profit from as wide a variety of modeling approaches as possible. These should include various hybrid approaches, e.g., between continuous and discrete models, and an optimum combination of spatio-temporal-chemical modeling for the systems that are stiff in temporal, spatial and chemical-concentration dimensions.

5. MODELS AND THE NONEQUILIBRIUM ORGANIZATION OF LIVING SYSTEMS

Of necessity, living systems are open and displaced from equilibrium. Self-organization, once thought to be dominant, appears to have a limited role, perhaps because biology needs to be robust and heritable. The expression of a robust genome appears to direct dissipative processes that may reinforce themselves through self-organization, such as through positive feedback loops. Mosaic nonequilibrium thermodynamics, kinetics, and metabolic control analysis are systems biology methods that are perhaps the natural successors (Westerhoff & Palsson, 2004) to the nonequilibrium thermodynamics that emphasized just self-organization (Glansdorff & Prigogine, 1971).

Two other topics that are highly relevant for systems biology and concern the organization of living systems have been addressed more at length in this book. Wolkenhauer and Ullah analyzed whether the types of models that are constructed at present can indeed capture all phenomena associated with living systems. They concluded that this is unlikely and that alternative approaches should also be considered. Indeed, systems biology is in need of new modeling methodologies. A second issue was addressed by Hofmeyr. He analyzed how the organization of cells may bring about self-replicating systems.

The related notion of complexity was also addressed by various authors. Some are of the opinion that cells are too complex to fully catch their behavior and that models serve the purpose of reducing the complexity (Wolkenhauer & Ullah), while others stress that it is not just complexity that matters for biology but organized forms of complexity (Keller) or functional organization in the light of context (Hofmeyr).

6. EMERGENT PROPERTIES

The concepts of mechanism and emergence have long been considered to be mutually exclusive. As soon as a mechanistic explanation could be given for some behavior, that behavior was not considered emergent, as it was to be expected, already implied by what was known. Conversely, it was thought that emergent properties could not be explained mechanistically. For some philosophers, this was a matter of definition: emergence was defined as the appearance of phenomena that could not be explained mechanistically (Kim, 1999). In this book, two different notions of emergence have surfaced. In their contribution, Westerhoff and Kell consider a weaker notion of emergence, taken by many scientists working with nonlinear systems, and note that it is relevant for science that in some systems properties appear that would not appear should the components have been isolated from each other. Such properties can differ qualitatively from any property that is present in the molecules in isolation or even be completely new. Noting that many of these properties are crucial aspects of life, they call these properties emergent. Then, on the basis of the fact that in the absence of vital forces, living systems also consist of nothing but a dynamic constellation of molecules that interact, they consider that, with the exception of cases of deterministic chaos, these emergent properties should be explainable mechanistically. They go even further by claiming that the construction of accurate computer replica enables the calculation of emergent properties. Thus emergence is perceived not only as being explainable, but also as being calculable on the basis of quantitative detailed kinetic models. The models are constructed on the basis of kinetic and binding constants measured for components in isolation but under the precise conditions that reign *in vivo*. The crux is whether the behavior *in vivo* will show up in the reconstruction of the system *in silico*. If so, the mechanistic model is also a reductive model, because only properties from the lower (component) level of organization suffice to reconstruct the system's behavior. The mechanistic model is not just a reductionist model, however, as the latter would maintain that the system properties are already present in the individual components, whereas Westerhoff and Kell argue that emergent system behavior arises in the nonlinear interactions between the molecules, as will be calculated in the silicon cell replicas. Even though the capability to interact is already present in the individual molecules, the interaction is not and the new properties do not arise in the abilities to interact but from the actual interactions. The system's emergent properties show up only when the interactions are actually happening. System's behavior is not taken for granted beforehand. Emergence is calculable and, in this view, eventually life would be calculable.

In the chapter of Richardson and Stephan, the concept of emergence is based exactly on such a failure of system behavior to be calculated from the component properties: system behavior should be called (strongly) emergent only if system behavior cannot be inferred or predicted from the behavior of components in isolation (or smaller subsystems). If one knows the behavior of components within the systemic context, then it is possible to give a mechanistic explanation of any system's behavior. This definition is stronger than Westerhoff and Kell's but follows the same type of reasoning.

In contrast, but consistent with the historical definition mentioned above, metaphysicians claim that properties (e.g., *qualia*) are emergent if they cannot be predicted even from the complete knowledge of components' behavior within the system. In order to approach reality better, knowledge should be gathered about component behavior in subsystems in isolation, encompassing a larger portion of cellular metabolism, or it may be required to study component behavior within the entire system. In the latter case, besides lower level component knowledge, system knowledge is required for the reconstruction of system behavior. We argue that this notion of emergence is not useful for natural sciences in general and for systems biology in particular, by virtue of the fact that such emergence cannot exist in the absence of vital forces or 'intelligent design'. There is no consistent evidence for such vital forces or design.

However, the ultimate resolution of these paradoxes may reside not so much in the definition of emergence but in the definition of what constitutes a mechanistic explanation. As also discussed by Westerhoff and Hofmeyr (2005), truly new properties do not emerge in any odd calculation of systems behavior. They arise only when that calculation is nonlinear in the sense that it accommodates that the properties of a component are codetermined by the properties of other components. The crux is that the definition of component properties is ambiguous in dynamic systems. The properties of components can be defined exclusive or inclusive of the effect of system reverberation. An example is the usual case where the control by an enzyme on the flux through itself is smaller than 1. This implies that if the enzyme 'decides' to become 10% more active, one may say that the component property of the enzyme is that its activity is 110%. However, its flux does not increase by 10% but by a percentage that is usually smaller and codetermined by the response of the other dynamic components in the system. If that percentage were only 3, then the second meaning of component property would be an activity of 103%. The actual percentage can be calculated from the interactive properties and network topology of the system, through metabolic control analysis. Now there would be a way to calculate the system properties without taking into account this mollification of the properties of the components by the dynamics of the system, i.e., assuming an activity of 110%. Such a calculation would not deliver the emergent properties and would not be consistent with a stable steady state. We surmise that emergent properties should be defined as properties that are not explainable/calculable through the latter, 'linear' (110%) method, which keeps component properties the same as those of the properties in isolation, i.e., does not calculate how the components properties change when the components are active in the system. It is here that the methodological and philosophical foundations of systems biology must be recognized and have strong implications in the context of emergence of unexpected properties.

7. THEORIES AND LAWS IN SYSTEMS BIOLOGY

Traditionally, in physics theories were considered to be coherent networks of natural laws. Laws were supposed to be universal, general, and necessary. However, even for physics, this picture of laws as being applicable throughout the universe, without exceptions and with effects that were deducible from causes, has been challenged (cf. Cartwright, 1983). In biology, many philosophers have raised their doubts about the existence of laws in the physical sense. Biological laws, if they exist, appear to be subject to spatial limitations, are never without exception, and have more often than not non-deducible cause–effect relationships. Some philosophers even completely reject the idea that biology has laws. In contrast, Shulman in his contribution claims that there is no need to abandon the philosophy of physics in biology, but his point of view seems to be an exception in this book. And Westerhoff and Kell pointed out that systems biology does have general, nomological laws, which are valid generally provided that the premises are defined as formally as is done in physics. After all, Boyle's law is never valid either, not even for Helium.

One concept of biological laws is linked to the unificationist idea of what counts as an explanation. The central dogma of molecular biology, i.e., information flows from DNA to RNA to protein, was considered by many to be a biological law that was global (but perhaps not universal), exceptionless, and necessary until the discovery of the retroviruses. Thereafter the old law was restricted, and the 'new law' claims that information flows from nucleic acids to protein, which has not been challenged seriously so far. This example seems prototypical of the fate of 'laws' in biology, but one may wonder if this is a problem at all. Generalizations with only limited applicability over a certain biological domain are still worthwhile to pursue. The theorems of metabolic control analysis are good examples in this respect. The conceptual dogma of the rate-limiting step (see Fell, this volume), which reigned sovereign for so many years in biochemistry textbooks, has been incorporated into the systems biological, much subtler notion of distributed control in biochemical pathways, where the rate-limiting step is still an explicit possibility, albeit it is acknowledged now that it only occurs in exceptional cases. Several additional useful theorems have been derived afterwards, although they all have limited applicability. The theorems of metabolic control analysis (see Fell; Shulman; Westerhoff & Kell, this volume) are also relevant with respect to the question whether the unificationist, the causal/mechanical, and the nomological modes of explanation can be linked in a useful manner. Biochemical pathways and their constituent enzymes are clear examples of a part–whole system that is composed of two adjacent levels of cellular organization, both of which feature in the quantitative mechanistic explanation. For, the 'local' role of each enzyme within the pathway and its level of control on the global steady-state pathway flux can be articulated in a quantitative way. At the same time, however, a systemic law can be formulated for the same part–whole system, i.e., the sum of all the flux control coefficients of the pathway should be equal to one. This result could be generalized to any arbitrary pathway, and it could therefore well be granted the status of a law. Also to more complicated nonlinear pathways – for example, pathways including branches and cycles – the same law applies. Thus metabolic control analysis enables a straightforward linkage of the two mainstream ideas about scientific explanations, the causal/mechanical and the unificationist type of explanation.

Systems biology should aim to find more of the above clarifying connections between particular part–whole mechanisms for behavior, which possess generalizable features at the systemic level.

8. EXPLANATORY PLURALISM: INTRALEVEL AND INTERLEVEL THEORIES

The importance of levels of organization, however difficult to define, cannot be underestimated in systems biology. As some levels of organization in living organisms are relatively autonomous, they are home to their own intralevel theories. These theories are valuable, but the inherent danger is that focusing too much on the autonomy of all levels of organization results in specialists who know a great deal about their own level, but quickly consider themselves nonspecialist outside their area of research. And of course one has to be modest in the light of the overwhelming body of knowledge that is available at each and every level of biological organization. This might explain why so many scientists stick to their own field of research, it seems. At the same time, it implies that there is more than enough space for connections and that it is worthwhile to seek such connections. In this sense, system biologists should be (become) generalists and should endorse a multidisciplinary approach. Connecting levels (e.g., via mechanisms) without eliminating the higher level (reduction without leveling) and without completely downplaying the lower level is very much possible and should be pursued in systems biology. In its focus on the interactions between the parts of a system, systems biology endeavors to find such connections. Several authors have stressed the importance of nonreductionist interlevel type of explanations (Fell; Richardson & Stephan; Shulman; Westerhoff & Kell). In short, we are in favor of an explanatory pluralism that incorporates intralevel as well as interlevel theorizing.

9. WHAT IS LIFE?

Erwin Schrödinger asked this question first in a now classical book (Schrödinger, 1944). Many biologists and philosophers of biology refrained from giving a definition of life, but instead gave lists of characteristics of life. In this book, several authors took up the issue. Hofmeyr did so by presenting his MCA (metabolism, construction, assembly) system (not to be confused with Metabolic Control Analysis), combining Rosen's MR system with Von Neumann's universal constructor. Hofmeyr sets out to make self-fabrication of cells formally intelligible in terms of the functional organization of a generic set of processes common to all organisms. Essential in his reasoning is the unassisted self-assembly of ribosomes that makes self-fabrication (i.e., life) possible. Moreno takes a different point of departure in his chapter. He analyses more deeply the role and the nature of the (self-) organization, self-maintenance, complexity, and autonomy of living systems. By giving a very in-depth analysis he manages to present a rich perspective on living systems laying bare many of the key characteristics of living systems and their relationships. As his analysis is so precise and encompassing he sketches a picture of living systems that is a lot more inspiring than giving a mere list of characteristics. Many of the views laid out by Hofmeyr come back in Moreno's contribution. Bechtel stressed the importance of the modes of organization required to explain life. Keller's exposition initially focuses more on self-organization, the history of the concept, its meaning, and its role in characterizing the nature of complex systems in general and living systems in particular. In the second part of her chapter, she investigates the typical organization of living systems as a functional organization and investigates the definition of function. Finally, she addresses how self-organized systems having a functional organization, in terms of organization, could have evolved in evolution. All in all, these three contributions give much food for thought and address in depth a major challenge for the philosophy of biology, and for the philosophy of systems biology in particular: What is life. We hope that systems biology and its philosophy shall tackle this fundamental biological question in the years to come. We anticipate that the recent field of evo-devo (evolutionary developmental biology) may offer the possibility for many interesting opportunities for synergetic interaction regarding this interesting and important topic (see Wimsatt, this volume).

10. CONCLUDING REMARKS

The science of systems biology appears to have much more philosophical consequences than molecular biology, which has been the biological science of the last decades. Molecular biology was largely driven by technology and its aim was reductionistic in essence and essentially lacking any philosophical consequences. It was about characterizing the molecular composition of living systems and not about coming to understand how the molecular components of living systems jointly bring about life. Nevertheless, the molecular biosciences proved to be one of the most successful scientific discipine. They paved the way for systems biology to emerge.

The topics addressed by systems biology seem to have many more philosophical consequences, such as mechanisms, emergence, self-organization, networks, self-maintenance, complexity, systems, control, modeling, function, and (interlevel) theory. Many of these topics have been discussed at length in this book. We hope that this book will be a spark for more systems biologists to engage in discussions about the philosophical aspects of their work, because by doing so we may come closer to an understanding (and perhaps a theory) of living systems. We are convinced that it will be necessary to view living systems from different perspectives in comparison to the perspectives physicists use to study complex systems in chemistry and physics; many of the reasons for this claim have been discussed in this book. One way to discusss those perspectives, their consequences, and relationships with evolutionary, molecular, and developmental biology is to engage in the philosophy of systems biology. We hope that many of you will do so. Systems biology is here to stay. Its aim is to offer an understanding of how molecular systems can be alive, which is the ultimate aim of biology. This is the challenge that systems biology and its philosophy is facing and currently taking up. We hope that this book has contributed to the development of systems biology.

REFERENCES

Alberghina L & Westerhoff HV, Eds. *Systems Biology: Definitions and Perspectives.* (Topics in Current Genetics), Springer-Verlag Berlin and Heidelberg GmbH & Co., 2005.

Alon U. *An introduction to systems biology.* Chapman & Hall/CRC, 2006.

Atkinson DE. *Cellular energy metabolism and its regulation.* Academic press, 1977.

Bakker BM, Mensonides FIC, Teusink B, van Hoek P, Michels PAM & Westerhoff HV. *Compartmentation protects trypanosomes from the dangerous design of glycolysis.* Proceedings of the National Acadamy of Sciences USA: 97(5), 2087–2092, 2000.

Bechtel W & Richardson RC. *Discovering complexity: decomposition and localization as strategies in scientific explanation.* Princeton university press, 1993.

Cartwright N. *How the laws of physics lie.* Oxford University Press, 1983.

Chance B, Garfinkel D, Higgins J & Hess B. *Metabolic control mechanisms. 5. A solution for the equations representing interaction between glycolysis and respiration in ascites tumor cells.* Journal of Biological Chemistry: 235, 2426–39, 1960.

Glansdorff P & Prigogine I. *Thermodynamic theory of structure, stability, and fluctuations.* New York, John Wiley & Sons, 1971.

Goldbeter A & Lefever R. *Dissipative structures for an allosteric model – application to glycolytic oscillations.* Biophysical Journal: 12(10), 1302–1315, 1972.

Goodwin BC. *Oscillatory behavior in enzymatic control processes.* Advances in Enzyme Regulation: 3, 425–38, 1965.

Heinrich R & Rapoport TA. *A linear steady-state treatment of enzymatic chains. General properties, control and effector strength.* European Journal of Biochemistry: 42(1), 89–95, 1974.

Hodgkin AL & Huxley AF. *A quantitative description of membrane current and its application to conduction and excitation in nerve.* Journal of Physiology: 117(4): 500–44, 1952.

Hull DL & Ruse M, Eds. *The philosophy of biology.* Oxford university press, 1998.

Kacser H & Burns JA. *The control of flux.* Symposium Society Experimental Biology: 27, 65–104, 1973.

Kauffman S. *Gene regulation networks: a theory for their global structure and behaviors.* Current Topics in Developmental Biology: 6(6), 145–82, 1971.

Kholodenko BN & Westerhoff HV, Eds. *Metabolic Engineering in the Post Genomic Era.* Horizon Bioscience S, Bios Scientific Publishers Ltd, 2004.

Kim J. *Making sense of emergence*. Philosophical Studies: 95, 3–36, 1999.
Kitano H, Ed. *Foundations of systems biology*. The MIT Press, 2001.
Klipp E, Herwig R, Kowald A, Wierling C & Lehrach H. *Systems biology in practice: concepts, implementation, and application*. John Wiley and Sons, 2005.
Kriete A & Eils R, Eds. *Computational systems biology*. Academic press, 2006.
Kuhn TS. *The structure of scientific revolutions*. University of Chicago press, third edition, 1996.
Mayr E. *Cause and effect in biology*. Science: 134, 1501–1506, 1961.
Mitchell P. *Coupling of phosphorylation to electron and hydrogen transfer by a chemi-osmotic type of mechanism*. Nature: 191, 144–8, 1961.
Nagel E. *The structure of science: problems in the logic of scientific explanation*. Hackett publishing company, second edition, 1979.
Palsson BO. *Systems Biology: properties of reconstructed networks*. Cambridge university press, 2006.
Price ND, Reed JL & Palsson BO. *Genome-scale models of microbial cells: evaluating the consequences of constraints*. Nature Reviews in Microbioly: 2(11), 886–97, 2004.
Rapoport TA, Heinrich R, Jacobasch G & Rapoport S. *Linear steady-state treatment of enzymatic chains – a mathematical model of glycolysis of human erythrocytes*. European Journal of Biochemistry: 42(1), 107–120, 1974.
Reich JG & Selkov EE. *Energy metabolism of the cell: a theoretical treatise*, Academic press, London, 1981.
Salmon WC. *Four decades of scientific explanation*. University of Pittsburgh press, 2006.
Savageau MA. *Biochemical systems theory*, Addison-Wesley, 1976.
Schaffner KF. *Discovery and explanation in biology and medicine*, University of Chicago press, 1994.
Schrödinger E. *What is Life? The Physical Aspect of the Living Cell*. Cambridge University Press, Cambridge, MA, 1944.
Sober E. *Philosophy of biology*, Westview press, 1999.
Sterelny K & Griffith PE. *Sex and death: an introduction to the philosophy of biology*, University of Chicago press, 1999.
Szallasi Z, Stelling J & Periwal V, Eds. *System modeling in cellular biology: from concepts to nuts and bolts*, The MIT press, 2006.
Teusink B, Walsh MC, Van Dam K & Westerhoff HV. *The danger of metabolic pathways with turbo design*. Trends in Biochemical Sciences: 23(5), 162–9, 1998.
Turing A. *The mathematical basis of morphogenesis*. Philosophical Transactions of the Royal Society B, Biological Sciences: 237, 37–47, 1952.
Von Bertalanffy L. *General Systems Theory*. George Braziller, revised edition, 1976.
Weber M. *Philosophy of experimental biology*, Cambridge university press, 2005.
Westerhoff HV & Hofmeyr JHS. *What is Systems Biology? From genes to function and back*. In: Systems Biology. Definitions and Perspectives (Eds.: Albergina L & Westeroff HV), 2005.
Westerhoff HV & Van Dam K. *Thermodynamics and control of biological free-energy transduction*. Amsterdam, Elsevier Science Publishers B.V. (Biomedical Division), 1987.
Westerhoff HV & Palsson BO. *The evolution of molecular biology into systems biology*. Nature Biotechnology: 22, 1249–1252, 2004.
Westerhoff HV & Chen YD. *How do enzyme activities control metabolite concentrations? An additional theorem in the theory of metabolic control*. European Journal of Biochemistry: 142(2), 425–30, 1984.
Wiener N. *Cybernetics*. The MIT Press, second edition, 1965.
Woodward J. *Making things happen: a theory of causal explanation*. Oxford university press, 2005.
Wouters A. Explanation without a cause. Ph.D. thesis, Utrecht University, The Netherlands, 1999.

Subject Index

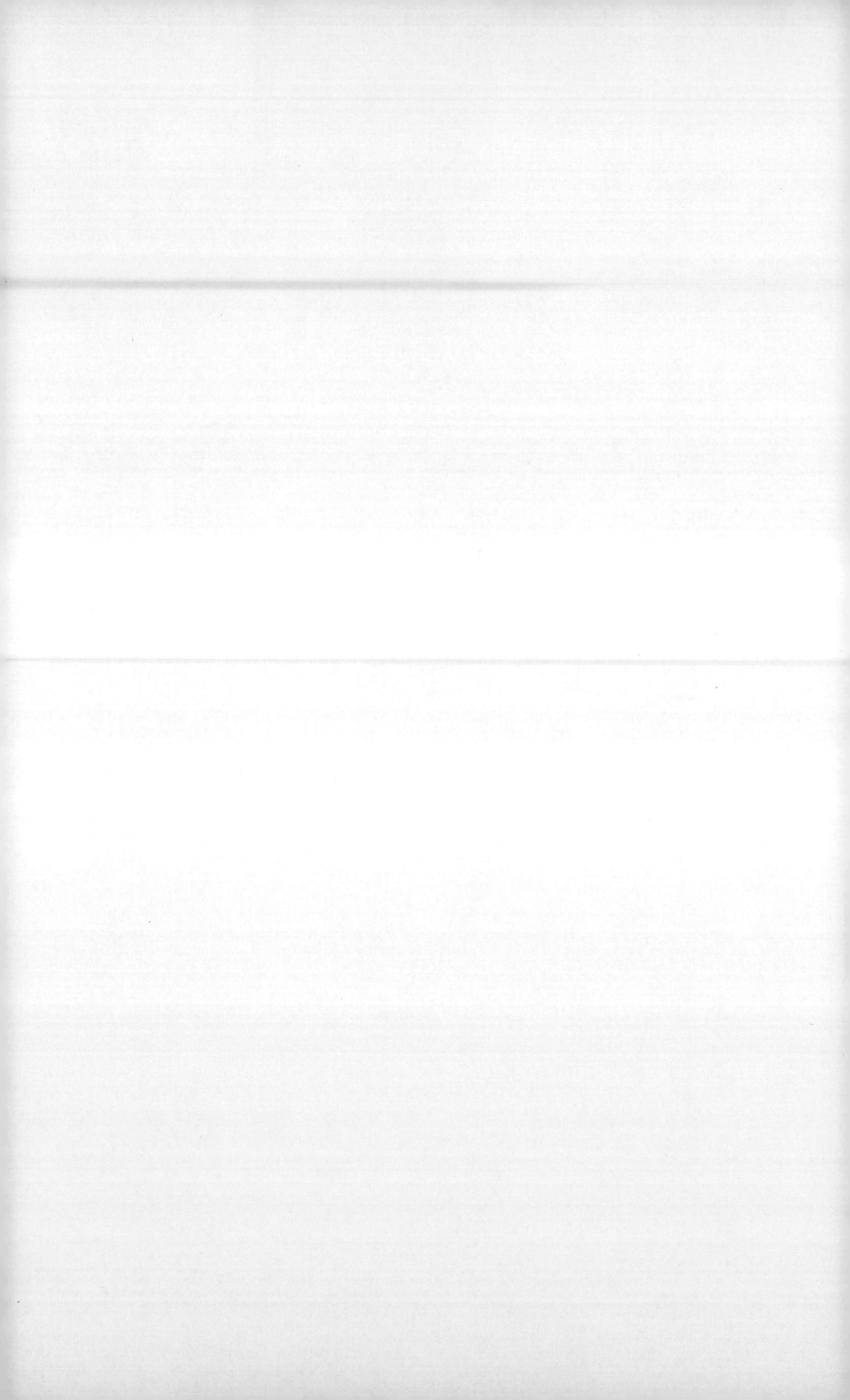

Printed and bound by CPI Group (UK) Ltd, Croydon, CR0 4YY

08/05/2025

01865008-0001